DISCRETE-PARAMETER MARTINGALES

North-Holland Mathematical Library

VOLUME 10

NORTH-HOLLAND PUBLISHING COMPANY–AMSTERDAM · OXFORD
AMERICAN ELSEVIER PUBLISHING COMPANY, INC. – NEW YORK

Discrete-Parameter Martingales

J. NEVEU

University of Paris
Paris, France

Translated by
T. P. Speed

1975

NORTH-HOLLAND PUBLISHING COMPANY–AMSTERDAM · OXFORD
AMERICAN ELSEVIER PUBLISHING COMPANY, INC. – NEW YORK

Library of Congress Catalog Card Number: 74-79241
North-Holland ISBN for the series 0 7204 2450 X
for this volume 0 7204 2810 6
American Elsevier ISBN: 0 444 10708 8

Translation and revised edition of:
MARTINGALES À TEMPS DISCRET
© Masson et Cie, Paris, 1972
Translated by T. P. Speed

Published by:
North-Holland Publishing Company – Amsterdam
North-Holland Publishing Company, Ltd. – Oxford

Sole distributors for the U.S.A. and Canada:
American Elsevier Publishing Company, Inc.
52 Vanderbilt Avenue
New York, N.Y. 10017

Printed in England

PREFACE

This volume presents a third cycle one-semester course first taught at the *Université de Paris* during 1970–71. It was published in French in 1972 by Masson et Cie. The present translation differs from the French text mainly in its last two chapters which have been rewritten in order to take into account the recent progresses of the theory.

Without any doubt the theory of martingales constitutes the mathematical technique at the base of modern probability. We have restricted ourselves to "discrete-time" martingales and give an exposition which, we hope, is fairly complete. At the outset the reader is assumed to be familiar with integration theory and to possess the basic notions of probability theory as taught in mathematics courses.

The first chapter gives a deep study of conditional expectations, without assuming of the reader any previous knowledge on this topic. Chapters II and IV then contain the basic results concerning discrete-time martingales, essentially the convergence and stopping theorems. Chapter III contains important applications, mainly to measure theory and Markov chains; Chapter V presents the various extensions of the notion of martingale (martingales with directed index set, vector-valued martingales, reversed martingales). Chapter VI is entirely devoted to an optimisation problem which has been of interest to numerous mathematicians over recent years. To finish up, Chapters VII and VIII present important theoretical results obtained in the last few years, and constitutes a first attempt at a synthesis of this material. Finally, a lengthy bibliography testifies to the intense research activity taking place in the theory of martingales; with some exceptions I have only cited those works on discrete-time martingales which have appeared since 1960.

I would like to thank my colleagues and students for their numerous remarks concerning the original text. I would also like to give particular thanks to Madame Baltzer and Mademoiselle Gillet for their excellent work in preparing the text for print. It is equally pleasant for me to be able to thank Dr. T. P. Speed for the care, patience and competence with which he carried out the translation and revision of the French text.

J. NEVEU

TABLE OF CONTENTS

CHAPTER I

PRELIMINARIES ON CONDITIONAL EXPECTATIONS

I-1. Sub-σ-fields of a probability space

Let $(\Omega, \mathscr{A}, P)$ be a probability space. The real vector space of equivalence classes of finite real-valued measurable functions defined on $(\Omega, \mathscr{A}, P)$ will be denoted by $L(\Omega, \mathscr{A}, P)$, or just by L if no confusion is possible. This vector space contains the classical Banach spaces

$$L^p(\Omega, \mathscr{A}, P) = L^p \qquad (1 \leqslant p \leqslant \infty).$$

A sub-σ-field $\mathscr{B}$ of $\mathscr{A}$ is said to be *complete* in the probability space $(\Omega, \mathscr{A}, P)$ if it contains all the $\mathscr{A}$-measurable subsets of Ω which have zero measure. For such a sub-σ-field $\mathscr{B}$, it is easy to verify that if two real-valued $\mathscr{A}$-measurable functions are equal a.s. (almost surely), then either both are $\mathscr{B}$-measurable or neither is $\mathscr{B}$-measurable. Consequently every equivalence class in L consists either of functions which are all $\mathscr{B}$-measurable, or of functions none of which is $\mathscr{B}$-measurable; in the former case we say that the equivalence class is $\mathscr{B}$-measurable. We will denote by $L(\mathscr{B})$ the subset of L formed by these $\mathscr{B}$-measurable equivalence classes; it is clear that $L(\mathscr{B})$ is a vector subspace of L.

If $\mathscr{B}$ is a sub-σ-field of $\mathscr{A}$ which is not complete in $(\Omega, \mathscr{A}, P)$, let us denote by $\tilde{\mathscr{B}}$ the complete σ-field which it generates, i.e. the sub-σ-field of $\mathscr{A}$ generated by $\mathscr{B}$ and the subsets of Ω in $\mathscr{A}$ of zero measure. We note that a real-valued $\mathscr{A}$-measurable function is $\tilde{\mathscr{B}}$-measurable if and only if there exists a $\mathscr{B}$-measurable function to which it is equal a.s. Consequently, if we agree to put $L(\mathscr{B}) = L(\tilde{\mathscr{B}})$, the space $L(\mathscr{B})$ consists of the equivalence classes in L containing *at least one* $\mathscr{B}$-measurable function.

More often than not in this section we will be considering complete sub-σ-fields of $\mathscr{A}$; on the other hand we will not find it natural to impose this condition in later sections.

We will denote by $L^p(\mathscr{B})$ the subspace $L(\mathscr{B}) \cap L^p$ of the Banach space L^p $(1 \leqslant p \leqslant \infty)$. Propositions **I-1**-1 and **I-1**-4 below provide intrinsic characterisations of the subspaces $L(\mathscr{B})$ and $L^p(\mathscr{B})$ of L and L^p respectively.

If as above we restrict consideration to complete sub-σ-fields, then it is easy to show that for every real number $p \in [1, \infty]$,

$$L(\mathscr{B}_1) \subset L(\mathscr{B}_2) \Leftrightarrow \mathscr{B}_1 \subset \mathscr{B}_2 \Leftrightarrow L^p(\mathscr{B}_1) \subset L^p(\mathscr{B}_2),$$

$$L(\mathscr{B}_1) = L(\mathscr{B}_2) \Leftrightarrow \mathscr{B}_1 = \mathscr{B}_2 \Leftrightarrow L^p(\mathscr{B}_1) = L^p(\mathscr{B}_2).$$

PROPOSITION I-**1**-1. *The spaces $L(\mathscr{B})$ (resp. $L^p(\mathscr{B})$, $p \in [1, \infty]$ fixed) obtained as $\mathscr{B}$ varies over complete sub-σ-fields of $\mathscr{A}$ are the vector subspaces of L (resp. L^p) which are lattice-ordered, contain the constant function* 1 *and which are closed under monotone limits (more precisely, where $\lim_n f_n \in L(\mathscr{B})$ or $L^p(\mathscr{B})$ for all monotone sequences $(f_n, n \geqslant 1)$ in $L(\mathscr{B})$ or $L^p(\mathscr{B})$ whose pointwise limit $\lim_n f_n$ belongs to L or L^p).*

When $p \in [1, \infty[$, the spaces $L^p(\mathscr{B})$ are also the vector subspaces of L^p which are closed, lattice-ordered and which contain the constant function 1.

To see that the second part of the proposition is not valid for $p = \infty$, it is enough to take the space $C[0,1]$ of all real-valued continuous functions defined on the interval $[0,1]$ and consider it (after passing to the equivalence classes) as a subspace of the space L^∞ constructed from the interval $[0,1]$ equipped with Lebesgue measure.

PROOF. The spaces $L(\mathscr{B})$ and $L^p(\mathscr{B})$ clearly have the properties indicated in the proposition. (To show that $L^p(\mathscr{B})$ is closed in L^p, i.e. that the limit f of every sequence $(f_n, n \geqslant 1)$ in $L^p(\mathscr{B})$ which converges in L^p also belongs to $L^p(\mathscr{B})$, one extracts a subsequence $(f_{n_k}, k \geqslant 1)$ which converges a.s. to f from the sequence $(f_n, n \geqslant 1)$ and notes that f, as the a.s. limit of $\mathscr{B}$-measurable functions, is then $\mathscr{B}$-measurable.)

If M is a subspace of L or of L^p satisfying the conditions in the first part of the proposition, the class $\mathscr{B} = \{B : B \in \mathscr{A}, 1_B \in M\}$ is a complete sub-σ-field of $\mathscr{A}$, as can easily be verified using the identities

$$1_{B^c} = 1 - 1_B, \qquad 1_{B_1 \cup B_2} = \sup(1_{B_1}, 1_{B_2}), \qquad 1_{\lim_n \uparrow B_n} = \lim_n \uparrow 1_{B_n}.$$

Further the class of functions $\{1_B : B \in \mathscr{B}\}$ generates $L(\mathscr{B})$ or $L^p(\mathscr{B})$ by forming linear combinations and monotone limits; thus the space $L(\mathscr{B})$ or $L^p(\mathscr{B})$ is contained in M. Conversely, if $f \in M$, the identity

$$1_{\{f > 0\}} = \lim_n \uparrow \min(nf^+, 1)$$

shows that $\{f > 0\} \in \mathscr{B}$ and thus $\{f > a\} = \{f - a > 0\} \in \mathscr{B}$ for all $a \in \mathbf{R}$; this shows that every $f \in M$ is $\mathscr{B}$-measurable and as a consequence $M \subset L(\mathscr{B})$ or $L^p(\mathscr{B})$.

To complete the proof of the proposition it only remains to remark that every closed subset of L^p is closed under monotone limits in L^p; indeed every monotone sequence $(f_n, n \geq 1)$ in L^p whose pointwise limit f belongs to L^p necessarily converges in L^p to its pointwise limit, for by the dominated convergence theorem $\int |f_n - f|^p \downarrow 0$. ■

Of the consequences of the preceding proposition, we begin with the following two corollaries:

Corollary I-1-2. *Let $p \in [1, \infty[$ be fixed. For every positive linear contraction (= of norm ≤ 1) U defined on the space L^p such that $U1 = 1$, the subspace $\{f : Uf = f\}$ of L^p formed by the invariant functions is of the form $L^p(\mathscr{B})$ for a complete sub-σ-field $\mathscr{B}$ of $\mathscr{A}$.*

Proof. The set $M = \{f : Uf = f\}$ is clearly a closed vector subspace of L^p which contains 1. Since U is a positive operator we have $U(f^+) \geq Uf$ and also $U(f^+) \geq 0$ for all $f \in L^p$; put in another way, the inequality $U(f^+) \geq (Uf)^+$ is satisfied for all $f \in L^p$. For all $f \in M$ we thus have $U(f^+) \geq f^+ \geq 0$; but since U is a contraction, this is only possible if $U(f^+) = f^+$, i.e. if $f^+ \in M$. This is enough to establish that M is lattice-ordered since for example

$$\sup(f_1, f_2) = f_1 + (f_2 - f_1)^+;$$

the preceding proposition then gives the result. ■

The second corollary below of Proposition I-1-1 could have an arbitrary increasing directed family of sub-σ-fields in its statement; its proof would be the same.

Corollary I-1-3. *Let $(\mathscr{B}_n, n \in \mathbf{N})$ be an increasing sequence of sub-σ-fields of $\mathscr{A}$ in the probability space $(\Omega, \mathscr{A}, P)$ and let $\mathscr{B}_\infty$ denote the σ-field generated by the sequence: $\mathscr{B}_\infty = \sigma(\bigcup_{\mathbf{N}} \mathscr{B}_n)$. Then for all finite real numbers $p \in [1, \infty[$, we have*

$$L^p(\mathscr{B}_\infty) = \overline{\bigcup_{\mathbf{N}} L^p(\mathscr{B}_n)}$$

in the Banach space L^p.

Proof. The fact that the sequence $(\mathscr{B}_n, n \in \mathbf{N})$ is increasing implies that $\bigcup_{\mathbf{N}} L^p(\mathscr{B}_n)$ is a vector subspace of the space L^p which contains the constant function 1. The closure $\overline{\bigcup_{\mathbf{N}} L^p(\mathscr{B}_n)}$ then also exhibits these properties; in fact this follows quite easily from the continuity of the mappings $f \to af$, $(f, g) \to f + g$ and $(f, g) \to f \vee g$, $(f, g) \to f \wedge g$. [For this we recall the inequality

$$\|f' \vee g' - f \vee g\|_p \leq \|f' - f\|_p + \|g' - g\|_p$$

valid if $f, f', g, g' \in L^p$, which follows from the analogous inequality for absolute values.]

By Proposition I-1-1 there exists a complete sub-σ-field $\mathscr{B}$ of $\mathscr{A}$ such that $\overline{\bigcup_{\mathbf{N}} L^p(\mathscr{B}_n)} = L^p(\mathscr{B})$. But it is then not hard to see that $L^p(\mathscr{B}) = L^p(\mathscr{B}_\infty)$. For on the one hand the inclusion $L^p(\mathscr{B}_n) \subset L^p(\mathscr{B})$, valid for all $n \in \mathbf{N}$, implies that $\mathscr{B}_n \subset \tilde{\mathscr{B}}_n \subset \mathscr{B}$ since $\mathscr{B}$ is complete; thus $\mathscr{B}_\infty \subset \mathscr{B}$ and hence $L^p(\mathscr{B}_\infty) \subset L^p(\mathscr{B})$. On the other hand, the inclusion $\mathscr{B}_n \subset \mathscr{B}_\infty$ implies that $L^p(\mathscr{B}_n) \subset L^p(\mathscr{B}_\infty)$ ($n \in \mathbf{N}$); thus

$$L^p(\mathscr{B}) = \overline{\bigcup_{\mathbf{N}} L^p(\mathscr{B}_n)} \subset L^p(\mathscr{B}_\infty).$$

(In terms of the σ-fields themselves, this result shows that $\mathscr{B}$ coincides with the completion $\tilde{\mathscr{B}}_\infty$ of $\mathscr{B}_\infty$.) ■

A result analogous to that in the preceding corollary holds for *decreasing* sequences of sub-σ-fields of $\mathscr{A}$. Its statement is, however, easier, and this time the proof is immediate from the definitions: for all $p \in [1, \infty]$, we have

$$\bigcap_{\mathbf{N}} L^p(\mathscr{B}_n) = L^p\Big(\bigcap_{\mathbf{N}} \mathscr{B}_n\Big)$$

if $(\mathscr{B}_n, n \in \mathbf{N})$ is decreasing sequence of sub-σ-fields of $\mathscr{A}$.

The following proposition provides a second characterisation of the subspaces of L^p of the form $L^p(\mathscr{B})$ $(1 \leq p < \infty)$, this one being based upon the algebraic structure of these spaces rather than on their lattice structure. The statement of the proposition is somewhat more complicated than that of the second part of Proposition I-1-1, since the L^p spaces $(1 \leq p < \infty)$ are not algebras!

PROPOSITION I-1-4. *Let p be a real number in $[1, \infty[$ and M a closed vector subspace of L^p containing the constants. If this space M contains a subalgebra Λ of L^∞ which is dense in M in the sense of L^p, then $M = L^p(\mathscr{B})$ for some complete sub-σ-field $\mathscr{B}$ of $\mathscr{A}$.*

Clearly the converse is true: for all complete sub-σ-fields $\mathscr{B}$ of $\mathscr{A}$, the space $L^p(\mathscr{B})$ contains the subalgebra $L^\infty(\mathscr{B})$ of L^∞ which is dense in $L^p(\mathscr{B})$ in the sense of L^p. On the other hand we note that the preceding proposition is false for $p = +\infty$: in general there exist L^∞-closed subalgebras of L^∞ which are not of the form $L^\infty(\mathscr{B})$. One such example is the algebra $C[0,1]$ of all real-valued continuous functions on $[0,1]$ embedded in the algebra L^∞ constructed from the interval $[0,1]$ equipped with Lebesgue measure.

We will return to the proof of Proposition I-1-4 after that of the following lemma:

LEMMA I-1-5. *Let Λ be a subalgebra of L^∞ containing the constants and let $\mathscr{B}$ be the complete sub-σ-field of $\mathscr{A}$ generated by Λ. Then for all $p \in [1, \infty[$ the closure $\bar{\Lambda}$ of Λ in L^p coincides with $L^p(\mathscr{B})$.*

PROOF. We fix p and show that $\bar{\Lambda}$ is lattice-ordered in L^p. Weierstrass's theorem implies that for any element $f \in \Lambda$ there is a sequence $(P_n, n \in \mathbf{N})$ of polynomials converging uniformly to the function x^+ on the bounded interval $\{x : |x| \leqslant \|f\|_\infty\}$. Since Λ is an algebra containing the constants and f, the sequence $(P_n(f), n \in \mathbf{N})$ is contained in Λ; further this sequence converges uniformly and hence in L^p to f^+. We have thus shown that $f^+ \in \bar{\Lambda}$ if $f \in \Lambda$; the continuity of the mapping $f \to f^+$ on L^p implies that $f^+ \in \bar{\Lambda}$ whenever $f \in \bar{\Lambda}$, and this shows that the vector space $\bar{\Lambda}$ is lattice-ordered.

Proposition I-1-1 shows that $\bar{\Lambda} = L^p(\mathscr{B}')$ for a complete sub-σ-field $\mathscr{B}'$ of $\mathscr{A}$. Although not needed for the proof of Proposition I-1-4, it is not hard to show that $\mathscr{B} = \mathscr{B}'$. In fact, since $\Lambda \subset L^p(\mathscr{B})$, it is clear that $L^p(\mathscr{B}') = \bar{\Lambda} \subset L^p(\mathscr{B})$ and thus that $\mathscr{B}' \subset \mathscr{B}$; conversely, since $\Lambda \subset L^p(\mathscr{B}')$, the functions in the equivalence classes of Λ are $\mathscr{B}'$-measurable, which implies that $\mathscr{B} \subset \mathscr{B}'$.

The lemma is thus proved.

We now prove the proposition. If the subalgebra Λ of L^∞ in the statement of the proposition contains the constants, an easy application of the lemma shows that the space $M = \bar{\Lambda}$ coincides with $L^p(\mathscr{B})$. If Λ does not contain the constants, consider the vector space Λ_1 generated by Λ and the function 1; it is also a subalgebra of L^∞, dense in M in the sense of L^p. We are thus back to the preceding case. ■

Lemma I-1-5 which we used to prove Proposition I-1-4 above is of independent interest. For example it implies that if $\mathscr{B}$ is the σ-field generated by a finite sequence $(f_0, f_1, \ldots, f_n)$ in L^∞, then the subalgebra of L^∞ formed by the polynomials $P(f_1, f_1, \ldots, f_n)$ of $f_0, f_1, \ldots f_n$ is a dense subspace of $L^p(\mathscr{B})$ for all $p \in [1, \infty[$. By considering the algebras of polynomials in a finite number of elements this result extends easily to infinite families in L^p.

I-2. Conditional expectations

We commence by studying conditional expectations on L^2 spaces; then we will extend them to all positive or integrable random variables.

Definition I-2-6. The orthogonal projection of the Hilbert space L^2 onto its closed vector subspace $L^2(\mathscr{B})$ is called the *conditional expectation with respect to the sub-σ-field $\mathscr{B}$ of $\mathscr{A}$* and is denoted by $E^{\mathscr{B}}$.

By the definition of an orthogonal projection, the conditional expectation $E^{\mathscr{B}}(f)$ of every $f \in L^2$ is characterised by the pair of properties:

(a) $E^{\mathscr{B}}(f) \in L^2(\mathscr{B})$,

(b) $\int E^{\mathscr{B}}(f) g \, dP = \int fg \, dP$ for all $g \in L^2(\mathscr{B})$.

Moreover, the equality (b) holds for all $g \in L^2(\mathscr{B})$ if it holds for all g belonging to a generating subset of $L^2(\mathscr{B})$; for example, $E^{\mathscr{B}}(f)$ is already the unique element of $L^2(\mathscr{B})$ such that

$$\int_B E^{\mathscr{B}}(f) dP = \int_B f dP \quad \text{for all } B \in \mathscr{B}.$$

The following proposition provides a simple characterisation of conditional expectations from amongst orthogonal projections.

PROPOSITION I-2-7. *For an orthogonal projection defined on the Hilbert space L^2 to be a conditional expectation, it is necessary and sufficient that it be positive and leave the constant function* 1 *invariant.*

PROOF. It is clear that $E^{\mathscr{B}}(1) = 1$. To show that $E^{\mathscr{B}}(f)$ is positive if f is, apply identity (b) above to the function

$$g = 1_{\{E^{\mathscr{B}}(f) < 0\}} \in L^2(\mathscr{B}),$$

we obtain

$$0 \leqslant \int f 1_{\{E^{\mathscr{B}}(f) < 0\}} dP = \int_{\{E^{\mathscr{B}}(f) < 0\}} E^{\mathscr{B}}(f) dP \leqslant 0.$$

The two integrals are thus zero; but then $\{E^{\mathscr{B}}(f) < 0\}$ is null and we have $E^{\mathscr{B}}(f) \geqslant 0$ in L^2.

Conversely, if U is a positive orthogonal projection on L^2 such that $U1 = 1$, Corollary I-1-2 shows that the space $\{f : Uf = f\}$ is of the form $L^2(\mathscr{B})$ for a complete sub-σ-field $\mathscr{B}$ of $\mathscr{A}$; but since U is the orthogonal projection onto $\{f : Uf = f\}$, we have $U = E^{\mathscr{B}}$. ■

REMARK. On a Hilbert space the orthogonal projections are exactly the idempotent ($U^2 = U$) linear operators of norm $\leqslant 1$. The preceding proposition thus allows us to characterise conditional expectations on a space $L^2(\Omega, \mathscr{A}, P)$ as those positive, idempotent linear operators of norm $\leqslant 1$ preserving the constants. ■

The following property of conditional expectations is equally fundamental; it is intimately connected with the property of $L^2(\mathscr{B})$ of being closed under multiplication by functions in $L^\infty(\mathscr{B})$.

PROPOSITION I-2-8. *The conditional expectation has the following "averaging property": For all* $f \in L^2$,

$$E^{\mathscr{B}}(hf) = hE^{\mathscr{B}}(f) \quad \text{if } h \in L^\infty(\mathscr{B}).$$

(The hypothesis of $\mathscr{B}$-measurability on h is essential here.)

PROOF. It suffices to show that $hE^{\mathscr{B}}(f)$ has the properties which characterise the conditional expectation $E^{\mathscr{B}}(hf)$ of hf. But $hE^{\mathscr{B}}(f) \in L^2(\mathscr{B})$ as it is a product of $\mathscr{B}$-measurable functions belonging to L^∞ and L^2 respectively. On the other hand, for all $g \in L^2(\mathscr{B})$ the function $hg \in L^2(\mathscr{B})$ for the same reason, and the definition of $E^{\mathscr{B}}(f)$ implies that

$$\int E^{\mathscr{B}}(f)hg \, \mathrm{d}P = \int fhg \, \mathrm{d}P;$$

in other words, $E^{\mathscr{B}}(f)h - fh$ is orthogonal to $L^2(\mathscr{B})$, and thus we have proved that $E^{\mathscr{B}}(fh) = E^{\mathscr{B}}(f)h$. ■

By virtue of their positivity, conditional expectations defined on the space L^2 can be extended to continuous linear operators on L^1. This result is formulated in Proposition I-2-10 below; we will be basing our proof on the following preliminary result.

LEMMA I-2-9. *Let* $\bar{L}_+$ *be the space of equivalence classes of measurable mappings of the probability space* $(\Omega, \mathscr{A}, P)$ *into* $\bar{\mathbf{R}}_+ = [0, \infty]$ *and, being given a sub-σ-field* $\mathscr{B}$ *of* $\mathscr{A}$, *let* $\bar{L}_+(\mathscr{B})$ *denote the subset of* $\mathscr{B}$*-measurable equivalence classes in* $\bar{L}_+$. *For all* $f \in \bar{L}_+$ *there exists a unique element* $E^{\mathscr{B}}(f)$ *in* $\bar{L}_+(\mathscr{B})$ *such that*

$$\int E^{\mathscr{B}}(f)\, g \, \mathrm{d}P = \int fg \, \mathrm{d}P$$

for all $g \in \bar{L}_+(\mathscr{B})$; *further,* $E^{\mathscr{B}}(f)$ *is already the unique element in* $\bar{L}_+(\mathscr{B})$ *such that*

$$\int_B E^{\mathscr{B}}(f) \, \mathrm{d}P = \int_B f \, \mathrm{d}P$$

for all $B \in \mathscr{B}$.

Also, for all $h \in \bar{L}_+(\mathscr{B})$ *we have*

$$E^{\mathscr{B}}(hf) = hE^{\mathscr{B}}(f) \quad \text{if } f \in \bar{L}_+.$$

It is not generally true that $E^{\mathscr{B}}(f)$ is finite if f is; this is why it is necessary to allow functions in the preceding lemma to take the value $+\infty$.

PROOF. If $f \in L^2_+$, the element $E^{\mathscr{B}}(f)$ of $L^2_+(\mathscr{B})$ defined in Definition I-2-6 and positive by Proposition I-2-7 satisfies the equality in the lemma for all $g \in L^2_+(\mathscr{B})$, hence also for all $g \in \bar{L}_+(\mathscr{B})$, as we can see by approximating g by the increasing sequence $(g \wedge n1,\ n \in \mathbf{N})$ of elements of $L^2_+(\mathscr{B})$. If f is an element of $\bar{L}_+$ which does not belong to L^2_+, the sequence $E^{\mathscr{B}}(f \wedge n1)$ is well-defined since $f \wedge n1 \in L^2_+$ $(n \in \mathbf{N})$ and increases in $L^2_+(\mathscr{B})$ with n since $E^{\mathscr{B}}$ is a positive linear, and hence monotone, operator on L^2. The pointwise limit of this increasing sequence, which we will denote by $E^{\mathscr{B}}(f)$, then satisfies the equality of the lemma for all $g \in \bar{L}_+(\mathscr{B})$, as can be seen by allowing $n \uparrow \infty$ in the equality

$$\int E^{\mathscr{B}}(f \wedge n1)g\,\mathrm{d}P = \int (f \wedge n1)g\,\mathrm{d}P \qquad (g \in \bar{L}_+(\mathscr{B})).$$

To establish the uniqueness of $E^{\mathscr{B}}(f)$, we remark that if h_1 and h_2 are two elements of $\bar{L}_+(\mathscr{B})$ such that $\int_B h_i \mathrm{d}P = \int_B f\,\mathrm{d}P$ $(i = 1, 2)$ for all $B \in \mathscr{B}$, then in particular we have

$$\int_{\{h_1 \leqslant a < b \leqslant h_2\}} h_1\,\mathrm{d}P = \int_{\{h_1 \leqslant a < b \leqslant h_2\}} h_2\,\mathrm{d}P$$

for all pairs a, b of real numbers with $0 \leqslant a < b$. But this equality is only possible if $P(h_1 \leqslant a < b \leqslant h_2) = 0$, because the left-hand side is bounded above by a times this probability and the right-hand side is bounded below by b times this probability. Since the set $\{h_1 < h_2\}$ is a *countable* union of sets $\{h_1 \leqslant a < b \leqslant h_2\}$ obtained as a and b vary over the rationals with $0 \leqslant a < b$, we see that $P(h_1 < h_2) = 0$. By symmetry we also have $P(h_1 > h_2) = 0$ and thus we have proved $h_1 = h_2$ a.s.

Finally, the product $hE^{\mathscr{B}}(f)$ belongs to $\bar{L}_+(\mathscr{B})$ for all $h \in \bar{L}_+(\mathscr{B})$ and all $f \in \bar{L}_+$ and, by virtue of the properties of $E^{\mathscr{B}}(f)$, it satisfies the identity

$$\int hE^{\mathscr{B}}(f)g\,\mathrm{d}P = \int hfg\,\mathrm{d}P \qquad (g \in \bar{L}_+(\mathscr{B}))$$

since $hg \in \bar{L}_+(\mathscr{B})$. By the uniqueness result we have just proved, this is only possible if

$$hE^{\mathscr{B}}(f) = E^{\mathscr{B}}(hf).$$

The final part of the proposition is thus proved. ■

From the definition of conditional expectation of positive functions and properties of the integral of such functions, it is easy to deduce the following:

COROLLARY I-2-10. *If $\mathscr{B}$ is a sub-σ-field of $\mathscr{A}$ in $(\Omega, \mathscr{A}, P)$ and if $(f_n, n \in \mathbf{N})$ is a sequence in $\bar{L}_+$, then:*

(a) *if the sequence is increasing,*

$$E^{\mathscr{B}}(\lim_n \uparrow f_n) = \lim_n \uparrow E^{\mathscr{B}}(f_n);$$

(b) *whatever the sequence,*

$$E^{\mathscr{B}}(\sum_{\mathbf{N}} f_n) = \sum_{\mathbf{N}} E^{\mathscr{B}}(f_n),$$

(c) *whatever the sequence,*

$$E^{\mathscr{B}}(\liminf_n f_n) \leqslant \liminf_n E^{\mathscr{B}}(f_n).$$

PROPOSITION I-2-11. *Let $\mathscr{B}$ be a sub-σ-field of $\mathscr{A}$ in the probability space $(\Omega, \mathscr{A}, P)$. For all $f \in L^1$ there exists a unique element $E^{\mathscr{B}}(f)$ in $L^1(\mathscr{B})$ such that*

$$\int E^{\mathscr{B}}(f) g \, \mathrm{d}P = \int fg \, \mathrm{d}P \quad \textit{for all } g \in L^\infty(\mathscr{B}).$$

The conditional expectation $E^{\mathscr{B}}$ thus defined on L^1 is a positive, idempotent, linear contraction such that $E^{\mathscr{B}}(1) = 1$. Further, for all real-valued $\mathscr{B}$-measurable functions h the identity

$$E^{\mathscr{B}}(hf) = hE^{\mathscr{B}}(f)$$

is valid whenever f and hf are integrable.

PROOF. For all $f \in L^1_+$ there exists, by Lemma I-2-9, an element $E^{\mathscr{B}}(f)$ of $\bar{L}_+(\mathscr{B})$ such that $\int E^{\mathscr{B}}(f) g \, \mathrm{d}P = \int fg \, \mathrm{d}P$ for all $g \in L^\infty_+(\mathscr{B})$. With $g = 1$, this equality shows that $E^{\mathscr{B}}(f)$ is integrable if f is, and the equality can then be extended to all $g \in L^\infty(\mathscr{B})$. Next, putting

$$E^{\mathscr{B}}(f) = E^{\mathscr{B}}(f^+) - E^{\mathscr{B}}(f^-)$$

for every $f \in L^1$, we define an element of $L^1(\mathscr{B})$ which also satisfies the above identity for all $g \in L^\infty(\mathscr{B})$. On the other hand, a standard argument analogous to that used at the end of Lemma I-2-9 shows that for all $f \in L^1$ the conditional expectation $E^{\mathscr{B}}(f)$ is already the unique element of $L^1(\mathscr{B})$ such that

$$\int_B E^{\mathscr{B}}(f) \, \mathrm{d}P = \int_B f \, \mathrm{d}P \quad \text{for all } B \in \mathscr{B}.$$

The above characterisation of $E^{\mathscr{B}}(f)$ implies that the mapping $f \to E^{\mathscr{B}}(f)$ from L^1 to $L^1(\mathscr{B})$ is linear; this mapping is also positive and such that $E^{\mathscr{B}}(1) = 1$. It contracts the norm on L^1 because the inequality

$$|E^{\mathscr{B}}(f)| \leqslant E^{\mathscr{B}}(f^+) + E^{\mathscr{B}}(f^-) = E^{\mathscr{B}}(|f|),$$

which follows from the definitions, implies that

$$\|E^{\mathscr{B}}(f)\|_1 = \int |E^{\mathscr{B}}(f)| \, \mathrm{d}P \leqslant \int E^{\mathscr{B}}(|f|) \, \mathrm{d}P = \int |f| \, \mathrm{d}P = \|f\|_1.$$

Finally, it is clear that $E^{\mathscr{B}}(f) = f$ if $f \in L^1(\mathscr{B})$; the operator $E^{\mathscr{B}}$ which takes its values in $L^1(\mathscr{B})$ is thus idempotent.

Lemma I-2-9 tells us that if h is $\mathscr{B}$-measurable,

$$E^{\mathscr{B}}(hf) = hE^{\mathscr{B}}(f)$$

whenever h and f are positive. Using the standard decomposition $f = f^+ - f^-$ and $h = h^+ - h^-$, it is then easy to see that this equality is also valid whenever f and hf are integrable. ■

The property "$E^{\mathscr{B}}(hf) = hE^{\mathscr{B}}(f)$ if h is $\mathscr{B}$-measurable" of conditional expectations allows us to "relativise" this expectation to every set $B \in \mathscr{B}$. Explicitly, if f_1 and f_2 are two positive or integrable real-valued functions, and if $f_1 = f_2$ on a set $B \in \mathscr{B}$, then $E^{\mathscr{B}}(f_1) = E^{\mathscr{B}}(f_2)$ on B for

$$1_B E^{\mathscr{B}}(f_1) = E^{\mathscr{B}}(1_B f_1) = E^{\mathscr{B}}(1_B f_2) = 1_B E^{\mathscr{B}}(f_2).$$

Consequently, if f is a real-valued measurable function defined only on a set $B \in \mathscr{B}$ and positive or integrable on B, the conditional expectation $E^{\mathscr{B}}(f)$ of f can be defined on B to be the common value on B of all the conditional expectations $E^{\mathscr{B}}(f^*)$ of functions f^*, positive or integrable, coinciding with f on B ($f_0 = f$ on B, $f_0 = 0$ on B^c gives one such function). The conditional expectation $E^{\mathscr{B}}$ thus relativised to a set B ($B \in \mathscr{B}$) clearly retains all the properties of this expectation!

The following inequality of Hölder allows us to study conditional expectations as operators on L^p spaces ($1 < p < \infty$).

Hölder's inequality. *Let p and q be two real numbers in $]1, \infty[$ conjugate in the sense that $p^{-1} + q^{-1} = 1$. The inequality*

$$|E^{\mathscr{B}}(fg)| \leqslant [E^{\mathscr{B}}(|f|^p)]^{1/p} [E^{\mathscr{B}}(|g|^q)]^{1/q}$$

is then true on Ω for all $f \in L^p$ and all $g \in L^q$.

Proof. We will use the following elementary inequality

$$xy \leqslant \frac{1}{p}x^p + \frac{1}{q}y^q \qquad (x, y \in \mathbf{R}_+),$$

which can be proved by calculating the maximum of the convex function $x \to xy - p^{-1}x^p$ defined on $\mathbf{R}_+$, for each fixed $y \in \mathbf{R}_+$. Further, the preceding inequality is strict for all pairs x, y such that $y \neq x^{p-1}$ ($\Leftrightarrow x \neq y^{q-1}$).

If f and g belong to L^p and L^q respectively, the functions $|f|^p$ and $|g|^q$ are integrable; then the function $|fg|$ is also, since it is dominated by $p^{-1}|f|^p + q^{-1}|g|^q$ using the preceding inequality. The conditional expectations $E^{\mathscr{B}}(fg)$, $E^{\mathscr{B}}(|f|^p)$ and $E^{\mathscr{B}}(|g|^q)$ are thus well-defined in L^1. The inequality above allows us to write

$$\frac{|f|}{[E^{\mathscr{B}}(|f|^p)]^{1/p}}\frac{|g|}{[E^{\mathscr{B}}(|g|^q)]^{1/q}} \leqslant \frac{|f|^p}{pE^{\mathscr{B}}(|f|^p)} + \frac{|g|^q}{qE^{\mathscr{B}}(|g|^q)}$$

at least on the set $B = \{E^{\mathscr{B}}(|f|^p) > 0, E^{\mathscr{B}}(|g|^q) > 0\}$. Since the set B belongs to the σ-field $\mathscr{B}$, we may take conditional expectations of both terms, finding

$$\frac{E^{\mathscr{B}}(|fg|)}{[E^{\mathscr{B}}(|f|^p)]^{1/p}[E^{\mathscr{B}}(|g|^q)]^{1/q}} \leqslant \frac{1}{p} + \frac{1}{q} = 1 \quad \text{on } B.$$

This proves Hölder's inequality on B since $|E^{\mathscr{B}}(fg)| \leqslant E^{\mathscr{B}}(|fg|)$. But the function f is zero on the set $B_f = \{E^{\mathscr{B}}(|f|^p) = 0\}$ which belongs to $\mathscr{B}$, since

$$\int_{B_f} |f|^p \, \mathrm{d}P = \int_{B_f} E^{\mathscr{B}}(|f|^p) \, \mathrm{d}P = 0;$$

we thus have $E^{\mathscr{B}}(fg) = 0$ on B_f since $fg = 0$ on B_f and $B_f \in \mathscr{B}$. Hölder's inequality reduces to $0 \leqslant 0$ on B_f; it is the same for B_g and since $B \cup B_f \cup B_g = \Omega$, Hölder's inequality is proved. ■

We will need a particular case of Hölder's inequality to prove the following proposition, which generalises the results for L^2 and L^1 contained in Propositions I-2-7 and I-2-11 to all the L^p spaces ($1 \leqslant p \leqslant \infty$).

PROPOSITION I-**2**-12. *Let p be a real number in* $[1, \infty]$ *and $\mathscr{B}$ a sub-σ-algebra of $\mathscr{A}$ in the probability space* $(\Omega, \mathscr{A}, P)$. *Then for all $f \in L^p$ the conditional expectation $E^{\mathscr{B}}(f)$ also belongs to L^p. The operator $E^{\mathscr{B}}$ on L^p is positive, idempotent and a linear contraction such that* $E^{\mathscr{B}}(1) = 1$. *It maps L^p onto $L^p(\mathscr{B})$ and satisfies the identity $E^{\mathscr{B}}(hf) = hE^{\mathscr{B}}(f)$ for all $f \in L^p$ and $h \in L^q(\mathscr{B})$, where q denotes the real number conjugate to p defined by $p^{-1} + q^{-1} = 1$.*

PROOF. Suppose initially that $p < \infty$. Since $L^p \subset L^1$, the conditional expectation $E^{\mathscr{B}}(f)$ as an element of L^1 is defined by Proposition I-2-11 for all $f \in L^p$. But Hölder's inequality applied to f and the constant function 1 shows that $|E^{\mathscr{B}}(f)|^p \leqslant E^{\mathscr{B}}(|f|^p)$; this implies that

$$\int |E^{\mathscr{B}}(f)|^p \, \mathrm{d}P \leqslant \int E^{\mathscr{B}}(|f|^p) \, \mathrm{d}P = \int |f|^p \, \mathrm{d}P.$$

Hence $E^{\mathscr{B}}(f) \in L^p$ and, more precisely, we have $\|E^{\mathscr{B}}(f)\|_p \leqslant \|f\|_p$.

In the case $p = \infty$ we can write

$$|E^{\mathscr{B}}(f)| \leqslant E^{\mathscr{B}}(|f|) \leqslant \|f\|_\infty E^{\mathscr{B}}(1) = \|f\|_\infty \quad \text{on } \Omega,$$

for all $f \in L^\infty$, and this shows that $\|E^{\mathscr{B}}(f)\|_\infty \leqslant \|f\|_\infty$.

The other properties in the proposition are immediate. We only note that in connection with the last property the two functions f and hf are integrable since $f \in L^p$ and $h \in L^q$. ∎

The remainder of this chapter is devoted to the study of two converses to the preceding proposition.

On every L^p space $(1 \leqslant p \leqslant \infty)$, every conditional expectation $E^{\mathscr{B}}$ defines a continuous linear operator with the following algebraic property:

$$T(f \cdot Tg) = T(f)T(g) \quad \text{if } f \in L^\infty \text{ and } g \in L^p.$$

In fact, $E^{\mathscr{B}}[L^p] = L^p(\mathscr{B})$ and the preceding identity is nothing more than a variant of the previously proved identity $E^{\mathscr{B}}(f \cdot h) = E^{\mathscr{B}}(f) \cdot h \, (f \in L^\infty, h \in L^p(\mathscr{B}))$. The following proposition shows that if $p < \infty$, the only operators having the preceding algebraic property and preserving the integral are conditional expectations; this characterisation of conditional expectations is indeed quite remarkable.

PROPOSITION I-2-13. *Let $p \geqslant 1$ be a finite real number. Every continuous linear operator T defined on L^p such that $\int Tf \, \mathrm{d}P = \int f \, \mathrm{d}P$ for all $f \in L^p$, satisfying the algebraic property*

$$T(f \cdot Tg) = T(f)T(g) \qquad (f \in L^\infty, g \in L^p)$$

is necessarily a conditional expectation $E^{\mathscr{B}}$ with respect to a complete sub-σ-field $\mathscr{B}$ of $\mathscr{A}$.

PROOF. (a) We begin by proving that $T(L^\infty) \subset L^\infty$. To this end, let us associate with every $f \in L^\infty$ a sequence $(f_n, n \geqslant 1)$ in L^p defined inductively by

$$f_1 = f, \qquad f_{n+1} = f \cdot T(f_n) \quad \text{if } n \geqslant 1.$$

It is clear that f_n belongs to L^p. The algebraic property of T implies that

$$Tf_{n+1} = T[f \cdot Tf_n] = T(f)\,T(f_n)$$

for all $n \geqslant 1$; hence $Tf_n = (Tf)^n$. All integral powers of Tf thus belong to L^p; it follows that $Tf \in L^r$ for all $r < \infty$. Further, the equality $Tf_{n+1} = T(f \cdot Tf_n)$ implies that

$$\|Tf_{n+1}\|_p \leqslant \|T\|\,\|f \cdot Tf_n\|_p \leqslant \|T\|\,\|f\|_\infty \|Tf_n\|_p,$$

where $\|T\|$ denotes the norm of the operator T. Since

$$\|Tf_1\|_p \leqslant \|T\|\,\|f\|_p \leqslant \|T\|\,\|f\|_\infty,$$

we can deduce that

$$\|(Tf)^n\|_p = \|Tf_n\|_p \leqslant [\|T\|\,\|f\|_\infty]^n.$$

But $\|(Tf)^n\|_p = \|Tf\|_{np}^n$; hence we have shown that $\|Tf\|_{np} \leqslant \|T\|\,\|f\|_\infty$ for every integer $n \geqslant 1$. But it is well known and easy to prove that for every function $g \in \bigcap_{r<\infty} L^r$, the family of norms $\|g\|_r$ increases with r towards $\|g\|_\infty$ or towards $+\infty$ according as the function g belongs to L^∞ or not; in our case this implies that $Tf \in L^\infty$ and that $\|Tf\|_\infty \leqslant \|T\|\,\|f\|_\infty$. We have thus proved that $T(L^\infty) \subset L^\infty$ (and also that the restriction of T to L^∞ is a continuous linear operator of norm not exceeding the norm of T on L^p).

(b) For every continuous linear operator T on L^p it can be immediately checked that

$$\Lambda = \{h : h \in L^\infty, T(fh) = T(f)\,h \quad \text{for all } f \in L^\infty\}$$

defines a subalgebra of L^∞ containing the constants. By Lemma I-1-5, the closure $\bar{\Lambda}$ of Λ in L^p is then of the form $L^p(\mathscr{B})$ for some complete sub-σ-algebra $\mathscr{B}$ of $\mathscr{A}$. But it is not difficult to extend the equality of definition of the functions in Λ to the functions in $\bar{\Lambda}$; we then have

$$T(fh) = T(f)\,h \qquad \text{for all } f \in L^\infty, \text{ if } h \in \bar{\Lambda} = L^p(\mathscr{B}).$$

[In fact, if $h \in L^p(\mathscr{B})$, there exists at least one sequence $(h_n, n \in N)$ in Λ converging in L^p to h; by continuity, $T(fh_n)$ then converges in L^p to $T(fh)$. By passing to a subsequence, we can suppose that $h_n \to h$ a.s. and that $T(fh_n) \to T(fh)$ a.s. when $n \uparrow \infty$; the equalities $T(fh_n) = Tf \cdot h_n$ then evidently imply that in the limit $T(fh) = Tf \cdot h$.]

The equality which we have just proved and the integral-preserving property of T imply that

$$\int fh \, \mathrm{d}P = \int T(fh) \, \mathrm{d}P = \int T(f) h \, \mathrm{d}P \quad \text{for all } f \in L^\infty \tag{1}$$

if $h \in L^p(\mathscr{B})$. Consequently, $E^{\mathscr{B}}(f) = E^{\mathscr{B}}(Tf)$ for all $f \in L^\infty$.

The algebraic property of T implies that $Tg \in \Lambda$ for all $g \in L^\infty$ since by the first part of the proof, $Tg \in L^\infty$. The inclusion $\Lambda \subset L^p(\mathscr{B})$ thus implies that Tg is $\mathscr{B}$-measurable for all $g \in L^\infty$; in other words, $E^{\mathscr{B}}(Tg) = Tg$. Comparing this equality with that proved in the previous paragraph, we see that $Tf = E^{\mathscr{B}}(f)$ for all $f \in L^\infty$; this equality extends by continuity to all $f \in L^p$. ■

Remark. Using Hölder's inequality, it is not difficult to show that if $u \in L^\infty$ (resp. if $E^{\mathscr{B}}(|u|^q) \in L^\infty$), the formula $Tf = E^{\mathscr{B}}(uf)$ defines a continuous linear operator on L^1 (resp. on L^p) of norm

$$\|T\| = \|u\|_\infty \quad (\text{resp. } \|T\| = \|E^{\mathscr{B}}(|u|^q)\|_\infty^{1/q}).$$

This operator also satisfies the algebraic identity of the previous proposition; on the other hand, it only preserves the integral if $u = 1$. A minor modification of the preceding proof allows us to show that all continuous linear operators on L^p such that $T(f \cdot Tg) = Tf \cdot Tg$ are of the form $E^{\mathscr{B}}(uf)$ for a function $u \in L^\infty$ if $p = 1$ or for a function u such that $E^{\mathscr{B}}(|u|^q) \in L^\infty$ if $1 < p < \infty$. This modification only consists in putting $u = T^*1$ (where T^* denotes the operator on L^q adjoint to T) and in replacing the first equality in the formula (1) above by $\int ufh \, \mathrm{d}P = \int T(fh) \, \mathrm{d}P$, which shows that $E^{\mathscr{B}}(Tf) = E^{\mathscr{B}}(uf)$ if $f \in L^\infty$. ■

Here is the second characterisation of conditional expectations which we have announced; we will only give part of the proof.

Proposition I-2-14. *Let p be a real number in the interval $[1, \infty[$ distinct from 2. Every idempotent linear contraction T defined on a space $L^p(\Omega, \mathscr{A}, P)$ such that $T1 = 1$ is necessarily the conditional expectation with respect to a complete sub-σ-field of $\mathscr{A}$.*

The above proposition is false for $p = 2$: every orthogonal projection on to a closed vector subspace H of $L^2(\Omega, \mathscr{A}, P)$ containing the constants satisfies the hypotheses of the proposition but is not necessarily of the form $L^2(\mathscr{B})$. (If there exists a function h in L^2 taking more than two values, the space H generated by h and the constants is not of the form $L^2(\mathscr{B})$.)

PROOF. (1) The proposition becomes easy to prove if, in addition to the assumptions we have already made, we also suppose that the operator T is positive and preserves the integral (with these further hypotheses the proposition is also true for $p = 2$).

Indeed, if T is positive, Corollary I-**1**-2 implies that the subspace $\{f : Tf = f\}$ of L^p consisting of the T-invariant functions is of the form $L^p(\mathscr{B})$ for a complete sub-σ-field $\mathscr{B}$ of $\mathscr{A}$. For all $B \in \mathscr{B}$ and all measurable functions $f : \Omega \to [0, 1]$ we thus have

$$T(f 1_B) = T(f)\, 1_B,$$

because the inequalities

$$0 \leqslant T(f 1_B) \leqslant T 1_B = 1_B, \qquad 0 \leqslant T(f 1_{B^c}) \leqslant T 1_{B^c} = 1_{B^c},$$

which follow from the positivity of T, imply that $T(f 1_B) = 0$ on B^c and that $T(f 1_{B^c}) = 0$ on B; this clearly proves the preceding equality since we also have $T(f 1_B) + T(f 1_{B^c}) = Tf$.

The above identity, proved for all measurable functions with values in $[0, 1]$, easily extends to all $f \in L^p$. It implies that for all $B \in \mathscr{B}$ and all $f \in L^p$,

$$\int_B f\, dP = \int f 1_B\, dP = \int T(f 1_B)\, dP = \int Tf \cdot 1_B\, dP = \int_B T(f)\, dP$$

since T preserves the integral; we have thus proved that $E^{\mathscr{B}}(f) = E^{\mathscr{B}}(Tf)$ for all $f \in L^p$.

We have not yet used the hypothesis of idempotence of the operator T. This implies that $Tf \in L^p(\mathscr{B})$ for all $f \in L^p$ since $T(Tf) = Tf$; the identity proved in the preceding paragraph then simplifies to $E^{\mathscr{B}}(f) = Tf\ (f \in L^p)$ and the proposition is thus proved under the extra conditions stated.

(2) We suppose that $p = 1$ and show that the hypotheses of the proposition imply that the operator T is positive and preserves the integral; to this end, it suffices to prove that $Tf \geqslant 0$ and $\int Tf\, dP = \int f\, dP$ for every real-valued measurable function f with values in $[0, 1]$. Now we note that all real numbers a satisfy the elementary inequality

$$|a| + |1 - a| \geqslant 1$$

and this inequality is an equality if and only if $0 \leqslant a \leqslant 1$. For a measurable function $f : \Omega \to [0, 1]$ we can then write

$$\begin{aligned} \int (|Tf| + |1 - Tf|)\, dP &= \int (|Tf| + |T(1 - f)|)\, dP \\ &\leqslant \int (|f| + |1 - f|)\, dP = 1, \end{aligned}$$

where we have used the fact that T is a contraction and $T1 = 1$; since $|Tf| + |1 - Tf| \geqslant 1$ follows from the preceding, the inequality just obtained can only be satisfied if $|Tf| + |1 - Tf| = 1$ a.s., that is, if $0 \leqslant Tf \leqslant 1$ a.s. On the other hand, the inequality between integrals is an equality only if $\int |Tf| \mathrm{d}P = \int |f| \mathrm{d}P$, that is, if $\int Tf \mathrm{d}P = \int f \mathrm{d}P$, since f and Tf are positive. Our assertion is thus proved.

(3) For the proof in the general case, which is more complicated, the reader is referred to [4]. ■

I-3. Supplement: Conditional expectations with respect to a σ-finite measure

A positive measure μ on the measurable space $(\Omega, \mathscr{A})$ is said to be *σ-finite* if there exists a countable partition $(A_n, n \in \mathbf{N})$ of Ω in $\mathscr{A}$ such that $\mu(A_n) < \infty$ for all $n \in \mathbf{N}$ or, equivalently, if there exists a real-valued measurable function on Ω which is finite and strictly positive, say $f: \Omega \to]0, \infty[$, such that $\int f \mathrm{d}\mu < \infty$. Multiplying f by a constant if necessary, we can suppose that $\int f \mathrm{d}\mu = 1$; then $P = f \cdot \mu$ becomes a probability measure having the same null sets in $(\Omega, \mathscr{A})$ as μ, and we will have $\mu = f^{-1} \cdot P$ on $(\Omega, \mathscr{A})$.

If $\mathscr{B}$ is a sub-σ-field of $\mathscr{A}$, the restriction of the measure μ to $\mathscr{B}$ is not generally σ-finite (for example, on the sub-σ-field of Borel subsets of $\mathbf{R}$ which are periodic with period 1, Lebesgue measure only takes on the values 0 or $+\infty$). But as $\mu = E^{\mathscr{B}}(f^{-1}) \cdot P$ on $\mathscr{B}$, the equivalence class $B_\mu = \{E^{\mathscr{B}}(f^{-1}) < \infty\}$, where the conditional expectation is taken on $(\Omega, \mathscr{A}, P)$, exhibits the following two properties which characterise it:

(a) There exists a partition $(B_n, n \in \mathbf{N})$ of B_μ in $\mathscr{B}$ such that $\mu(B_n) < \infty$ for all $n \in \mathbf{N}$, whereas every set $B \in \mathscr{B}$ disjoint from B_μ has measure equal to 0 or $+\infty$ (for example, put $B_n = \{n \leqslant E^{\mathscr{B}}(f^{-1}) < n + 1\}$ for all $n \in \mathbf{N}$).

(b) Every real-valued, μ-integrable and $\mathscr{B}$-measurable function is a.s. zero outside B_μ; there exists at least one real-valued μ-integrable $\mathscr{B}$-measurable function which is strictly positive a.s. on B_μ.

We call B_μ the *set of σ-finiteness* of μ relative to the sub-σ-field $\mathscr{B}$. All the spaces $L^p(\mathscr{B}, \mu)$ $(1 \leqslant p < \infty)$ are formed from equivalence classes which are zero outside this set B_μ of σ-finiteness of μ relative to $\mathscr{B}$.

Let us denote by $E_\mu^{\mathscr{B}}$ the orthogonal projection of the space $L^2(\Omega, \mathscr{A}, \mu)$ on to its closed linear subspace $L^2(\mathscr{B}, \mu)$. It is not difficult to show that this projection can be expressed in terms of the conditional expectation $E^{\mathscr{B}}$ (taken relative to P) by the formula

$$E_\mu^{\mathscr{B}}(g) = \begin{cases} \dfrac{E^{\mathscr{B}}(f^{-1}g)}{E^{\mathscr{B}}(f^{-1})} & \text{on } B_\mu, \\ 0 & \text{outside } B_\mu. \end{cases}$$

(Note that on B_μ we have $E^{\mathscr{B}}(f^{-1}) > 0$ a.s. and $E^{\mathscr{B}}(f^{-1}|g|) < \infty$ a.s. if $g \in L^2(\mu)$.) The preceding formula can be used to extend the domain of definition of $E_\mu^{\mathscr{B}}$ to positive functions and then to all functions in

$$\bigcup_{1 \leqslant p < \infty} L^p(\Omega, \mathscr{A}, \mu);$$

for example for all positive measurable functions g, the function $E_\mu^{\mathscr{B}}(g)$ is the unique $\mathscr{B}$-measurable function zero outside B_μ such that

$$\int_B E_\mu^{\mathscr{B}}(g)\,\mathrm{d}\mu = \int_B g\,\mathrm{d}\mu \quad \text{for all } B \in \mathscr{B},\ B \subset B_\mu.$$

[In this definition it is not possible to omit the set B_μ!] On the other hand, it is easy to prove that the operator $E_\mu^{\mathscr{B}}$ on $L^p(\Omega, \mathscr{A}, \mu)$ is linear, positive, idempotent, a contraction, and maps L^p onto $L^p(\mathscr{B}, \mu)$.

CHAPTER II

POSITIVE MARTINGALES AND SUPERMARTINGALES

Probability theory begins with the definition of probability spaces. However, as soon as one studies random phenomena evolving in time, it is also important to introduce the σ-fields of events prior to various possible times, for these σ-fields are basic to all of the important later definitions. In this book we will be restricting ourselves to the case of discrete time represented by the half-line $\mathbf{N}$ of non-negative integers. We are thus led to the following definitions.

In all that follows we will denote by $(\Omega, \mathscr{A}, P)$ a probability space and by $(\mathscr{B}_n, n \in \mathbf{N})$ an increasing sequence of sub-σ-fields of $\mathscr{A}$ which we will fix once and for all. We will denote by $\mathscr{B}_\infty = \bigvee_{\mathbf{N}} \mathscr{B}_n$ the sub-σ-field of $\mathscr{A}$ generated by the $\mathscr{B}_n$. In applications, the sequence $(\mathscr{B}_n, n \in \mathbf{N})$ will frequently be the increasing sequence

$$\mathscr{B}_n = \sigma(Y_m, m \leqslant n) \qquad (n \in \mathbf{N})$$

associated with a given sequence $(Y_n, n \in \mathbf{N})$ of r.v.'s; in this case the σ-field $\mathscr{B}_\infty$ is the σ-field generated by the infinite sequence $(Y_n, n \in \mathbf{N})$. In all cases, the events in $\mathscr{B}_n$ are said to be *prior* to n.

A sequence $(X_n, n \in \mathbf{N})$ of random variables defined on $[\Omega, \mathscr{A}, P; (\mathscr{B}_n, n \in \mathbf{N})]$ is said to be *adapted* if for all $n \in \mathbf{N}$, the r.v. X_n is $\mathscr{B}_n$-measurable. If the sequence $(X_n, n \in \mathbf{N})$ of r.v.'s is adapted, the same is true of every sequence

$$(X_n' = f_n(X_0, X_1, \ldots, X_n), n \in \mathbf{N})$$

constructed using measurable functions f_n $(n \in \mathbf{N})$. On the other hand, a sequence $(X_n, n \in \bar{\mathbf{N}})$ also defined for the value $n = +\infty$ will be called adapted if the r.v.'s X_n are $\mathscr{B}_n$-measurable even for the value $n = +\infty$.

The r.v.'s introduced in what follows only enter into the discussion through their equivalence classes. Thus all the equalities and inequalities which we write are to be taken in the a.s. sense, without this being explicitly stated. We also say that $E^{\mathscr{B}}(X)$ is a $\mathscr{B}$-measurable r.v., thus replacing the equivalence class defining the conditional expectation $E^{\mathscr{B}}(X)$ by one of the $\mathscr{B}$-measurable r.v.'s in this class; it is the indeterminateness of this r.v. which obliges us to interpret the relations only a.s.

The notion of stopping time is one of the most fruitful in probability theory. The following paragraph is devoted to it.

II-1. Stopping times

The term "stopping time" is an expression from gambling. A game of chance which evolves in time (for example an infinite sequence of coin tosses) can be adequately represented by a space $[\Omega, \mathscr{A}, P; (\mathscr{B}_n, n \in \mathbf{N})]$, the sub-$\sigma$ fields $\mathscr{B}_n$ giving the information on the results of the game available to the player at time n. A stopping rule for the player thus consists of giving a rule for leaving the game, based at each time n ($n \in \mathbf{N}$) on information at his disposal at that time (by this definition we exclude from our considerations "dishonest" players who decide to leave the game at time n already knowing certain subsequent outcomes of the game). Denoting by v the time of stopping the game given by such a rule, we are led to the definition below; we also admit stopping times taking the value $+\infty$, corresponding to the case where the game is not stopped.

Often in probability, concepts rendered intuitive by the ideas of gambling have a much wider significance; we will show that this is the case with the concept of stopping time.

Definition II-1-1. A mapping $v:\Omega \to \bar{\mathbf{N}}$ is called a *stopping time* if

$$\{v = n\} \in \mathscr{B}_n \qquad \text{for all } n \in \mathbf{N}.$$

We associate with it the σ-field $\mathscr{B}_v$ of subsets of Ω defined by

$$\mathscr{B}_v = \{B : B \in \mathscr{B}_\infty, B \cap \{v = n\} \in \mathscr{B}_n \text{ for all } n \in \mathbf{N}\}$$

and we say that the events in $\mathscr{B}_v$ are *prior to* v. [It is immediate that $\mathscr{B}_v$ is in fact a σ-field and thus a sub-σ-field of $\mathscr{B}_\infty$.]

Here is the first important example of a stopping time. For every adapted sequence $(X_n, n \in \mathbf{N})$ of r.v.'s with values in an arbitrary measurable space $(E, \mathscr{F})$, the *hitting time* v_F of a subset F ($F \in \mathscr{F}$) by the sequence which is defined by

$$v_F(\omega) = \begin{cases} \inf\{n : X_n(\omega) \in F\} & \text{if } \omega \in \bigcup_{\mathbf{N}} \{X_n \in F\}, \\ +\infty & \text{otherwise} \end{cases}$$

is a stopping time; more precisely, for all $n \in \mathbf{N}$ we have

$$\{\nu_F = n\} = \bigcap_{m<n} \{X_m \notin F\} \cap \{X_n \in F\} \in \mathscr{B}_n$$

since the r.v.'s X_m ($m \leqslant n$) are $\mathscr{B}_n$-measurable.

Every constant map of Ω into $\bar{\mathbf{N}}$, say $\nu \equiv p$ for a fixed $p \in \bar{\mathbf{N}}$, is clearly a stopping time, and in this case the σ-field $\mathscr{B}_\nu$ associated with ν coincides with $\mathscr{B}_p$; thus our notations are consistent.

For any stopping time ν, the set $\{\nu = \infty\}$ belongs to $\mathscr{B}_\infty$, for it is the complement of the union of the events $\{\nu = n\}$ ($n \in \mathbf{N}$) which all belong to $\mathscr{B}_\infty$. It follows that $B \cap \{\nu = \infty\} \in \mathscr{B}_\infty$ for all $B \in \mathscr{B}_\nu$. It also follows that $\nu : \Omega \to \bar{\mathbf{N}}$ is $\mathscr{B}_\infty$-measurable.

LEMMA II-1-2. *For a mapping $\nu : \Omega \to \bar{\mathbf{N}}$ to be a stopping time, it is necessary and sufficient that $\{\nu \leqslant n\} \in \mathscr{B}_n$ for all $n \in \mathbf{N}$. For an event $B \in \mathscr{B}_\infty$ to belong to the σ-field $\mathscr{B}_\nu$ associated with a stopping time ν, it is necessary and sufficient that $B \cap \{\nu \leqslant n\} \in \mathscr{B}_n$ for all $n \in \mathbf{N}$.*

PROOF. The proof of this lemma is immediate because of the formulae

$$\{\nu \leqslant n\} = \bigcup_{m \leqslant n} \{\nu = m\}, \qquad \{\nu = n\} = \{\nu \leqslant n\} \cap \{\nu \leqslant n-1\}^c$$

and because the sequence of σ-fields $(\mathscr{B}_n, n \in \mathbf{N})$ is increasing. ■

Let us observe that although for any stopping time ν it is true that $\{\nu > n\} \in \mathscr{B}_n$ for all $n \in \mathbf{N}$, it is false in general that $B \cap \{\nu > n\} \in \mathscr{B}_n$ if $B \in \mathscr{B}_\nu$ and $n \in \mathbf{N}$!

It follows easily from the preceding lemma that the minimum $\bigwedge_k \nu_k$ and the maximum $\bigvee_k \nu_k$ of a finite or infinite sequence of stopping times is again a stopping time, since the sets

$$\{\bigwedge_k \nu_k \leqslant n\} = \bigcup_k \{\nu_k \leqslant n\}, \qquad \{\bigvee_k \nu_k \leqslant n\} = \bigcap_k \{\nu_k \leqslant n\}$$

belong to $\mathscr{B}_n$ for all $n \in \mathbf{N}$. In particular the increasing sequence $\nu \wedge p$ ($p \in \mathbf{N}$) of stopping times which can be associated with a stopping time ν will play an important technical role in the sequel. We will also meet many instances of stopping times being defined as a limit $\lim\uparrow_k \nu_k$ of an increasing sequence of stopping times.

The following proposition and its corollary describe more fully the σ-field $\mathscr{B}_\nu$ associated with a stopping time ν. In what follows it will be convenient for us to say that a mapping X only defined on a subset Ω' of a probability space $(\Omega, \mathscr{A}, P)$ is a *random variable defined on* Ω' (resp. a $\mathscr{B}$-measurable r.v. defined on Ω') if it is measurable with respect to the trace σ-field $\mathscr{A} \cap \Omega'$ (resp. $\mathscr{B} \cap \Omega'$). Recall that a trace σ-field $\mathscr{B} \cap \Omega'$ is defined to be the class $\{B \cap \Omega'$ where B varies over $\mathscr{B}\}$ of subsets of Ω', and is such that if $\Omega' \in \mathscr{B}$, it coincides with the σ-field $\{B: B \in \mathscr{B}, B \subset \Omega'\}$.

PROPOSITION II-1-3. *Let ν be a stopping time. In order that a $\mathscr{B}_\infty$-measurable real-valued function $f: \Omega \to \bar{R}$ be measurable with respect to the σ-field $\mathscr{B}_\nu$, it is necessary and sufficient that for all $n \in \mathbf{N}$ the restriction of f to $\{\nu = n\}$ be measurable with respect to $\mathscr{B}_n$.*

For every positive or integrable real-valued measurable function g defined on the probability space $(\Omega, \mathscr{A}, P)$, the conditional expectation $E^{\mathscr{B}_\nu}(g)$ is given by the formula

$$E^{\mathscr{B}_\nu}(g) = E^{\mathscr{B}_n}(g) \quad \text{on } \{\nu = n\} \qquad (n \in \bar{\mathbf{N}}).$$

PROOF. If f is an indicator function, say $f = 1_B$ where $B \in \mathscr{B}_\infty$, then the first part of the proposition follows from the definition of $\mathscr{B}_\nu$. The extension to an arbitrary $\mathscr{B}_\infty$-measurable function is then immediate.

Let us suppose that g is positive. By what we have just proved, the right-hand side of the above formula defines a positive $\mathscr{B}_\nu$-measurable function; further the integral of this function over a set $B \in \mathscr{B}_\nu$ is

$$\sum_{\bar{\mathbf{N}}} \int_{B \cap \{\nu = n\}} E^{\mathscr{B}_n}(g)\, \mathrm{d}P = \sum_{\bar{\mathbf{N}}} \int_{B \cap \{\nu = n\}} g \,\mathrm{d}P = \int_B g \,\mathrm{d}P = \int_B E^{\mathscr{B}_\nu}(g)\, \mathrm{d}P$$

since $B \cap \{\nu = n\} \in \mathscr{B}_n$ for all $n \in \bar{\mathbf{N}}$. This suffices to prove that this function is equal to $E^{\mathscr{B}_\nu}(g)$. The extension to an integrable function of arbitrary sign is immediate. ■

The following corollary is apparent.

COROLLARY II-1-4. *For every adapted sequence $(X_n, n \in \mathbf{N})$ of r.v.'s and for every stopping time ν, the r.v. X_ν defined on $\{\nu < \infty\}$ by $X_\nu(\omega) = X_{\nu(\omega)}(\omega)$ or, equivalently, by*

$$X_\nu = X_n \quad \text{on } \{\nu = n\} \qquad (n \in \mathbf{N}),$$

is $\mathscr{B}_\nu$-measurable.

An analogous result is valid for an adapted sequence $(X_n, n \in \bar{\mathbf{N}})$ whose index n can take the value $+\infty$; in this case the r.v. X_ν is defined on the entire space Ω. In particular, taking $X_n = n$, we see that the stopping time ν is itself $\mathscr{B}_\nu$-measurable on $(\Omega, \mathscr{A}, P)$.

PROPOSITION II-1-5. *For every pair* v, v' *of stopping times, the events* $\{v < v'\}$, $\{v = v'\}$ *and* $\{v \leqslant v'\}$ *belong to* $\mathscr{B}_v$ *and to* $\mathscr{B}_{v'}$. *On the other hand,*

$$B \in \mathscr{B}_v \Rightarrow B \cap \{v \leqslant v'\} \in \mathscr{B}_{v'}.$$

It follows that $\mathscr{B}_v \subset \mathscr{B}_{v'}$ *for every pair of stopping times such that* $v \leqslant v'$ *on all of the space* Ω.

PROOF. Since both v and v' are necessarily $\mathscr{B}_\infty$-measurable, it is clear that all the sets under consideration belong to $\mathscr{B}_\infty$. On the other hand, as

$$\{v < v'\} \cap \{v = n\} = \{v' > n\} \cap \{v = n\} \in \mathscr{B}_n,$$

$$\{v = v'\} \cap \{v = n\} = \{v' = n\} \cap \{v = n\} \in \mathscr{B}_n,$$

for all $n \in \mathbf{N}$, it is clear that the two events $\{v < v'\}$ and $\{v = v'\}$ belong to $\mathscr{B}_v$; thus their union $\{v \leqslant v'\}$ also belongs to $\mathscr{B}_v$. By symmetry we see that $\{v = v'\}$ belongs to $\mathscr{B}_{v'}$; finally, taking complements of $\{v < v'\}$ and $\{v \leqslant v'\}$ and reversing the roles of v and v', we find that the same events belong to $\mathscr{B}_{v'}$.

If $B \in \mathscr{B}_v$, then

$$B \cap \{v \leqslant v'\} \cap \{v' = n\} = B \cap \{v \leqslant n\} \cap \{v' = n\} \in \mathscr{B}_n$$

for all $n \in \mathbf{N}$, and as a consequence the event $B \cap \{v \leqslant v'\}$, which obviously belongs to $\mathscr{B}_\infty$, actually belongs to $\mathscr{B}_{v'}$. ∎

Finally note that by taking the intersection of $B \cap \{v \leqslant v'\}$ with $\{v = v'\}$ or $\{v < v'\}$, we find that $B \cap \{v = v'\}$ and $B \cap \{v < v'\} \in \mathscr{B}_{v'}$ whenever $B \in \mathscr{B}_v$.

II-2. Positive supermartingales

We begin with a definition in which we recall that we have taken once and for all a probability space $(\Omega, \mathscr{A}, P)$ and an increasing sequence $(\mathscr{B}_n, n \in \mathbf{N})$ of sub-σ-fields of $\mathscr{A}$.

Definition II-2-6. An adapted sequence $(X_n, n \in \mathbf{N})$ of positive real-valued r.v.'s is called a *positive supermartingale* if the a.s. inequality

$$X_n \geqslant E^{\mathscr{B}_n}(X_{n+1})$$

is satisfied for all $n \in \mathbf{N}$. The sequence is called a *positive martingale* if the inequality is replaced by an equality.

This definition makes sense as the conditional expectation of an arbitrary positive real-valued r.v. was defined in the first chapter (Lemma I-2-9).

A supermartingale is by definition a sequence of r.v.'s which "decrease in conditional mean". For a sequence $(X_n, n \in \mathbf{N})$ of positive r.v.'s denoting the sequence of values of the fortune of a gambler, the supermartingale condition expresses the property that at each play the game is unfavourable to the player in conditional mean. On the other hand, a martingale "remains constant in conditional mean" and, for the gambler, corresponds to a game which is on average fair.

We note that the inequality defining a supermartingale implies that

$$X_m \geqslant E^{\mathscr{B}_m}(X_p)$$

for every pair of positive integers $m < p$, and not only for pairs of consecutive integers; indeed the inequality $X_n \geqslant E^{\mathscr{B}_n}(X_{n+1})$ implies that

$$E^{\mathscr{B}_m}(X_n) \geqslant E^{\mathscr{B}_m} E^{\mathscr{B}_n}(X_{n+1}) = E^{\mathscr{B}_m}(X_{n+1})$$

if $m \leqslant n$; thus the sequence $(E^{\mathscr{B}_m}(X_n),\ n \geqslant m)$ is decreasing and so $X_m = E^{\mathscr{B}_m}(X_m) \geqslant E^{\mathscr{B}_m}(X_p)$ if $p > m$.

The following proposition is important and its proof foreshadows that of Theorem II-**2**-9, which is the basic result of this section.

PROPOSITION II-**2**-7. *For every positive supermartingale* $(X_n, n \in \mathbf{N})$, *the r.v.* $\sup_{\mathbf{N}} X_n$ *is a.s. finite on the set* $\{X_0 < \infty\}$, *and, more precisely, satisfies the following inequality*

$$P^{\mathscr{B}_0}(\sup_{\mathbf{N}} X_n \geqslant a) \leqslant \min\left(\frac{X_0}{a}, 1\right)$$

for all constants $a > 0$.

This inequality is called the *maximal inequality* or the maximal lemma (for positive supermartingales). We often write $P^{\mathscr{B}}(A)$ instead of $E^{\mathscr{B}}(1_A)$ and will frequently use the identity

$$\int_B P^{\mathscr{B}}(A)\,\mathrm{d}P = P(AB)$$

valid when $B \in \mathscr{B}$.

PROOF. We rely upon an auxiliary lemma which constitutes a switching principle for supermartingales.

LEMMA II-2-8. *Given two positive supermartingales* $(X_n^{(i)}, n \in \mathbf{N})$ $(i = 1,2)$ *and a stopping time* ν *such that* $X_\nu^{(1)} \geqslant X_\nu^{(2)}$ *on* $\{\nu < \infty\}$, the formula

$$X_n(\omega) = \begin{cases} X_n^{(1)}(\omega) & \text{if } n < \nu(\omega) \\ X_n^{(2)}(\omega) & \text{if } n \geqslant \nu(\omega) \end{cases} \qquad (n \in \mathbf{N})$$

defines a new positive supermartingale.

Indeed, the defining formula of the X_n $(n \in \mathbf{N})$ can also be written

$$X_n = 1_{\{\nu > n\}} X_n^{(1)} + 1_{\{\nu \leqslant n\}} X_n^{(2)}$$

and it is then clear that X_n is $\mathscr{B}_n$-measurable for all $n \in \mathbf{N}$. The supermartingale property of the $X_{\bullet}^{(i)}$ allows us to write

$$\begin{aligned} X_n &= 1_{\{\nu > n\}} X_n^{(1)} + 1_{\{\nu \leqslant n\}} X_n^{(2)} \\ &\geqslant 1_{\{\nu > n\}} E^{\mathscr{B}_n}(X_{n+1}^{(1)}) + 1_{\{\nu \leqslant n\}} E^{\mathscr{B}_n}(X_{n+1}^{(2)}) \\ &= E^{\mathscr{B}_n}[1_{\{\nu > n\}} X_{n+1}^{(1)} + 1_{\{\nu \leqslant n\}} X_{n+1}^{(2)}]. \end{aligned}$$

But the assumption $X_\nu^{(1)} \geqslant X_\nu^{(2)}$ on $\{\nu < \infty\}$ implies that $X_{n+1}^{(1)} \geqslant X_{n+1}^{(2)}$ on $\{\nu = n + 1\}$; it then follows that

$$1_{\{\nu > n\}} X_{n+1}^{(1)} + 1_{\{\nu \leqslant n\}} X_{n+1}^{(2)} \geqslant 1_{\{\nu > n+1\}} X_{n+1}^{(1)} + 1_{\{\nu \leqslant n+1\}} X_{n+1}^{(2)} = X_{n+1}.$$

This proves that $X_n \geqslant E^{\mathscr{B}_n}(X_{n+1})$.

The lemma thus being proved, let us associate with the positive supermartingale $(X_n, n \in \mathbf{N})$ of the proposition the stopping time defined by

$$\nu_a = \begin{cases} \min(n : X_n > a), & \text{if } \sup_{\mathbf{N}} X_n > a \\ \infty & \text{if } \sup_{\mathbf{N}} X_n \leqslant a \end{cases} \qquad (a > 0).$$

Since $X_{\nu_a} > a$ on $\{\nu_a < \infty\}$ and since the constant a can be considered a supermartingale, the formula

$$Y_n = \begin{cases} X_n & \text{if } n < \nu_a \\ a & \text{if } n \geqslant \nu_a \end{cases} \qquad (n \in \mathbf{N})$$

defines a new supermartingale by Lemma II-2-8. Then we may write $Y_0 \geqslant E^{\mathscr{B}_0}(Y_n)$; since Y_0 takes the value X_0 or a according as $X_0 \leqslant a$ or $X_0 > a$, and since $Y_n \geqslant a 1_{\{\nu_a \leqslant n\}}$, the preceding inequality implies that

$$aP^{\mathscr{B}_0}(\nu_a \leqslant n) \leqslant \min(X_0, a).$$

Letting $n \uparrow \infty$ and dividing by a, we obtain

$$P^{\mathscr{B}_0}(\sup_{\mathbf{N}} X_n > a) \leqslant \min\left(\frac{X_0}{a}, 1\right)$$

since $\{\nu_a < \infty\} = \{\sup_{\mathbf{N}} X_n > a\}$. This inequality implies that of the proposition; it suffices to replace a by $a(1 - k^{-1})$ in the inequality above and then let $k \uparrow \infty$ (k an integer $\geqslant 2$) to obtain the same inequality with $\geqslant$ instead of $>$ on the left-hand side.

Finally, let us integrate both sides of this inequality over the event $\{X_0 < \infty\}$, which belongs to $\mathscr{B}_0$; we find that

$$P(X_0 < \infty, \sup_{\mathbf{N}} X_n > a) \leqslant \int_{\{X_0 < \infty\}} \min\left(\frac{X_0}{a}, 1\right) \mathrm{d}P.$$

When $a \uparrow \infty$, the right-hand side tends to zero by the dominated convergence theorem and we have thus proved that

$$P(X_0 < \infty, \sup_{\mathbf{N}} X_n = \infty) = 0,$$

which completes the proof of Proposition II-2-7. ■

(We see in the latter part of the proof the simplification achieved by writing the upper bound of $P^{\mathscr{B}_0}(\sup_{\mathbf{N}} X_n > a)$ in the form $\min(X_0/a, 1)$ rather than just X_0/a.)

REMARK. The preceding proof is valid without any change if we replace the constant a by an arbitrary positive $\mathscr{B}_0$-measurable r.v. A. We would then obtain

$$P^{\mathscr{B}_0}(\sup_{\mathbf{N}} X_n \geqslant A) \leqslant \min\left(\frac{X_0}{A}, 1\right) \quad \text{on } \{A > 0\}$$

if A is $\mathscr{B}_0$-measurable and $\geqslant 0$. This inequality has the following consequence: for every positive supermartingale $(X_n, n \in \mathbf{N})$, every $\mathscr{B}_0$-measurable r.r.v. $A \geqslant 0$ satisfies the implication

$$A \leqslant \sup_{\mathbf{N}} X_n \Rightarrow A \leqslant X_0;$$

put another way, X_0 is the largest $\mathscr{B}_0$-measurable r.v. dominated by $\sup_{\mathbf{N}} X_n$. Furthermore, more generally, every $\mathscr{B}_p$-measurable r.r.v. $A \geqslant 0$ such that

$A \leqslant \sup_{\mathbf{N}} X_n$ satisfies the inequality $A \leqslant \sup_{n \leqslant p} X_n$, as can be seen by applying the preceding result to the supermartingale $(\sup_{n \leqslant p} X_n, X_{p+1}, X_{p+2}, \ldots)$ adapted to the sequence $(\mathscr{B}_p, \mathscr{B}_{p+1}, \mathscr{B}_{p+2}, \ldots)$ of σ-fields. ■

The main result of this paragraph is then:

THEOREM II-2-9. *Every positive supermartingale $(X_n, n \in \mathbf{N})$ converges almost surely. Furthermore, the limit $X_\infty = \lim_{n \to \infty}$ a.s. X_n satisfies the inequalities*

$$E^{\mathscr{B}_n}(X_\infty) \leqslant X_n \qquad (n \in N).$$

PROOF. (1) We begin by proving a criterion for convergence of sequences of real numbers.

Given a sequence $(x_n, n \in \mathbf{N})$ in $\overline{\mathbf{R}}$ and a pair $a < b$ of finite real numbers, let us define integers ν_k $(k \geqslant 1)$ inductively by

$$\begin{aligned} \nu_1 &= \min(n : n \geqslant 0, x_n \leqslant a), \\ \nu_2 &= \min(n : n \geqslant \nu_1, x_n \geqslant b), \\ \nu_3 &= \min(n : n \geqslant \nu_2, x_n \leqslant a), \\ &\vdots \end{aligned}$$

alternating the inequalities $x_n \leqslant a$ and $x_n \geqslant b$. If one of the indices ν_k is not defined (for example ν_1 is not defined if $x_n > a$ for all $n \in \mathbf{N}$), it is convenient to put it equal to $+\infty$ and similarly for all the subsequent indices. We will denote by $\beta_{a,b}$ the largest integer p for which ν_{2p} is finite, and put $\beta_{a,b} = \infty$ if all of the ν_k are finite; this number $\beta_{a,b}$ represents the number of times that the sequence $(x_n, n \in \mathbf{N})$ "upcrosses the interval $[a, b]$". It is then easy to establish the following implications

$$\varliminf_n x_n < a < b < \varlimsup_n x_n \Rightarrow \beta_{a,b} = +\infty \Rightarrow \varliminf_n x_n \leqslant a < b \leqslant \varlimsup_n x_n,$$

from which we deduce that a sequence $(x_n, n \in \mathbf{N})$ in $\overline{\mathbf{R}}$ is convergent if and only if $\beta_{a,b} < \infty$ for every pair $a < b$ in $\mathbf{R}$ (or, equivalently, in $\mathbf{Q}$).

Now let us consider a sequence $(X_n, n \in \mathbf{N})$ of real-valued r.v.'s. The indices $\nu_k(\omega)$ which are associated as above with each of the sequences $(X_n(\omega), n \in \mathbf{N})$ are then real-valued random variables (i.e. measurable functions of ω); this is easily proved inductively by writing

$$\{\nu_{2p} = n\} = \sum_{m<n} \{\nu_{2p-1} = m;\ X_{m+1} < b, \ldots, X_{n-1} < b,\ X_n \geqslant b\}$$

with an analogous formula in the case of an odd index. Since $\{\beta_{a,b} \geqslant p\} = \{\nu_{2p} < \infty\}$, we see that $\beta_{a,b}$ is also a random variable. The convergence criterion enunciated above then implies that

$$\{X_n \to\} = \bigcap_{\substack{a<b \\ a,b \in \mathbf{R}}} \{\beta_{a,b} < \infty\} = \bigcap_{\substack{a<b \\ a,b \in \mathbf{Q}}} \{\beta_{a,b} < \infty\}$$

and this allows us to formulate the following lemma, which furthermore is useful beyond the theory of martingales.

LEMMA II-2-10. *For a sequence* $(X_n, n \in \mathbf{N})$ *of real-valued r.v.'s to converge a.s., it is necessary and sufficient that the upcrossing numbers* $\beta_{a,b}$ *are finite a.s. for all* $a < b$ *in* $\mathbf{R}$ *or in* $\mathbf{Q}$.

(2) The proof of Theorem II-2-9 thus consists of proving $\beta_{a,b} < \infty$ a.s. for every pair $a < b$; since the r.v.'s X_n are positive, we need only consider pairs $a < b$ of positive numbers (as we can see for example using a homeomorphism of $\overline{\mathbf{R}}$ onto $[0, \infty]$). To this end we will establish an extremely precise result.

Dubins' inequalities. *For every positive supermartingale* $(X_n, n \in \mathbf{N})$, *the upcrossing numbers are r.v.'s satisfying the inequalities*

$$P^{\mathscr{B}_0}(\beta_{a,b} \geqslant k) \leqslant \left(\frac{a}{b}\right)^k \min\left(\frac{X_0}{a}, 1\right)$$

whatever the integers $k \geqslant 1$ *and real numbers* $0 < a < b < \infty$. *Thus the r.v.'s* $\beta_{a,b}$ *are a.s. finite.*

The indices $\nu_k(\omega)$ $(k \geqslant 1)$ associated with the sequences $(X_n(\omega), n \in \mathbf{N})$ define stopping times, since each of the sets $\{\omega : \nu_k(\omega) = n\}$ clearly depends only on $X_0(\omega), \ldots, X_n(\omega)$ and thus belongs to the σ-field $\mathscr{B}_n$ $(n \in \mathbf{N})$. This being so, repeated application of Lemma II-2-8 shows that for $k \geqslant 1$ a fixed integer, the following formulae define a positive supermartingale

$$\begin{aligned} Y_n &= 1 && \text{if } 0 \leqslant n < \nu_1, \\ &= \frac{X_n}{a} && \text{if } \nu_1 \leqslant n < \nu_2, \\ &= \frac{b}{a} \cdot 1 && \text{if } \nu_2 \leqslant n < \nu_3, \end{aligned}$$

$$= \frac{b}{a} \cdot \frac{X_n}{a} \qquad \text{if } \nu_3 \leqslant n < \nu_4,$$

$$\vdots$$

$$= \left(\frac{b}{a}\right)^{k-1} \frac{X_n}{a} \qquad \text{if } \nu_{2k-1} \leqslant n < \nu_{2k},$$

$$= \left(\frac{b}{a}\right)^{k} \qquad \text{if } \nu_{2k} \leqslant n.$$

Indeed, the constant sequences $(b/a)^j$ $(0 \leqslant j \leqslant k)$ and the sequences $((b/a)^j X_n/a, n \in \mathbf{N})$ proportional to $(X_n, n \in \mathbf{N})$ are positive supermartingales satisfying the inequalities

$$1 \geqslant \frac{X_{\nu_1}}{a}, \frac{X_{\nu_2}}{a} \geqslant \frac{b}{a}, \ldots, \left(\frac{b}{a}\right)^{k-1} \frac{X_{\nu_{2k}}}{a} \geqslant \left(\frac{b}{a}\right)^k,$$

at the times ν_j, at least on the sets on which the respective stopping times are finite.

Let us remark that $Y_0 = \min(1, X_0/a)$, for according as X_0/a is less than or greater than 1, we have $\nu_1 = 0$ or $\nu_1 > 0$. On the other hand, we have the inequality $Y_n \geqslant (b/a)^k 1_{\{\nu_{2k} \geqslant n\}}$, and, since $Y_0 \geqslant E^{\mathscr{B}_0}(Y_n)$ because $(Y_n, n \in \mathbf{N})$ is a supermartingale, we find that

$$\left(\frac{b}{a}\right)^k P^{\mathscr{B}_0}(\nu_{2k} \leqslant n) \leqslant \min\left(\frac{X_0}{a}, 1\right).$$

It only remains to let $n \uparrow \infty$ and remark that

$$\{\nu_{2k} < \infty\} = \{\beta_{a,b} \geqslant k\}$$

in order to see that the r.v. $\beta_{a,b}$ satisfies Dubins' inequality and hence is a.s. finite. With the help of Lemma II-2-10 this implies the a.s. convergence of the positive supermartingale $(X_n, n \in \mathbf{N})$.

(3) Let X_∞ be the a.s. limit of the sequence $(X_n, n \in \mathbf{N})$. The inequality

$$E^{\mathscr{B}_p}(\inf_{m \geqslant n} X_m) \leqslant E^{\mathscr{B}_p}(X_n) \leqslant X_p$$

valid if $n > p$ shows when we let $n \uparrow \infty$ that

$$E^{\mathscr{B}_p}(X_\infty) \leqslant X_p \quad \text{for all } p \in \mathbf{N},$$

since X_∞ is the limit as $n \uparrow \infty$ of the increasing sequence $(\inf_{m \geqslant n} X_m, n \in \mathbf{N})$ of positive r.v.'s and the conditional expectation has the increasing continuity property on positive r.v.'s (see Corollary I-2-10). ■

REMARK. The inequality $E^{\mathscr{B}_n}(X_\infty) \leqslant X_n$ $(n \in \mathbf{N})$ which we have just established implies that

$$X_\infty < \infty \text{ a.s.} \quad \text{outside } \{X_n = +\infty \text{ for all } n \in \mathbf{N}\}$$

because for every n, the r.v. X_∞ is integrable on each of the events

$$\{E^{\mathscr{B}_n}(X_\infty) \leqslant a\} \qquad (a \in \mathbf{R}_+),$$

and hence is finite on $\{E^{\mathscr{B}_n}(X_\infty) < \infty\}$. In fact this result already follows from Proposition II-2-7, which, in showing that $\sup_{\mathbf{N}} X_n < \infty$ on $\{X_0 < \infty\}$, also implies that $X_\infty < \infty$ on $\{X_0 < \infty\}$; applying this result to the supermartingale $(X_n, n \geqslant p)$ adapted to the sequence $(\mathscr{B}_n, n \geqslant p)$, we find that $X_\infty < \infty$ a.s. on $\{X_p < \infty\}$ for all $p \in \mathbf{N}$, which is equivalent to the above result. ■

The inequality $E^{\mathscr{B}_n}(X_\infty) \leqslant X_n$ $(n \in \mathbf{N})$ for the positive supermartingale $(X_n, n \in \mathbf{N})$ implies that $X_\infty \in L^1$ whenever $X_n \in L^1$, but it is not possible to conclude that the sequence $(X_n, n \in \mathbf{N})$ converges in L^1 to X_∞ when the supermartingale $(X_n, n \in \mathbf{N})$ is integrable. We will see numerous counterexamples in later paragraphs (see especially Section III-1). For a positive martingale $(X_n, n \in \mathbf{N})$, which of course satisfies the equalities $X_m = E^{\mathscr{B}_m}(X_n)$ if $m \leqslant n$, not only is it untrue that the inequality $E^{\mathscr{B}_n}(X_\infty) \leqslant X_n$ automatically becomes an equality, but there even exist many positive martingales not a.s. zero which converge a.s. to zero, and for such martingales the equality $E^{\mathscr{B}_n}(X_\infty) = X_n$ is clearly impossible (see also Section III-1).

Nevertheless, the possibilities which we have just indicated cannot arise for the martingales $(E^{\mathscr{B}_n}(Z), n \in \mathbf{N})$ of the following proposition. This proposition furnishes an important class of positive martingales whose limits can be identified.

PROPOSITION II-2-11. *Let p be a real number in $[1, \infty[$. For every $Z \in L^p_+$ the sequence $(Z_n = E^{\mathscr{B}_n}(Z), n \in \mathbf{N})$ is a positive martingale which converges a.s. and in L^p to the r.v. $Z_\infty = E^{\mathscr{B}_\infty}(Z)$.*

(This positive martingale $(Z_n, n \in \mathbf{N})$ and its limit Z_∞ thus satisfy the equalities $Z_n = E^{\mathscr{B}_n}(Z_\infty)$ $(n \in \mathbf{N})$ since $E^{\mathscr{B}_n} E^{\mathscr{B}_\infty}(Z) = E^{\mathscr{B}_n}(Z)$.)

This proposition implies that martingales of the form $(E^{\mathscr{B}_n}(Z),\ n \in \mathbf{N})$ where $Z \in L^p_+$ are precisely those positive martingales which converge in L^p as $n \uparrow \infty$; in fact, if $(Z_n, n \in \mathbf{N})$ is a positive martingale converging in L^p, the equality $Z_n = E^{\mathscr{B}_n}(Z_r)$ valid if $n \leqslant r$ and the continuity of conditional expectations on L^p imply that $Z_n = E^{\mathscr{B}_n}(Z_\infty)$ for all $n \in \mathbf{N}$.

The usual decomposition $Z = Z^+ - Z^-$ allows an immediate extension of the preceding proposition to an arbitrary $Z \in L^p$.

Proof. The sequence $(E^{\mathscr{B}_n}(Z),\ n \in \mathbf{N})$ is a martingale since

$$E^{\mathscr{B}_n}(E^{\mathscr{B}_{n+1}}(Z)) = E^{\mathscr{B}_n}(Z) \quad \text{for all } n \in \mathbf{N}.$$

By Theorem II-2-9 above, this sequence converges a.s. to a limit which we will denote by Z; this limit is clearly $\mathscr{B}_\infty$-measurable.

If the r.v. Z is a.s. bounded by a finite constant a, the conditional expectations $E^{\mathscr{B}_n}(Z)$ are similarly bounded and the dominated convergence theorem then implies that

$$\int_A E^{\mathscr{B}_n}(Z)\,\mathrm{d}P \to \int_A Z_\infty\,\mathrm{d}P \qquad (n \to \infty)$$

for all $A \in \mathscr{A}$. But if $A \in \mathscr{B}_m$ for some $m \in \mathbf{N}$, the integrals on the left-hand side are all equal to $\int_A Z\mathrm{d}P$ when $n \geqslant m$, and we thus conclude that

$$\int_A Z\,\mathrm{d}P = \int_A Z_\infty\,\mathrm{d}P \quad \text{for all } A \in \bigcup_{\mathbf{N}} \mathscr{B}_m.$$

This identity remains true if $A \in \mathscr{B}_\infty$ for by a standard result of measure theory, two positive *finite* measures, here $Z \cdot P$ and $Z_\infty \cdot P$, cannot coincide on a Boolean algebra without coinciding on the σ-algebra it generates. It then follows that $Z_\infty = E^{\mathscr{B}_\infty}(Z)$ since Z_∞ is $\mathscr{B}_\infty$-measurable. Further, the convergence of the martingale $(E^{\mathscr{B}_n}(Z),\ n \in \mathbf{N})$ to $E^{\mathscr{B}_n}(Z)$ also takes place in all the L^p-spaces $(1 \leqslant p < \infty)$ by the dominated convergence theorem (since $E^{\mathscr{B}_n}(Z) \leqslant a$ for all $n \in \mathbf{N}$).

It is not hard to see that the convergence of $E^{\mathscr{B}_n}(Z)$ to $E^{\mathscr{B}_n}(Z)$ in L^p remains pth power, the martingale $(E^{\mathscr{B}_n}(Z),\ n \in \mathbf{N})$ converges a.s. to a limit Z_∞, and In fact, the decomposition

$$Z = Z \wedge a + (Z - a)^+$$

allows us to write

$$\|E^{\mathscr{B}_n}(Z) - E^{\mathscr{B}_\infty}(Z)\|_p \leqslant \|E^{\mathscr{B}_n}(Z \wedge a) - E^{\mathscr{B}_\infty}(Z \wedge a)\|_p + 2\|(Z - a)^+\|_p$$

since conditional expectations decrease the L^p-norm. But by the foregoing, the first term on the right-hand side tends to zero as $n \uparrow \infty$, while the second term decreases by dominated convergence to zero as $a \uparrow \infty$; thus letting $n \uparrow \infty$ and then $a \uparrow \infty$ in the above inequality, we find that $\|E^{\mathscr{B}_n}(Z) - E^{\mathscr{B}_\infty}(Z)\|_p \to 0$ when $n \uparrow \infty$. We have thus proved that for an arbitrary r.v. $Z \geqslant 0$ with integrable pth power, the martingale $(E^{\mathscr{B}_n}(Z), n \in \mathbf{N})$ converges a.s. to a limit Z_∞, and converges in L^p to $E^{\mathscr{B}_\infty}(Z)$; this implies that $Z_\infty = E^{\mathscr{B}_\infty}(Z)$ and the proposition is proved. ■

COROLLARY II-2-12. *For any positive r.r.v. Z we have*

$$E^{\mathscr{B}_n}(Z) \to E^{\mathscr{B}_\infty}(Z)\ a.s. \quad outside\ \{E^{\mathscr{B}_n}(Z) = \infty\ for\ all\ n\}.$$

PROOF. For all fixed $m \in \mathbf{N}$ and $a \in \mathbf{R}_+$, the positive random variable $Z' = Z 1_{\{E^{\mathscr{B}_m}(Z) \leqslant a\}}$ is integrable since

$$E(Z') = E(E^{\mathscr{B}_m}(Z)\, 1_{\{E^{\mathscr{B}_m}(Z) \leqslant a\}}) \leqslant a.$$

The preceding proposition then implies that $E^{\mathscr{B}_n}(Z') \to E^{\mathscr{B}_\infty}(Z')$ a.s. when $n \uparrow \infty$. But $E^{\mathscr{B}_n}(Z') = E^{\mathscr{B}_n}(Z)$ on $\{E^{\mathscr{B}_m}(Z) \leqslant a\}$ if $m \leqslant n \leqslant +\infty$; consequently, on $\{E^{\mathscr{B}_m}(Z) \leqslant a\}$ the sequence $(E^{\mathscr{B}_n}(Z), n \in \mathbf{N})$ also converges a.s. to $E^{\mathscr{B}_\infty}(Z)$. To obtain the corollary it only remains to let m run over $\mathbf{N}$ and let $a \uparrow \infty$. ■

REMARK. The preceding corollary cannot be improved. Here is an example of an a.s. *finite* $\mathscr{B}_\infty$-measurable r.v. Z such that $E^{\mathscr{B}_n}(Z) = +\infty$ a.s. for all $n \in \mathbf{N}$ (for such a r.v. the convergence $E^{\mathscr{B}_n}(Z) \to E^{\mathscr{B}_\infty}(Z)$ can thus only occur on a null set!). In the probability space obtained by equipping the unit interval $[0, 1[$ with Lebesgue measure, let us consider the increasing sequence $\mathscr{B}_n$ of sub-σ-fields generated by the dyadic partitions $\{[2^{-n}k, 2^{-n}(k+1)[, 0 \leqslant k < 2^n\}$ $(n \in \mathbf{N})$. For each $n \in \mathbf{N}$ let us then choose a positive measurable function $f_n: [0, 1[\to \mathbf{R}_+$ of period 2^{-n} such that $\int f_n dx = 1$ and $\int 1_{\{f_n > 0\}}\, dx = 2^{-n}$. The positive random variable $Z = \sum_{\mathbf{N}} f_n$ is then a.s. finite, for the sum $\sum_{\mathbf{N}} 1_{\{f_n > 0\}}$ is integrable, and thus a.s. finite, which shows that the series $\sum_{\mathbf{N}} f_n(\omega)$ contains no more than a finite number of non-zero terms for almost all ω. On the other hand, for all $n \in \mathbf{N}$ and all k $(0 \leqslant k < 2^n)$,

$$\int_{2^{-n}k}^{2^{-n}(k+1)} Z\, dx \geqslant \sum_{p \geqslant n} \int_{2^{-n}k}^{2^{-n}(k+1)} f_p\, dx = \frac{1}{2^n} \sum_{p \geqslant n} \int_0^1 f_p\, dx = +\infty$$

by the periodicity of f_p, and this shows that $E^{\mathscr{B}_n}(Z) = +\infty$ a.s. for all $n \in \mathbf{N}$. ■

We will finish this paragraph with an important result which shows that the inequality defining supermartingales remains true when written with a pair of stopping times and not just for a pair of constant times.

THEOREM II-2-13. *Let* $(X_n, n \in \mathbf{N})$ *be a positive supermartingale whose a.s. limit will be denoted by* X_∞. *Then for any pair* v_1, v_2 *of stopping times, we have*

$$X_{v_1} \geqslant E^{\mathscr{B}_{v_1}}(X_{v_2}) \text{ a.s.} \quad \text{on } \{v_1 \leqslant v_2\}.$$

PROOF. To prove this inequality on $\{v_1 = n\}$ it suffices, by Proposition II-1-3, to show that

$$X_n \geqslant E^{\mathscr{B}_n}(X_v) \text{ a.s. on } \{v \geqslant n\}$$

for all stopping times v. But this result is an easy consequence of Theorem II-2-9 and the lemma below which states that for all stopping times v the sequence $(X_{v \wedge n}, n \in \mathbf{N})$ is a supermartingale; indeed, as this supermartingale converges a.s. to X_v when $n \uparrow \infty$, the stated theorem shows that $X_{v \wedge n} \geqslant E^{\mathscr{B}_n}(X_v)$ a.s. and in particular $X_n \geqslant E^{\mathscr{B}_n}(X_v)$ a.s. on $\{v \geqslant n\}$.

LEMMA II-2-14. *For a positive supermartingale* $(X_n, n \in \mathbf{N})$, *the sequence* $(X_{v \wedge n}, n \in \mathbf{N})$ *"stopped at the stopping time* v*" is also a positive supermartingale.*

In fact, the r.v. $X_{v \wedge n}$ is $\mathscr{B}_n$-measurable since

$$X_{v \wedge n} = \sum_{m<n} X_m 1_{\{v=m\}} + X_n 1_{\{v \geqslant n\}}.$$

Upon taking the conditional expectation $E^{\mathscr{B}_{n-1}}$ of both sides, it follows that for $n \geqslant 1$,

$$\begin{aligned} E^{\mathscr{B}_{n-1}}(X_{v \wedge n}) &= \sum_{m<n} X_m 1_{\{v=m\}} + E^{\mathscr{B}_{n-1}}(X_n) 1_{\{v \geqslant n\}} \\ &\leqslant \sum_{m<n} X_m 1_{\{v=m\}} + X_{n-1} 1_{\{v \geqslant n\}} = X_{v \wedge (n-1)}. \blacksquare \end{aligned}$$

For a positive martingale $(X_n, n \in \mathbf{N})$, the sequence $(X_{v \wedge n}, n \in \mathbf{N})$ stopped at a stopping time is likewise a positive martingale; this is immediately verified in the preceding proof. On the other hand, the inequality

$$X_{v_1} \geqslant E^{\mathscr{B}_{v_1}}(X_{v_2})$$

of Theorem II-2-13 does not in general become an equality for a positive martingale; we will study the case where there is effective equality in Chapter IV. (This will be an important study for applications).

REMARK. For a positive supermartingale $(X_n, n \in \mathbf{N})$, all the r.v.'s X_ν associated with a stopping time are integrable whenever the initial r.v. X_0 of the supermartingale is integrable. Indeed this follows immediately by integrating the inequality $X_0 \geqslant E^{\mathscr{B}_0}(X_\nu)$ which we have already established. ■

II-3. Exercises

II-1. Let ν be a stopping time relative to the increasing sequence $(\mathscr{B}_n, n \in \mathbf{N})$ of sub-σ-fields of $\mathscr{A}$ in the probability space $(\Omega, \mathscr{A}, P)$. For all $n \in \mathbf{N}$, denote by $\phi(n)$ the smallest integer p such that $\{\nu = n\} \in \mathscr{B}_p$. Show that $\phi(\nu)$ is a stopping time dominated by ν.

II-2. Dubins' inequalities cannot be improved: for every pair a, b of real numbers such that $0 < a < b < \infty$, there does in fact exist a positive supermartingale for which Dubins' inequalities

$$P^{\mathscr{B}_0}(\beta_{a,b} \geqslant k) \leqslant \left(\frac{a}{b}\right)^k \min\left(\frac{X_0}{a}, 1\right)$$

become equalities for all integers $k \geqslant 1$. Here is an example: Take a probability space supporting a sequence $(\xi_n, n \in \mathbf{N})$ of independent and identically distributed random variables taking the values 1 and 0 with probabilities a/b and $1 - a/b$ respectively; take $\mathscr{B}_0$ to be the trivial σ-field, and the σ-field generated by $\xi_0, \ldots, \xi_k$ for the σ-fields $\mathscr{B}_{2k+1} = \mathscr{B}_{2k+2}$ $(k \in \mathbf{N})$. The formulae $X_0 = a$ and

$$X_{2k+1} = b\xi_0 \cdots \xi_k, \qquad X_{2k+2} = a\xi_0 \cdots \xi_k \qquad (k \in \mathbf{N})$$

then define a supermartingale of the desired type.

II-3. On the space $[\Omega, \mathscr{A}, P; (\mathscr{B}_n, n \in \mathbf{N})]$, let $(Z_n, n \in \mathbf{N})$ be an adapted sequence of r.r.v.'s such that $E(\sup_{\mathbf{N}} |Z_n|) < \infty$, and let ν be an a.s. finite stopping time. Show that the formula

$$\nu^* = \inf(n : n \in \mathbf{N}, Z_n \geqslant E^{\mathscr{B}_n}(Z_\nu))$$

defines an a.s. finite stopping time such that $\nu^* \leqslant \nu$, and that

$$Z_n < E^{\mathscr{B}_n}(Z_{\nu^*}) \text{ a.s. } \quad \text{on } \{\nu^* > n\} \qquad (n \in \mathbf{N}).$$

II-4. Let $(X_n, n \in \mathbf{N})$, $(\beta_n, n \in \mathbf{N})$ and $(Y_n, n \in \mathbf{N})$ be three adapted sequences of finite positive r.r.v.'s defined on $[\Omega, \mathscr{A}, P; (\mathscr{B}_n, n \in \mathbf{N})]$ such that

$$E^{\mathscr{B}_n}(X_{n+1}) \leqslant (1 + \beta_n) X_n + Y_n \qquad (n \in \mathbf{N}).$$

This relation expresses the fact that $(X_n, n \in \mathbf{N})$ is "almost" a supermartingale.

Show that the limit $\lim_{n\to\infty} X_n$ exists and is finite a.s. on the event

$$A = \left\{ \sum_{\mathbf{N}} \beta_n < \infty,\ \sum_{\mathbf{N}} Y_n < \infty \right\}.$$

[To this end, one should consider the sequence $(U_n, n \in \mathbf{N})$ defined by

$$U_n = X_n' - \sum_{m<n} Y_n' \qquad (n \in \mathbf{N})$$

where

$$X_n' = X_n/(1+\beta_1)\ldots(1+\beta_{n-1}), \qquad Y_n' = Y_n/(1+\beta_1)\ldots(1+\beta_{n-1})$$

and also the stopping times

$$\nu_a = \min\left(n\colon \sum_{m\leqslant n} Y_m/(1+\beta_1)\ldots(1+\beta_{m-1}) > a \right).$$

One then observes that $(a + U_{\nu_a \wedge n}, n \in \mathbf{N})$ is a finite positive supermartingale.]

II-5. If $(X_n, n \in \mathbf{N})$ is a positive supermartingale and ν a stopping time on $[\Omega, \mathscr{A}, P; (\mathscr{B}_n, n \in \mathbf{N})]$, show that the sequence $(X_n', n \in \mathbf{N})$ defined by

$$X_n' = E^{\mathscr{B}_n}(X_{\nu \vee n}) \qquad (n \in \mathbf{N})$$

is again a positive supermartingale. [One can proceed directly or use the switching Lemma II-2-8.]

II-6. If $(X_n, n \in \mathbf{N})$ is a positive supermartingale on the space $[\Omega, \mathscr{A}, P; (\mathscr{B}_n, n \in \mathbf{N})]$ and if ν is a stopping time such that $X_\nu \geqslant X_{\nu-1}$ on $\{0 < \nu < \infty\}$, show that the formula

$$Y_n = \begin{cases} X_{(\nu-1)\wedge n} & \text{if } \nu \geqslant 1, \\ 0 & \text{if } \nu = 0 \end{cases}$$

defines a new positive supermartingale. In particular, the stopping time

$$\nu_a = \inf(n\colon X_n > a)$$

satisfies the hypotheses for all $a \in \mathbf{R}_+$ and the corresponding supermartingale $(Y_n, n \in \mathbf{N})$, which coincides with $(X_n, n \in \mathbf{N})$ on $\{\nu_a = \infty\}$, is then such that $0 \leqslant Y_n \leqslant a$ $(n \in \mathbf{N})$.

II-7. Let Z be an integrable r.r.v. defined on the probability space $(\Omega, \mathscr{A}, P)$. Beginning with the trivial σ-field $\mathscr{B}_0$, let us define inductively a sequence $(\mathscr{B}_n, n \in \mathbf{N})$ of sub-σ-fields of $\mathscr{A}$: the σ-field $\mathscr{B}_{n+1}$ is generated by $\mathscr{B}_n$ and the event $\{Z > E^{\mathscr{B}_n}(Z)\}$.

Show that the r.v. Z is measurable with respect to the σ-field $\bar{\mathscr{B}}_\infty$ generated by the σ-fields $\mathscr{B}_n$ $(n \in \mathbf{N})$ and the null sets. To this end one should show that on the event $A = \limsup_n \{Z > E^{\mathscr{B}_n}(Z)\}$, which belongs to $\mathscr{B}_\infty$, one has $Z \geqslant E^{\mathscr{B}_\infty}(Z)$, whilst on the complementary event one has $Z \leqslant E^{\mathscr{B}_\infty}(Z)$; one then concludes that $Z = E^{\mathscr{B}_\infty}(Z)$.

CHAPTER III

APPLICATIONS

III-1. Positive martingales and set functions. The Lebesgue and Radon–Nikodym theorems on the decomposition of measures

The following proposition merits careful study. It associates a set function with every positive integrable martingale and connects the σ-additivity properties of this set function with the properties at ∞ of the martingale.

PROPOSITION III-1-1. *For every positive integrable martingale* $(f_n, n \in \mathbf{N})$, *the formula*

$$Q(A) = \int_A f_n \, \mathrm{d}P \qquad (A \in \mathscr{B}_n, n \in \mathbf{N})$$

unambiguously defines a set function $Q : \bigcup_{\mathbf{N}} \mathscr{B}_n \to \mathbf{R}_+$ *whose restriction to each* $\mathscr{B}_n$ *is* σ*-additive* $(n \in \mathbf{N})$. *The a.s. limit* $f_\infty = \lim_n f_n$ *of the martingale* $(f_n, n \in \mathbf{N})$ *then furnishes the largest* $\mathscr{B}_\infty$*-measurable function (up to P-equivalence) such that* $\int_A f_\infty \mathrm{d}P \leqslant Q(A)$ *for all* $A \in \bigcup_{\mathbf{N}} \mathscr{B}_n$.

For the set function Q *to be* σ*-additive on* $\bigcup_{\mathbf{N}} \mathscr{B}_n$ *(i.e. for* $Q(A) = \sum_{\mathbf{N}} Q(A_p)$ *for any partition of a set* $A \in \bigcup_{\mathbf{N}} \mathscr{B}_n$ *into a countable disjoint union of sets in the same class), it is necessary and sufficient that* $\int_\Omega f_\nu \mathrm{d}P = Q(\Omega)$ *for all finite stopping times* ν, *or again, that* $E^{\mathscr{B}_n}(f_\nu) = f_{\nu \wedge n}$ *for all* $n \in \mathbf{N}$, *for all finite stopping, times* ν. *In this case the set function* Q *admits a unique* σ*-additive extension to the* σ*-field* $\mathscr{B}_\infty = \sigma(\bigcup_{\mathbf{N}} \mathscr{B}_n)$, *and for all finite stopping times* ν *the measure* Q *thus obtained is such that* $Q = f_\nu \cdot P$ *on* $\mathscr{B}_\nu$.

For the set function Q *to extend to a measure on the* σ*-field* $\mathscr{B}_\infty$ *of the form* $g \cdot P$ *where* $g \in L^1_+(P)$, *it is necessary and sufficient that* $\int_\Omega f_\infty \mathrm{d}P = Q(\Omega)$, *or again, that* $E^{\mathscr{B}_n}(f_\infty) = f_n$ *for all* $n \in \mathbf{N}$.

PROOF. (1) If $A \in \mathscr{B}_n$, the martingale identity $f_n = E^{\mathscr{B}_n}(f_p)$, where $p > n$, implies that

$$\int_A f_n \, \mathrm{d}P = \int_A f_p \, \mathrm{d}P,$$

and this shows that the indefinite integrals $\int f_n \mathrm{d}P$, defined respectively on the σ-fields B_n $(n \in \mathbf{N})$, extend one another as n increases. The existence of the set function Q on $\bigcup_{\mathbf{N}} \mathscr{B}_n$ is thus assured and clearly $Q = f_n \cdot P$ on $\mathscr{B}_n$.

Fatou's lemma then implies that

$$\int_A f_\infty \, \mathrm{d}P \leqslant \liminf_{p\to\infty} \int_A f_p \, \mathrm{d}P = Q(A)$$

for all $A \in \bigcup_{\mathbf{N}} \mathscr{B}_n$, since for $A \in \mathscr{B}_n$ we have the equality $\int_A f_p \mathrm{d}P = Q(A)$ whenever $p \geqslant n$. Conversely, if g is a positive $\mathscr{B}_\infty$-measurable function such that $\int_A g \, \mathrm{d}P \leqslant Q(A)$ for all $A \in \bigcup_{\mathbf{N}} \mathscr{B}_n$, we have

$$\int_A E^{\mathscr{B}_n}(g) \, \mathrm{d}P = \int_A g \, \mathrm{d}P \leqslant Q(A) = \int_A f_n \, \mathrm{d}P \quad \text{if } A \in \mathscr{B}_n;$$

we then deduce that $E^{\mathscr{B}_n}(g) \leqslant f_n$ since both functions are $\mathscr{B}_n$-measurable (consider $A = \{E^{\mathscr{B}_n}(g) > f_n\}$). As the function g is integrable, Proposition II-**2**-11 shows that

$$g = \lim_n E^{\mathscr{B}_n}(g) \leqslant \lim_n f_n = f_\infty.$$

The first part of the proposition being proved, we pass now to the second part of the proof.

(2) If Q is σ-additive on $\bigcup_{\mathbf{N}} \mathscr{B}_n$ and ν is a finite stopping time on Ω, then $Q(\Omega) = \sum_{\mathbf{N}} Q(\{\nu = n\})$, and consequently

$$Q(\Omega) = \sum_{\mathbf{N}} \int_{\{\nu=n\}} f_n \, \mathrm{d}P = \int_\Omega f_\nu \, \mathrm{d}P.$$

Conversely, if ν is a finite stopping time and if $\int_\Omega f_\nu \mathrm{d}P = Q(\Omega)$, the inequality $E^{\mathscr{B}_n}(f_\nu) \leqslant f_{\nu \wedge n}$ proved in Proposition II-**2**-13 can only be an equality, since the two sides have the same finite integral. [Note that $(f_{\nu \wedge n}, n \in \mathbf{N})$ is a martingale with initial term $f_{\nu \wedge 0} = f_0$ and consequently

$$\int f_{\nu\wedge n} \, \mathrm{d}P = \int f_0 \, \mathrm{d}P = Q(\Omega).]$$

Finally, let us suppose that the equality $E^{\mathscr{B}_n}(f_\nu) = f_{\nu \wedge n}$ is true for all finite stopping times and let us proceed to show that Q is σ-additive on $\bigcup_{\mathbf{N}} \mathscr{B}_n$; given a partition $A = \sum_{\mathbf{N}} A_p$ with $A \in \mathscr{B}_n$ and $A_p \in \mathscr{B}_{n_p}$, let us consider the finite stopping time ν defined by

$$\nu = \begin{cases} n & \text{on } A^c, \\ n \vee n_p & \text{on } A_p (p \in \mathbf{N}) \end{cases}$$

and write

$$\sum_{\mathbf{N}} Q(A_p) = \sum_{\mathbf{N}} \int_{A_p} f_{n \vee n_p} \mathrm{d}P = \int_A f_\nu \, \mathrm{d}P.$$

The a.s. equality $E^{\mathscr{B}_n}(f_\nu)=f_n$ valid on Ω since $\nu \geqslant n$ a.s. then implies that

$$\sum_{\mathbf{N}} Q(A_p) = \int_A f_\nu \,\mathrm{d}P = \int_A f_n \,\mathrm{d}P = Q(A)$$

since $A \in \mathscr{B}_n$.

If Q is σ-additive on the Boolean algebra $\bigcup_{\mathbf{N}} \mathscr{B}_n$, a theorem of Carathéodory shows that Q admits a unique σ-additive extension to the σ-field $\mathscr{B}_\infty = \sigma(\bigcup_{\mathbf{N}} \mathscr{B}_n)$ (see for example [223]). Then for every stopping time ν and every $B \in \mathscr{B}_\nu$, we have

$$Q(B \cap \{\nu < \infty\}) = \sum_{\mathbf{N}} Q(B \cap \{\nu = n\})$$

$$= \sum_{\mathbf{N}} \int_{B\cap\{\nu=n\}} f_n \,\mathrm{d}P = \int_{B\cap\{\nu<\infty\}} f_\nu \,\mathrm{d}P$$

because $B \cap \{\nu = n\} \in \mathscr{B}_n$ for all $n \in \mathbf{N}$; this shows that $Q = f_\nu \cdot P$ on $\mathscr{B}$ whenever ν is a finite stopping time.

(3) If Q admits an extension to the σ-field $\mathscr{B}_\infty$ of the form $g \cdot P$, we have $f_n = E^{\mathscr{B}_n}(g)$, since

$$\int_A f_n \,\mathrm{d}P = Q(A) = \int_A g \,\mathrm{d}P \quad \text{for all } A \in \mathscr{B}_n.$$

The a.s. limit f_∞ of the martingale $(f_n,\ n \in \mathbf{N})$ is then equal to $E^{\mathscr{B}_\infty}(g)$ by Proposition II-2-11, and consequently $f_n = E^{\mathscr{B}_n}(f_\infty)$ for all $n \in \mathbf{N}$. Conversely, if $f_n = E^{\mathscr{B}_n}(f_\infty)$ for all $n \in \mathbf{N}$, Proposition II-2-11 shows that f_∞ is also the limit in L^1 of the sequence $(f_n,\ n \in \mathbf{N})$; as a result,

$$\int_A f_\infty \,\mathrm{d}P = \lim_{p\to\infty} \int_A f_p \,\mathrm{d}P = Q(A) \quad \text{for all } A \in \bigcup_{\mathbf{N}} \mathscr{B}_n$$

and the measure $f_\infty \cdot P$ thus extends the set function Q to the σ-field $\mathscr{B}_\infty$. ■

Here is a simple example showing that the conditions of the second part of the previous proposition are not always satisfied. In Proposition III-2-6 to follow we will study a class of examples for which the conditions of the third part of the preceding proposition may or may not be fulfilled.

EXAMPLE. Let us take $\mathbf{N}$ equipped with the σ-field of all its subsets and a strictly positive probability $P = (p(n),\ n \in \mathbf{N})$ as the space $(\Omega, \mathscr{A}, P)$; for $\mathscr{B}_n$ take the σ-field generated by the partition $(\{0\}, \{1\}, \ldots, \{n-1\}, [n, \infty[)$ of $\mathbf{N}$. The sequence $(\mathscr{B}_n,\ n \in \mathbf{N})$ is increasing and the class $\bigcup_{\mathbf{N}} \mathscr{B}_n$ consists of all finite subsets of $\mathbf{N}$ and their complements, whilst $\mathscr{B}_\infty = \mathscr{P}(\mathbf{N})$. The set function Q

defined on $\bigcup_{\mathbf{N}} \mathscr{B}_n$ by $Q(A) = 0$ or 1 according as A or A^c is finite, is associated with the positive integrable martingale $(f_n, n \in \mathbf{N})$ defined by

$$f_n = \begin{cases} 0 & \text{on } [0, n[, \\ [\sum_{m \geq n} p(m)]^{-1} & \text{on } [n, \infty[. \end{cases}$$

It is clear that the set function Q is not σ-additive on $\bigcup_{\mathbf{N}} \mathscr{B}_n$ since $Q(\mathbf{N}) = 1$ and $\sum_{\mathbf{N}} Q(\{n\}) = 0$; on the other hand, the finite stopping time v defined on $\mathbf{N}$ by $v(\omega) = \omega + 1$ is such that $f_v = 0$ and thus fails to satisfy the equality $Q(\Omega) = \int f_v \mathrm{d}P$ considered in Proposition III-**1**-1. ■

At the beginning of this section we reduced certain asymptotic properties of a martingale to properties of a set function which we associated with it beforehand. For the remainder of this section we adopt the reverse viewpoint: we shall use martingales to study certain decompositions of measures.

We begin by proving Lebesgue's theorem on the decomposition of measures, whose statement is given below.

PROPOSITION III-**1**-2. *Let $(\Omega, \mathscr{A}, P)$ be a probability space and let Q be a second positive finite measure defined on the measurable space $(\Omega, \mathscr{A})$. There exists a positive integrable function f on $(\Omega, \mathscr{A}, P)$ and a subset $N \in \mathscr{A}$ of Ω of zero P-measure such that*

$$Q(A) = \int_A f \mathrm{d}P + Q(AN) \quad \textit{for all } A \in \mathscr{A}.$$

Furthermore, this decomposition is essentially unique: the function f is determined up to P-equivalence and the set N up to $(P+Q)$-equivalence. Moreover, the P-equivalence class of the function f is characterised by the property of being the largest class such that $f \cdot P \leq Q$ on $\mathscr{A}$.

PROOF. We will be relying on the following lemma, which is essentially a special case of the proposition.

LEMMA III-**1**-3. *Let μ and v be two positive finite measures on the measurable space $(\Omega, \mathscr{A})$ such that $\mu \leq v$ on $\mathscr{A}$. Then there exists a real-valued measurable function h with values in* [0, 1] *such that $\mu = h \cdot v$ on $\mathscr{A}$.*

We will prove this lemma by remarking that the mapping $g \to \int g \, \mathrm{d}\mu$ defines a continuous linear functional on the Hilbert space $L^2(v)$; Schwartz's inequality implies that for all $g \in L^2(\nu)$,

$$\left(\int g \, d\mu\right)^2 \leqslant \mu(\Omega) \int g^2 \, d\mu \leqslant \mu(\Omega) \int g^2 \, d\nu = \mu(\Omega) \|g\|^2_{L^2(\nu)}$$

since $\mu \leqslant \nu$. By the general form of continuous linear functionals on a real Hilbert space there thus exists an element $h \in L^2(\nu)$ such that

$$\int g \, d\mu = \int gh \, d\nu \quad \text{for all } g \in L^2(\nu).$$

To prove the lemma it remains to prove that $0 \leqslant h \leqslant 1$ a.s. ν on Ω.

But putting $g = 1_{\{h \leqslant -\varepsilon\}}$ in the preceding formula, we find that

$$0 \leqslant \mu(h \leqslant -\varepsilon) = \int_{\{h \leqslant -\varepsilon\}} h \, d\nu \leqslant -\varepsilon \, \nu(h \leqslant -\varepsilon) \leqslant 0,$$

which for $\varepsilon > 0$ is only possible if $\nu(h \leqslant -\varepsilon) = 0$; it is thus clear that $h \geqslant 0$ ν-a.s. On the other hand, putting $g = 1_{\{h \geqslant 1+\varepsilon\}}$ in the above formula we find

$$\mu(h \geqslant 1 + \varepsilon) = \int_{\{h \geqslant 1+\varepsilon\}} h \, d\nu \geqslant (1 + \varepsilon) \, \nu(h \geqslant 1 + \varepsilon);$$

but the left-hand side is bounded above by $\nu(h \geqslant 1 + \varepsilon)$ since $\mu \leqslant \nu$ on $\mathscr{A}$, and the preceding inequality is thus only possible if

$$\nu(h \geqslant 1 + \varepsilon) = 0;$$

as a consequence, $h \leqslant 1$ ν-a.s.

To prove Proposition III-1-2 we will apply the lemma just proved to $\mu = Q$ and $\nu = P + Q$; the hypotheses of the lemma clearly being satisfied, there thus exists a measurable function $h : \Omega \to [0, 1]$ such that $Q = h \cdot (P + Q)$ on $\mathscr{A}$. This formula establishes the equality between positive measures:

$$(1 - h) \cdot Q = h \cdot P \quad \text{on } \mathscr{A}.$$

Then put $N = \{h = 1\}$; this set is P-null since by the preceding equality,

$$P(N) = \int_N h \, dP = \int_N (1 - h) \, dQ = 0$$

because $h = 1$ on N. On the other hand, on N^c we can write

$$\frac{1}{1 - h} \cdot (1 - h) = 1$$

so that for all $A \in \mathscr{A}$,

$$Q(AN^c) = \int_{AN^c} \frac{1}{1-h} \cdot (1-h)\,\mathrm{d}Q = \int_{AN^c} \frac{1}{1-h} h\,\mathrm{d}P$$

which can also be written $Q(AN^c) = \int_A f\,\mathrm{d}P$ if we put $f = h/(1-h)$ on Ω (since $P(N) = 0$). We have thus proved the formula

$$Q(A) = Q(AN^c) + Q(AN) = \int_A f\mathrm{d}P + Q(AN) \qquad (A \in \mathscr{A}).$$

The second part of Proposition III-1-2 is very easily proved. If

$$Q(A) = \int_A f'\,\mathrm{d}P + Q(AN') \qquad (A \in \mathscr{A})$$

is a second decomposition of Q with $P(N') = 0$, then for all $A \in \mathscr{A}$,

$$\int_A f'\,\mathrm{d}P = \int_{A \cap (NN')^c} f'\,\mathrm{d}P = Q(A \cap (NN')^c) = \int_{A \cap (NN')^c} f\mathrm{d}P = \int_A f\mathrm{d}P$$

for $P(NN') = 0$, and this implies that $f = f'$ P-a.s. On the other hand,

$$Q(NN'^c) = \int_N f'\,\mathrm{d}P = 0$$

since $P(N) = 0$, and symmetrically $Q(N^c N') = 0$, whence N and N' are Q-equivalent; since these sets are P-null, they are also $(P + Q)$-equivalent.

Finally, if g is a positive measurable function such that $\int_A g\,\mathrm{d}P \leqslant Q(A)$ for all $A \in \mathscr{A}$, we also have

$$\int_A g\,\mathrm{d}P = \int_{AN^c} g\,\mathrm{d}P \leqslant Q(AN^c) = \int_{AN^c} f\mathrm{d}P = \int_A f\mathrm{d}P$$

for all $A \in \mathscr{A}$, and this implies that $g \leqslant f$ P-a.s. It is thus clear that the P-equivalence class of f is the largest satisfying the inequality $f \cdot P \leqslant Q$ on $(\Omega, \mathscr{A})$. ■

The Radon–Nikodym theorem stated in the corollary below is one of the more interesting particular cases of Lebesgue's decomposition theorem of measures.

COROLLARY III-1-4. *Let $(\Omega, \mathscr{A}, P)$ be a probability space. For every positive finite measure Q defined on $(\Omega, \mathscr{A})$, the following conditions are equivalent:*

(a) $Q(A) = 0$ *for all $A \in \mathscr{A}$ with probability $P(A) = 0$.*

(b) $Q = f \cdot P$ *for a positive integrable* f *on* $(\Omega, \mathscr{A})$.

(c) *For every real number* $\varepsilon > 0$, *there exists a real number* $\delta(\varepsilon) > 0$ *such that* $Q(A) \leqslant \varepsilon$ *if* $P(A) \leqslant \delta(\varepsilon)$ (*the condition called absolute continuity*).

Proof. The preceding proposition has in particular established the implication (a) $\Rightarrow$ (b), since the condition (a) implies that $Q(N) = 0$ because $P(N) = 0$. To prove the implication (b) $\Rightarrow$ (c), let us associate with every $\varepsilon > 0$ a real number a such that $Q(f > a) \leqslant \frac{1}{2}\varepsilon$, which is possible since $f < \infty$ Q-a.s.; for all $A \in \mathscr{A}$ we can then write

$$Q(A) \leqslant Q(f > a) + Q(A \cap \{f \leqslant a\}) = Q(f > a) + \int_{A \cap \{f \leqslant a\}} f \, \mathrm{d}P$$

$$\leqslant Q(f > a) + aP(A).$$

The inequality $Q(A) \leqslant \varepsilon$ will thus be satisfied whenever $P(A) \leqslant \varepsilon/2a$ and it remains to put $\delta(\varepsilon) = \varepsilon/2a$ for all $\varepsilon > 0$. Finally, the implication (c) $\Rightarrow$ (a) is clear. ■

Theorem II-2-9 for positive supermartingales allows us to immediately study the behaviour of Lebesgue's decomposition (Proposition III-2-4) when the base σ-field $\mathscr{A}$ varies.

Proposition III-1-5. *Let* $(\Omega, \mathscr{A}, P)$ *be a probability space and let* $(\mathscr{B}_n, n \in \mathbf{N})$ *be an increasing sequence of sub-*σ*-fields of* $\mathscr{A}$, *generating the sub-*σ*-field* $\mathscr{B}_\infty$ *of* $\mathscr{A}$. *Let* Q *be a positive finite measure on* $(\Omega, \mathscr{A})$ *whose Lebesgue decompositions on* $\mathscr{B}_n$ *can be written*

$$Q = f_n \cdot P + Q(\cdot \cap N_n) \quad \text{on } \mathscr{B}_n \quad (n \in \mathbf{N} \text{ or } n = \infty).$$

Then the sequence $(f_n, n \in \mathbf{N})$ *is a positive supermartingale which converges* P*-a.s. to the function* f_∞.

If the restrictions of the measure Q *to the* σ*-fields* $\mathscr{B}_n$ $(n \in \mathbf{N})$ *are absolutely continuous with respect to the corresponding restrictions of the probability* P *so that* $Q = f_n \cdot P$ *on* $\mathscr{B}_n$ $(n \in \mathbf{N})$, *the sequence* $(f_n, n \in \mathbf{N})$ *is a positive martingale. Furthermore, for the measure* Q *to be absolutely continuous with respect to* P *on the* σ*-field* $\mathscr{B}_\infty$, *it is necessary and sufficient that the a.s. convergence* $f = \lim_{n \to \infty} f_n$ *also takes place in the* L^1 *sense.*

Proof. The inequality $f_{n+1} \cdot P \leqslant Q$ on $\mathscr{B}_{n+1}$ implies that

$$\int_A E^{\mathscr{B}_n}(f_{n+1}) \, \mathrm{d}P = \int_A f_{n+1} \, \mathrm{d}P \leqslant Q(A)$$

for all $A \in \mathscr{B}_n$; but by Proposition III-1-2, f_n is, up to P-equivalence, the largest $\mathscr{B}_n$-measurable function satisfying the above inequality on $\mathscr{B}_n$. It then follows that $E^{\mathscr{B}_n}(f_{n+1}) \leqslant f_n$ for all $n \in \mathbf{N}$. We will denote by f the a.s. limit of the positive supermartingale $(f_n, n \in \mathbf{N})$, the existence of which is guaranteed by Theorem II-2-9.

The inequality $f_\infty \cdot P \leqslant Q$ on $\mathscr{B}_\infty$ implies as above that $E^{\mathscr{B}_n}(f_\infty) \leqslant f_n$ for all $n \in \mathbf{N}$; letting $n \uparrow \infty$ we obtain $f_\infty \leqslant f$ by Proposition II-2-11. On the other hand, Fatou's lemma shows that

$$\int_A f \, \mathrm{d}P \leqslant \liminf_{n\to\infty} \int_A f_n \, \mathrm{d}P \leqslant Q(A) \quad \text{if } A \in \bigcup_{\mathbf{N}} \mathscr{B}_n$$

since $\int_A f_n \, \mathrm{d}P \leqslant Q(A)$ if $A \in \mathscr{B}_p$ and $n > p$; the preceding inequality extends to the σ-field $\mathscr{B}_\infty$ generated by $\bigcup_{\mathbf{N}} \mathscr{B}_n$ and the maximality of f_∞ in Proposition III-1-2 implies that $f \leqslant f_\infty$. We have thus proved that $f = f_\infty$.

If $Q = f_n \cdot P$ on $\mathscr{B}_n$ $(n \in \mathbf{N})$, then it is easy to check that the sequence $(f_n, n \in \mathbf{N})$ is a positive martingale and not just a positive supermartingale. Now let us suppose that Q is absolutely continuous with respect to P on $\mathscr{B}_\infty$. By the preceding this is equivalent to supposing that $Q = f_\infty \cdot P$ on $\mathscr{B}_\infty$, where $f_\infty = \lim_{n\to\infty} f_n$ P-a.s. From the identity

$$|f_n - f_\infty| = f_n + f_\infty - 2(f_n \wedge f_\infty) \quad (n \in \mathbf{N})$$

we find that

$$\int |f_n - f_\infty| \, \mathrm{d}P = 2 - 2\int (f_n \wedge f_\infty) \, \mathrm{d}P \to 0 \quad \text{as } n \uparrow \infty,$$

the last step following from the dominated convergence theorem since $0 \leqslant f_n \wedge f_\infty \leqslant f_\infty$ $(n \in \mathbf{N})$ and $f_n \wedge f_\infty \to f_\infty$ a.s. as $n \uparrow \infty$. (This argument constitutes Scheffé's lemma.)

Finally, let us suppose that $f_n \to f_\infty$ a.s. *and* in L^1. Then by the remark following Proposition II-2-11 we have $E^{\mathscr{B}_n}(f_\infty) = f_n$ a.s. $(n \in \mathbf{N})$. Thus for $A \in \mathscr{B}_n$ $(n \in \mathbf{N})$ we have

$$\int_A f_\infty \, \mathrm{d}P = \int_A E^{\mathscr{B}_n}(f_\infty) \, \mathrm{d}P = \int_A f_n \, \mathrm{d}P = Q(A),$$

whence Q and $f_n \cdot P$ agree on $\bigcup_{n \in \mathbf{N}} \mathscr{B}_n$. Applying Carathéodory's theorem as above, we deduce that $Q = f_\infty \cdot P$ on $\mathscr{B}_\infty$, and the proposition is completely proved. ■

The sets N_n ($n \in \mathbf{N}$ or $n = \infty$) of the previous proposition increase with n (up to Q-equivalence) because

$$Q(N_n \cap N_p^c) = \int_{N_n} f_p \, \mathrm{d}P = 0$$

when $n \leqslant p \leqslant \infty$. But it is not generally true that N_∞ is the limit of the N_n when $n \uparrow \infty$; for example, when the hypotheses of case (b) of Proposition III-**2**-6 below are satisfied, the sets N_n can be taken to be empty for all n whilst $Q(\Omega \setminus N_\infty) = 0$.

To conclude this section, here is a concrete example of the application of the preceding proposition. Suppose that the σ-field $\mathscr{A}$ of the space $(\Omega, \mathscr{A}, P)$ can be generated by a countable sequence of sets (perhaps up to sets of measure zero); such is the case with the Borel σ-field of the interval $[0, 1[$ which we equip with Lebesgue measure. Moreover, let us choose a sequence $(A_n^i, i \in I_n)$ of increasingly finer countable partitions of Ω in $\mathscr{A}$ which generate the σ-field $\mathscr{A}$; then the σ-fields $(\mathscr{B}_n, n \in \mathbf{N})$ associated with these partitions increase with n and generate $\mathscr{A}$. In the case of $[0, 1[$, we can take the dyadic partitions $([2^{-n}k, 2^{-n}(k+1)[, 0 \leqslant k < 2^n)$. For simplicity we will suppose that the partitions have been chosen such that $P(A_n^i) > 0$ for all $i \in I_n$ and all $n \in \mathbf{N}$; this is always possible and is seen to be the case with the dyadic partitions of $[0, 1[$ and Lebesgue measure.

If Q is a second positive measure, finite on $(\Omega, \mathscr{A})$, by the Radon–Nikodym theorem there exist $\mathscr{B}_n$-measurable positive integrable functions f_n such that $Q = f_n \cdot P$ on $\mathscr{B}_n$ ($n \in \mathbf{N}$); by hypothesis the set $\emptyset$ is the only $\mathscr{B}_n$-measurable set of zero P-measure, so that Corollary III-**1**-4 is applicable. However, since the σ-fields $\mathscr{B}_n$ are generated by a countable partition, it is not necessary to appeal to the Radon–Nikodym theorem to see that $Q = f_n \cdot P$ on $\mathscr{B}_n$ if we put

$$f_n(\omega) = \frac{Q(A_n^i)}{P(A_n^i)} \quad \text{if } \omega \in A_n^i \qquad (i \in I_n, n \in \mathbf{N});$$

the functions f_n are thus explicitly determined.

By Proposition III-**1**-5 this sequence $(f_n, n \in \mathbf{N})$ converges P-a.s. to the function f in the Lebesgue decomposition of Q with respect to P relative to the σ-field $\mathscr{B}_\infty$; in other words, for P-almost all ω, we have

$$\lim_{n \to \infty} \frac{Q[A_n(\omega)]}{P[A_n(\omega)]} = f(\omega)$$

if $A_n(\omega)$ denotes that element of the partition $(A_n^i, i \in I_n)$ which contains ω.

Thus the martingale theorem in this case gives us an explicit procedure for constructing the function f in the Lebesgue decomposition of Q with respect to P on $\mathscr{B}_\infty$.

III-2. Applications to statistical tests

Using the results of Section III-1 above we now introduce a class of martingales which are important in statistics and which are also of interest in themselves. In this class we shall find some martingales which satisfy the conditions of the third part of Proposition III-1-1, and others which do not; on the other hand, by construction the hypotheses of the second part of this proposition will always be fulfilled.

PROPOSITION III-2-6. *Let* $P = \Pi_{\mathbf{N}^*} P_n$ *and* $Q = \Pi_{\mathbf{N}^*} Q_n$ *be two probabilities on* $\mathbf{R}^{\mathbf{N}^*}$ *defined as infinite products of two sequences* $(P_n, n \in \mathbf{N}^*)$ *and* $(Q_n, n \in \mathbf{N}^*)$ *of probabilities on* $\mathbf{R}$. *We suppose moreover that for all* $n \in \mathbf{N}^*$, *the probability* Q_n *possesses a density with respect to* P_n, *which is written* $Q_n = g_n \cdot P_n$ *on* $\mathbf{R}$. *Then if* Y_n $(n \in \mathbf{N}^*)$ *denote the coordinate mappings of* $\mathbf{R}^{\mathbf{N}^*}$ *onto* $\mathbf{R}$ *and if we put* $\mathscr{B}_n = \sigma(Y_1, \ldots, Y_n)$ *for* $n \in \bar{\mathbf{N}}$, *the formula*

$$f_n = \prod_1^n g_m \circ Y_m \qquad (n \in \mathbf{N})$$

defines a positive integrable martingale on the space $[\mathbf{R}^{\mathbf{N}^*}, \mathscr{B}_\infty, P; (\mathscr{B}_n, n \in \mathbf{N})]$ *with initial term* $f_0 = 1$, *such that*

$$Q = f_n \cdot P \quad \text{on } \mathscr{B}_n$$

for all $n \in \mathbf{N}$. *More generally, for all stopping times* $\nu : \mathbf{R}^{\mathbf{N}^*} \to \bar{\mathbf{N}}$ *which are a.s. finite for* BOTH *probabilities* P *and* Q, *we have* $Q = f_\nu \cdot P$ *on the* σ*-field* $\mathscr{B}_\nu$.

Moreover, two cases are possible: Either

(a) *the infinite product*

$$\prod_{\mathbf{N}^*} \int_{\mathbf{R}} \sqrt{g_n}\, \mathrm{d}P_n$$

of the integrals $\int_{\mathbf{R}} \sqrt{(g_n)}\, \mathrm{d}P_n$ *which are* $\leqslant 1$, *is* > 0; *in this case the martingale* $(f_n, n \in \mathbf{N})$ *converges a.s. and in* L^1 *to a r.v.* f_∞; *furthermore,* $Q = f_\infty \cdot P$ *on* $\mathscr{B}_\infty$;

or

(b) *the infinite product above is* 0; *in this case the martingale* $(f_n, n \in \mathbf{N})$ *converges a.s.* P *to* 0 *and the same sequence converges a.s.* Q *to* $+\infty$. *The two probabilities* P *and* Q *are then mutually singular on* $\mathscr{B}_\infty$ *(in the sense that there exists a set* $B \in \mathscr{B}_\infty$ *such that* $P(B^c) = 0 = Q(B)$).

PROOF. (1) We begin by showing that $Q = f_n \cdot P$ on $\mathscr{B}_n$ for all $n \in \mathbf{N}$. If $B_1, \ldots, B_n$ are Borel subsets of $\mathbf{R}$, we have

$$Q[B_1 \times \ldots \times B_n \times \mathbf{R} \times \mathbf{R} \ldots] = \prod_1^n Q_m(B_m) = \prod_1^n \int_{B_m} g_m \, \mathrm{d}P_m$$

$$= \int_{B_1 \times \ldots \times B_n \times \mathbf{R} \times \mathbf{R} \ldots} \prod_1^n g_m \circ Y_m \, \mathrm{d}P;$$

this shows that the probabilities Q and $f_n \cdot P$ agree on the class of all sets of the form $B_1 \times \ldots \times B_n \times \mathbf{R} \times \mathbf{R} \times \ldots$ which is closed under intersection and which contains Ω; thus they agree on the σ-field $\mathscr{B}_n$ generated by these sets.

The equality $Q = f_\nu \cdot P$ on $\mathscr{B}_\nu$ for all stopping times ν which are a.s. finite for both P and Q is then proved from the equalities $Q = f_n \cdot P$ on $\mathscr{B}_n$ ($n \in \mathbf{N}$) along the lines of the proof of Proposition III-1-1: for all stopping times ν and every $B \in \mathscr{B}_\nu$ the following equalities hold:

$$Q(B \cap \{\nu < \infty\}) = \sum_{\mathbf{N}} Q(B \cap \{\nu = n\}) = \sum_{\mathbf{N}} \int_{B \cap \{\nu = n\}} f_n \, \mathrm{d}P$$

$$= \sum_{\mathbf{N}} \int_{B \cap \{\nu = n\}} f_\nu \, \mathrm{d}P = \int_{B \cap \{\nu < \infty\}} f_\nu \, \mathrm{d}P.$$

If ν is a.s. finite for both P and Q, then we have $Q = f_\nu \cdot P$ on $\mathscr{B}_\nu$.

(2) The equalities $Q = f_n \cdot P$ on $\mathscr{B}_n$ ($n \in \mathbf{N}$) which we have already proved imply that $(f_n, n \in \mathbf{N})$ is a positive martingale with expectation 1 on the space $\mathbf{R}^{\mathbf{N}^*}$ equipped with the probability P and the sequence $(\mathscr{B}_n, n \in \mathbf{N})$. For each $n \in \mathbf{N}$, the r.v. f_n is indeed $\mathscr{B}_n$-measurable and such that

$$\int_B f_n \, \mathrm{d}P = Q(B) = \int_B f_{n+1} \, \mathrm{d}P \quad \text{if } B \in \mathscr{B}_n,$$

so that $f_n = E^{\mathscr{B}_n}(f_{n+1})$. By Proposition III-1-5 the martingale $(f_n, n \in \mathbf{N})$ thus converges a.s. to a limit f_∞ which is such that $Q = f_\infty \cdot P + Q(\cdot \cap N)$ for a P-null set N; furthermore, $\mathbf{N} = \emptyset$ if and only if the martingale $(f_n, n \in \mathbf{N})$ converges in L^1.

On the other hand, Schwartz's inequality implies that

$$\int_R \sqrt{g_n} \, \mathrm{d}P_n \leqslant 1$$

since $\int_R g_n \, \mathrm{d}P_n = 1$; hence the infinite product of the $\sqrt{g_n}$ always makes sense. If this product is strictly positive, the sequence $(\sqrt{f_n},\, n \in \mathbf{N})$ converges in L^2 for

$$\int_{\mathbf{R}^{\mathbf{N}^*}} \sqrt{f_n f_{n+p}} \, \mathrm{d}P = \int_{\mathbf{R}^{\mathbf{N}^*}} \prod_1^n g_m \circ Y_m \prod_{n+1}^{n+p} \sqrt{g_m \circ Y_m} \, \mathrm{d}P$$

$$= \prod_{n+1}^{n+p} \int_{\mathbf{R}} \sqrt{g_m} \, \mathrm{d}P_m$$

which implies that

$$\|\sqrt{f_{n+p}} - \sqrt{f_n}\|_2^2 = \int_{\mathbf{R}^{\mathbf{N}^*}} (f_{n+p} + f_n - 2\sqrt{f_n f_{n+p}}) \, \mathrm{d}P$$

$$= 2\left(1 - \prod_{n+1}^{n+p} \int_{\mathbf{R}} \sqrt{g_m} \, \mathrm{d}P_m\right) \to 0$$

when $n + p \geqslant n \to \infty$. Then the sequence $(f_n,\, n \in \mathbf{N})$ converges in L^1, because

$$\|f_{n+p} - f_n\|_1 = \int_{\mathbf{R}^{\mathbf{N}^*}} |\sqrt{f_{n+p}} - \sqrt{f_n}| \, (\sqrt{f_{n+p}} + \sqrt{f_n}) \, \mathrm{d}P$$

$$\leqslant \|\sqrt{f_{n+p}} - \sqrt{f_n}\|_2 \, (\|\sqrt{f_{n+p}}\|_2 + \|\sqrt{f_n}\|_2)$$

$$\to 0 \quad \text{as } n + p \geqslant n \to \infty;$$

by the result recalled above we then have $Q = f_\infty \cdot P$ on $\mathscr{B}_\infty$.

On the other hand, if the infinite product of integrals of the $\sqrt{g_n}$ is zero, the relations

$$\int_{\mathbf{R}^{\mathbf{N}^*}} \sqrt{f_n} \, \mathrm{d}P = \prod_1^n \int_R \sqrt{g_m} \, \mathrm{d}P_m \downarrow 0 \quad \text{as } n \uparrow \infty$$

and Fatou's lemma imply that the positive r.v. $f_\infty = \lim_{n \to \infty}$ a.s. f_n is a.s. zero. By the result above the probability Q is then carried by a P-null set N. It only remains to show that $\lim_{n \to \infty} f_n$ exists and is infinite a.s. Q.

To this end let us note that $(1/\sqrt{f_n},\, n \in \mathbf{N})$ is a positive supermartingale on the space $\mathbf{R}^{\mathbf{N}^*}$, when the latter is given the probability Q. In fact for all $B \in \mathscr{B}_n$ we can write

$$\int_B \frac{1}{\sqrt{f_n}} \mathrm{d}Q = \int_{B \cap \{f_n \neq 0\}} \frac{1}{\sqrt{f_n}} f_n \, \mathrm{d}P$$

because $Q = f_n \cdot P$ on $\mathscr{B}_n$ and $Q(\{f_n = 0\}) = 0$; hence

$$\int_B \frac{1}{\sqrt{f_n}} \mathrm{d}Q = \int_{B \cap \{f_n \neq 0\}} \sqrt{f_n} \mathrm{d}P = \int_B \sqrt{f_n} \mathrm{d}P$$

(the introduction of the set $\{f_n \neq 0\}$ is unavoidable since the expression $f_n / \sqrt{f_n}$ is undefined on $\{f_n = 0\}$). The supermartingale inequality

$$\int_B \sqrt{f_n} \mathrm{d}P \geqslant \int_B \sqrt{f_{n+1}} \mathrm{d}P \qquad (B \in \mathscr{B}_n, n \in \mathbf{N})$$

then implies that

$$\int_B \frac{1}{\sqrt{f_n}} \mathrm{d}Q \geqslant \int_B \frac{1}{\sqrt{f_{n+1}}} \mathrm{d}Q \qquad (B \in \mathscr{B}_n, n \in \mathbf{N})$$

as we wished to show. The positive supermartingale $(1/\sqrt{f_n}, n \in \mathbf{N})$ converges Q-a.s. by Theorem II-2-9 and converges to 0 in $L^1(Q)$ because

$$\int_{\mathbf{R}^{\mathbf{N}^*}} \frac{1}{\sqrt{f_n}} \mathrm{d}Q = \int_{\mathbf{R}^{\mathbf{N}^*}} \sqrt{f_n} \mathrm{d}P \downarrow 0 \quad \text{when } n \uparrow \infty$$

by the assumption made. We have thus proved $1/\sqrt{f_n} \to 0$ Q-a.s. or, equivalently, that $f_n \to \infty$ Q-a.s. ■

COROLLARY III-2-7. *Let P_0 and Q_0 be two distinct probabilities on $(\mathbf{R}, \mathscr{R})$ and let us suppose that $Q_0 = g \cdot P_0$ for a probability density g on $(\mathbf{R}, \mathscr{R}, P_0)$. Then if on the space $\mathbf{R}^{\mathbf{N}^*}$ equipped with the σ-field $\mathscr{B}_\infty$, $P = P_0^{\mathbf{N}^*}$ and $Q = Q_0^{\mathbf{N}^*}$ denote the infinite product probabilities of P_0 and Q_0 respectively, generated by the coordinates $Y_n : \mathbf{R}^{\mathbf{N}^*} \to \mathbf{R}$ $(n \in \mathbf{N}^*)$, we have*

$$\prod_1^n g \circ Y_m \begin{cases} \to 0 & P\text{-a.s.} \\ \to +\infty & Q\text{-a.s.} \end{cases} \quad \text{on } \mathbf{R}^{\mathbf{N}^*},$$

and the probabilities P and Q are thus mutually singular on $(\mathbf{R}^{\mathbf{N}^*}, \mathscr{B}_\infty)$.

PROOF. It suffices to apply the preceding proposition in the case where the densities g_n are all equal to the same density g. Since $Q_0 = g \cdot P_0$ is distinct from P_0, the function g is not P_0-a.s. equal to 1 and Schwartz's inequality $\int \sqrt{g} \, \mathrm{d}P_0 < 1$ is thus strict: we are in the second case considered in the last part of the proposition. ■

REMARK. An alternative proof of the preceding corollary consists of using the law of large numbers (which is itself a martingale theorem!). The law of large numbers shows that when $n \to \infty$,

$$\frac{1}{n}\sum_{1}^{n} \log(g \circ Y_m) \to \int_{\mathbf{R}^{\mathbf{N}^*}} \log(g \circ Y_1)\,dP = \int_{\mathbf{R}} \log(g)\,dP_0$$

in the sense of a.s. convergence for the probability P, since the sequence $(\log(g \circ Y_m),\ m \in \mathbf{N}^*)$ is, like the sequence $(Y_m,\ m \in \mathbf{N}^*)$, independent and identically distributed on $(\mathbf{R}^{\mathbf{N}^*}, \mathscr{B}_\infty, P)$; but by the strict convexity of the logarithm function, we have

$$\int_R \log(g)\,dP_0 < \log\left(\int_R g\,dP_0\right) = 0$$

since g is not P_0-a.s. equal to 1. The convergence above implies, after multiplying the left-hand side by n, that

$$\sum_{1}^{n} \log(g \circ Y_m) \to -\infty \quad P\text{-a.s.} \quad \text{as } n \to \infty.$$

The first convergence in the corollary is thus proved; the second can be proved similarly. (In fact, for the law of large numbers to be applicable in the above, it is necessary to suppose that the function $\log g$ is P_0-integrable; however, if this was not the case, it would suffice to replace the function $\log g$ by the larger function $\sup(\log g, -a)$ which is integrable since it is between $-a$ and g and thus its integral with respect to P_0 could be made negative if a was chosen sufficiently large.) ■

The preceding results are applicable to the study of statistical tests between two simple hypotheses. We begin with the basic ideas.

Let us consider a random experiment described by a measurable space $(\Omega, \mathscr{B})$ and one or the other of two probabilities P and Q defined on this space. To *test* which of these two probabilities P or Q governs the random experiment based on a single outcome ω of this experiment, the statistician adopts the following *strategy*: He takes an event $D \in \mathscr{B}$ called the *critical region* of the test beforehand and decides that the experiment is governed by Q (resp. P) if the result ω of the observation belongs to D (resp. D^c). By adopting such a strategy the statistician makes an incorrect decision with probability $P(D)$ (resp. $Q(D^c)$) if P (resp. Q) is the true probability governing the experiment. He thus wishes to make these two error probabilities $P(D)$ and $Q(D^c)$ as small as possible, but as these probabilities are increasing and decreasing functions, respectively, of the critical region D, the way to minimise them is not clear! We propose to minimise a linear combination with positive coefficients of the two errors, say $aP(D)+bQ(D^c)$ where a and b are two strictly positive real numbers. Suppose now that the probability Q can be written in the form $Q=f\cdot P$ on the space $(\Omega, \mathscr{B})$. It is then easy to establish that

$$\inf_{D\in\mathscr{B}} [aP(D)+bQ(D^c)] = \int_\Omega \min(a, bf)\,\mathrm{d}P$$

and that this lower bound is attained exactly on the sets D^* satisfying the double inclusion

$$\{a<bf\}\subset D^*\subset\{a\leqslant bf\}.$$

(When $P(\{f=a/b\})=0$, this double inclusion determines D^* uniquely up to P-equivalence; on the other hand, the probability $P(\{f=c\})$ cannot be different from zero for more than a countable set of values of c ($c\in\mathbf{R}_+$).)

After recalling these general notions we go on to study a specific problem which counts as one of the most classical in mathematical statistics. Take a sequence $(Y_n, n\in\mathbf{N}^*)$ of independent and identically distributed real-valued r.v.'s governed by one or the other of the two distinct probability laws P_0 and Q_0 on $\mathbf{R}$; more precisely, we will suppose further that, as in the hypotheses of Corollary III-**2**-7, the $(Y_n, n\in\mathbf{N}^*)$ are the coordinate mappings of $\mathbf{R}^{\mathbf{N}^*}$ and that this space is equipped with probabilities $P=P_0^{\mathbf{N}^*}$ and $Q=Q_0^{\mathbf{N}^*}$ which makes $(Y_n, n\in\mathbf{N}^*)$ a sequence of independent and identically distributed r.r.v's with laws P_0 and Q_0 respectively. We shall also suppose that $Q_0=g\cdot P_0$ for a P_0-density g defined on $\mathbf{R}$.

By Corollary III-**2**-7, the probabilities P and Q are mutually singular on $(\mathbf{R}^{\mathbf{N}^*}, \mathscr{B}_\infty)$, and it is thus quite impossible that Q should admit a density with respect to P on this measure space. But in practice the statistician never has all the information in the σ-field $\mathscr{B}_\infty$ generated by the infinite sequence $(Y_n, n\in\mathbf{N}^*)$ at his disposal; rather it is the σ-fields $\mathscr{B}_n$ corresponding to the information carried by the finite sequences $(Y_1, \ldots, Y_n)$ which should be considered. But for all finite n we know that

$$Q=f_n\cdot P \quad \text{on } \mathscr{B}_n \quad \text{if } f_n=\prod_1^n g\circ Y_m;$$

by the above, the statistician will thus use tests based on the critical regions defined by

$$D_{n,c}=\left\{\prod_1^n g\circ Y_m\geqslant c\right\} \qquad (n\in\mathbf{N}^*, c>0).$$

For a fixed value of n and for $c = b/a$, the test based on $D_{n,c}$ will give a minimum weighted error probability $aP(D) + bQ(D^c)$ equal to

$$I_{n;a,b} = \int_{\mathbf{R}^{\mathbf{N}^*}} \min(a, bf_n)\,\mathrm{d}P.$$

We know that $(f_n, n \in \mathbf{N})$ is a positive martingale on $\mathbf{R}^{\mathbf{N}^*}$ for the probability P; hence the sequence $(\min(a, bf_n), n \in \mathbf{N})$ is a positive supermartingale for P since the conditional expectation $E^{\mathscr{B}_n}(\min(a, bf_{n+1}))$ is dominated by both $E^{\mathscr{B}_n}(a) = a$ and by $E^{\mathscr{B}_n}(bf_{n+1}) = bf_n$. It follows that the integrals $I_{n;a,b}$ of the terms in this supermartingale decrease with n, which corresponds to the fact that the minimum weighted errors $I_{n;a,b}$ decrease when the amount of information $\mathscr{B}_n$ at the disposal of the statistician increases. Further when $n \uparrow \infty$, the dominated convergence theorem shows that the error $I_{n;a,b} \downarrow 0$ since $f_n \to 0$ P-a.s. (Corollary III-2-7); this result should be compared with the fundamental theorem of statistics which permits, given the observation of an infinite sequence of independent and identically distributed r.r.v's, the recovery a.s. of their common probability law.

The test based on the critical region $D_{n,c} = \{f_n \geqslant c\}$ consists of "accepting" the probability law P_0 (resp. Q_0) when the probability density f_n of Q with respect to P on $\mathscr{B}_n$ is small (resp. large); this is natural since a priori (before the experiment) the probability Q assigns high probability to the region $D_{n,c}$ and low probability to its complement, in comparison with P. As a matter of fact, the latter argument is only convincing for the subset $D_{n,c}$ where f_n is very much larger than c and for the subset $D_{n,c}^c$ where f_n is very much smaller than c; in the intermediate region where $f_n \sim c$, the statistician might prefer, if he has the opportunity, to continue observing the sequence $(Y_m, m \in \mathbf{N}^*)$ beyond the time n. We are thus led to a study of tests based upon a random number, say ν, of observations (for example, we might take $\nu = n$ if $f_n \geqslant 10c$ or $f_n \leqslant \frac{1}{10}c$ and $\nu = 2n$ otherwise). This random number ν cannot be arbitrary: the "decision" $\{\nu = n\}$ to observe exactly n random variables Y_m must be based solely upon information in $\mathscr{B}_n$; put another way, ν should be a stopping time. The σ-field $\mathscr{B}_\nu$ will then represent the information available to the statistician adopting this "rule for terminating sampling".

For a fixed stopping time ν, which we will suppose *finite* (since this will be the case in practice!), we know that $Q = f_\nu \cdot P$ on $\mathscr{B}_\nu$ (with $f_\nu = f_n$ on $\{\nu = n\}$!). When such a stopping time ν is fixed, the best critical regions are of the form

$$\{f_\nu \geqslant c\} = \sum_{n \in \mathbf{N}^*} \{\nu = n\} \cap D_{n,c}$$

and correspond with minimal weighted errors $aP(D)+bQ(D^c)$ taking the values

$$I_{\nu;a,b} = \int_{\mathbf{R}^{\mathbf{N}^*}} \min(a, bf_\nu)\,\mathrm{d}P$$

if $c=b/a$. Let us also note that the inequality $I_{\nu_1;a,b} \geqslant I_{\nu_2;a,b}$ holds whenever $\nu_1 \leqslant \nu_2$ as a consequence of Theorem II-2-13 and admits an immediate statistical interpretation.

The problem of the choice of the stopping time ν in this statistical problem will be studied in Section VI-5.

III-3. Haar systems and bases

Definition III-3-8. In a probability space $(\Omega, \mathscr{A}, P)$, an increasing sequence $(\mathscr{B}_n, n \in \mathbf{N})$ of sub-σ-fields of $\mathscr{A}$ which generates $\mathscr{A}$ is called a *Haar system* if for all $n \in \mathbf{N}$ the σ-field $\mathscr{B}_n$ is generated by a partition $(B_0^{(n)}, \ldots, B_n^{(n)})$ of Ω into $n+1$ events of strictly positive probability.

The σ-field $\mathscr{B}_0$ is thus the minimum σ-field $\{\emptyset, \Omega\}$. The property that each σ-field $\mathscr{B}_n$ is generated by a partition of exactly $n+1$ sets is crucial; together with the increasing property of the sequence $(\mathscr{B}_n, n \in \mathbf{N})$ it implies that for all $n \in \mathbf{N}$, the partition $(B_i^{(n+1)}, 0 \leqslant i \leqslant n+1)$ is obtained by dividing one of the events $B_j^{(n)}$ $(0 \leqslant j \leqslant n)$ into two non-null events and keeping the other events $B_j^{(n)}$ intact. We remark in passing that amongst all the strictly increasing sequences of sub-σ-fields of $\mathscr{A}$, the Haar systems are the minimal sequences.

With every Haar system $(\mathscr{B}_n, n \in \mathbf{N})$ we will associate an adapted sequence $(U_n, n \in \mathbf{N})$ of r.r.v.'s as follows: firstly we will put $U_0 = 1$ and for all $n \in \mathbf{N}$, we define

$$U_{n+1} = \begin{cases} a & \text{on } B_k^{(n+1)}, \\ b & \text{on } B_l^{(n+1)}, \\ 0 & \text{otherwise,} \end{cases}$$

denoting by $B_k^{(n+1)}$, $B_l^{(n+1)}$ the only two elements of the partition generating $\mathscr{B}_{n+1}$ which did not already belong to $\mathscr{B}_n$, and choosing the real numbers a and b in such a way that

$$E(U_{n+1}) = 0, \qquad E(U_{n+1}^2) = 1.$$

Each of the r.r.v.'s U_{n+1} is thus defined uniquely up to a change of sign and it is then easy to check that:

(1) $E^{\mathscr{B}_n}(U_{n+1}) = 0$ *for all* $n \in \mathbf{N}$,

(2) *for all* $n \in \mathbf{N}$, *the sequence* $(U_0, \ldots, U_n)$ *is an orthonormal basis for* $L^2(\mathscr{B}_n)$ *so that* $(U_n, n \in \mathbf{N})$ *is an orthonormal basis for the space* $L^2(\Omega, \mathscr{A}, P)$.

In what follows we shall call every sequence $(U_n, n \in \mathbf{N})$ in $L^2(\Omega, \mathscr{A}, P)$ constructed from a Haar system $(\mathscr{B}_n, n \in N)$ in the preceding manner a *Haar basis.*

The importance of Haar bases is connected with the following result.

PROPOSITION III-3-9. *For every Haar basis $(U_n, n \in \mathbf{N})$ and every function $f \in L^2$, the series $\sum_{\mathbf{N}} \hat{f}_n U_n$,* where $\hat{f}_n = \int f U_n \,\mathrm{d}P$, *converges a.s. (and in L^2) to the function f. More generally, for all $p \in [1, \infty[$ and for all $f \in L^p$, the preceding coefficients $\hat{f}_n$ $(n \in \mathbf{N})$ remain well-defined and the series $\sum_{\mathbf{N}} \hat{f}_n U_n$ converges a.s. and in L^p to the function f.*

PROOF. If $f \in L^2$, we have

$$E^{\mathscr{B}_n}(f) = \sum_{m \leqslant n} \hat{f}_m U_m \qquad (n \in \mathbf{N})$$

as $(U_0, \ldots, U_n)$ is an orthonormal basis of $L^2(\mathscr{B}_n)$; this formula remains valid for all $f \in L^1$ (and a fortiori for all $f \in L^p$) by continuity on L^1. Proposition II-2-11 then gives the result. ■

EXAMPLE. Every independent sequence $(A_n, n \geqslant 1)$ of events in a probability space $(\Omega, \mathscr{A}, P)$ of probabilities different from 0 and 1 allows us to construct a Haar system in the following way: the σ-fields $\mathscr{B}_2{}^n$ are generated by the sequences $(A_1, \ldots, A_n)$, i.e. by the partition

$$\pi_n = (A_1 \ldots A_{n-1} A_n, A_1 \ldots A_{n-1} A_n^c, A_1 \ldots A_{n-1}^c A_n, A_1 \ldots A_{n-1}^c A_n^c, \ldots)$$

whilst the intermediate σ-fields $\mathscr{B}_{2^n+k}$ $(0 \leqslant k < 2^n, n \in \mathbf{N})$ are obtained by intersecting the first k sets of π_n with A_{n+1} and A_{n+1}^c (which gives $2k$ sets) and then taking the remaining $2^n - k$ other sets of π_n intact. Note that all the events thus defined are non-null by virtue of the formula

$$P(A_1' \ldots A_n') = P(A_1') \ldots P(A_n') > 0$$

for all $n \in \mathbf{N}$ and $A_m' = A_m$ or A_m^c $(1 \leqslant m \leqslant n)$. In particular when we take the sequence

$$A_n = \sum_{0 \leqslant k < 2^{n-1}} \left[\frac{2k}{2^n}, \frac{2k+1}{2^n}\right[$$

in the real interval $[0, 1[$ equipped with Lebesgue measure for the sequence $(A_n, n \in \mathbf{N})$, the partitions π_n are the dyadic partitions $([2^{-n}k, 2^{-n}(k+1)[$, $0 \leqslant k < 2^n)$ and the functions $(U_n, n \in \mathbf{N})$ are the classical Haar functions U_{2^n+k} equalling 0 outside the interval $[2^{-n}k, 2^{-n}(k+1)[$ and equalling $\pm 2^{n/2}$ on the first and second half of this interval, respectively $(U_0 = 1)$.

The following proposition gives two easy but interesting characterisations of Haar bases.

PROPOSITION III-3-10. (a) *Every orthonormal basis $(U_n, n \in \mathbf{N})$ of a space $L^2(\Omega, \mathscr{A}, P)$ such that $U_0 = 1$ and for all $n \in \mathbf{N}$ we have $E^{\mathscr{B}_n}(U_{n+1}) = 0$, where $\mathscr{B}_n$ denotes the σ-field generated by $U_0, U_1, \ldots, U_n$, is necessarily a Haar basis.*

(b) *Every orthonormal basis $(U_n, n \in \mathbf{N})$ of a space $L^2(\Omega, \mathscr{A}, P)$ such that $U_0 = 1$ and $\sum_{m \leqslant n} U_m \otimes U_m \geqslant 0$ a.s on $(\Omega, \mathscr{A}, P)^{2\otimes}$ is necessarily a Haar basis. (By definition, $U_m \otimes U_m$ denotes the function defined on Ω^2 by $U_m \otimes U_m(\omega_1, \omega_2) = U_m(\omega_1) U_m(\omega_2)$.)*

PROOF. To establish the first of these properties, let us consider a function $f \in L^2$ that we will expand as $f = \sum_{\mathbf{N}} \hat{f}_n U_n$ in L^2 in terms of the orthonormal basis $(U_n, n \in \mathbf{N})$. The hypothesis $E^{\mathscr{B}_n}(U_{n+1}) = 0$ then implies that $E^{\mathscr{B}_n}(f) = \sum_{m \leqslant n} \hat{f}_m U_m$. It is then easy to deduce that $(U_0, \ldots, U_n)$ is an orthonormal basis of $L^2(\mathscr{B}_n)$; this space is hence of dimension $n + 1$ and this implies that the σ-field is generated by a partition of $n + 1$ non-null events. It is then easy to see that the sequence $(U_n, n \in \mathbf{N})$ can only be the Haar basis associated with the Haar system $(\mathscr{B}_n, n \in \mathbf{N})$.

To establish the second property above, we remark that the function $D_n = \sum_{m \leqslant n} U_m \otimes U_m$ is the kernel of the orthogonal projection of L^2 onto the vector subspace Λ_n generated by $U_0, \ldots, U_n$; as $1 = U_0 \in \Lambda_n$, this projection can only be positive if it is a conditional expectation (Proposition I-2-7). It is then easy to conclude with an argument similar to the above. ∎

III-4. Gaussian spaces

A closed vector subspace H, say, of the space $L^2(\Omega, \mathscr{A}, P)$ is called a gaussian space if the r.r.v.'s which comprise H are all centred and gaussian. These spaces exhibit the following remarkable property, whose proof can be found in [221]:

For every closed vector subspace H_1 of H, the conditional expectation $E^{\mathscr{B}(H_1)}$ associated with the completed σ-field $\mathscr{B}(H_1)$ generated by the r.v.'s in H_1 coincides with the orthogonal projection E^{H_1} on the space H.

Suppose that the gaussian space H which we have been given is separable and infinite dimensional; the orthonormal bases of H will then be sequences. The following proposition is then an easy corollary of Proposition II-2-11.

PROPOSITION III-4-11. *For every orthonormal basis* $(X_n, n \in \mathbf{N})$ *of the gaussian space* H *and for every r.v.* Y *of this space, the orthogonal expansion of* Y *in terms of the basis* $(X_n, n \in N)$, say

$$\sum_{\mathbf{N}} E(YX_n)\, X_n$$

converges a.s. to Y.

PROOF. For all $n \in \mathbf{N}$, let us denote by H_n the finite-dimensional vector subspace of H generated by the r.v.'s X_m $(m \leqslant n)$; then

$$E^{H_n}(Y) = \sum_{m \leqslant n} \cdot E(YX_m) X_m.$$

But by the property recalled above, we have $E^{H_n}(Y) = E^{\mathscr{B}(H_n)}(Y)$, so that the sequence $(\sum_{m \leqslant n} E(YX_m)\, X_m, n \in \mathbf{N})$ coincides with the sequence

$$(E^{\mathscr{B}(H_n)}(Y), n \in \mathbf{N}),$$

which, since the σ-fields $\mathscr{B}(H_n)$ obviously increase with n, is a martingale. The series $\sum_{\mathbf{N}} E(YX_m)\, X_m$, which converges to Y in L^2 by elementary Hilbert space theory, thus converges a.s. to the same limit by the martingale theorem. ∎

III-5. Application to Markov chains

The relationship between the theory of martingales and the theory of Markov chains is very deep; we will limit ourselves here to the consideration of Markov chains in discrete time defined on a countable state space.

Let E be a countable set, whose points will be the "states" of the chain. Let $\Omega = E^{\mathbf{N}}$ be the space of all infinite sequences $(x_n, n \in \mathbf{N})$ of states which will be called "paths" of the chain, and let $X_n : \Omega \to E$ $(n \in \mathbf{N})$ be the coordinate mappings defined on the product space $\Omega = E^{\mathbf{N}}$. We will denote by $\mathscr{B}$ the product σ-field on Ω, i.e. the σ-field generated by the X_n $(n \in \mathbf{N})$, and for all $p \in \mathbf{N}$ we denote by $\mathscr{B}_p$ the σ-field of subsets of Ω generated by the r.v.'s $X_0, X_1, \ldots, X_p$, which is also the σ-field generated by the countable partition

$$(\{X_0 = x_0, \ldots, X_p = x_p\}; x_0, \ldots, x_p \in E)$$

of Ω. If P denotes a Markov matrix on E, i.e. a matrix $P = (P(x, y); x, y \in E)$ such that $P(x, \cdot)$ is a probability on E for each $x \in E$, Kolmogorov's theorem shows that there exists a family $(\mathbf{P}_x; x \in E)$ of probabilities on $(\Omega, \mathscr{B})$ characterised by the formula

$$\mathbf{P}_x(X_0 = x_0, \ldots, X_p = x_p) = \varepsilon_x(x_0) P(x_0, x_1) \ldots P(x_{p-1}, x_p),$$

where $p \in \mathbf{N}$ and $x_0, \ldots, x_p \in E$. When the space $(\Omega, \mathscr{B})$ is equipped with the probability $\mathbf{P}_x$, the sequence $(X_n, n \in \mathbf{N})$ is said to be the canonical Markov chain with initial state x and transition matrix P. Let us recall also that a sequence

$$X_n^*: \Omega^* \to E \qquad (n \in \mathbf{N})$$

of r.v.'s defined on an arbitrary probability space $(\Omega^*, \mathscr{B}^*, \mathbf{P}^*)$ is called a Markov chain with initial state x and transition matrix P if the probability $\mathbf{P}^*(X_0^* = x_0, \ldots, X_p^* = x_p)$ is given by the above expression for all $p \in \mathbf{N}$ and $x_0, \ldots, x_p \in E$; furthermore, by means of the measurable map $X^* = (X_n^*, n \in \mathbf{N})$ of Ω^* into $E^{\mathbf{N}} = \Omega$, any problem relating to the sequence $(X_n^*, n \in \mathbf{N})$ of random variables can be reduced to a problem concerning the canonical Markov chain $(X_n, n \in \mathbf{N})$. In what follows we will also make use of the coordinate shift operators $\theta_p : \Omega \to \Omega$ which are defined for all $p \in \mathbf{N}$ by the property $X_n \circ \theta_p = X_{n+p}$ $(n \in \mathbf{N})$.

On the space $(\Omega, \mathscr{B})$ equipped with the probabilities $\mathbf{P}_x$ $(x \in E)$ and the increasing sequence $(\mathscr{B}_n, n \in \mathbf{N})$ of sub-σ-fields, the positive supermartingales which can be written in the form $(f(X_n), n \in \mathbf{N})$ for a function f defined on E play an important role in the study of the chain $(X_n, n \in \mathbf{N})$. In the sequel we will only consider finite supermartingales for these are the only really interesting ones to be found in the theory of Markov chains. Let us begin by characterising those positive supermartingales of the form just mentioned.

LEMMA III-5-12. *For every function $f : E \to \mathbf{R}^+$, the following two conditions are equivalent:*

(a) *The function f is superharmonic* (*resp. harmonic*) *for the matrix P, i.e., satisfies the inequality $f \geq Pf$ on E* (*resp. the equality $f = Pf$ on E*).

(b) *The sequence $(f(X_n), n \in N)$ is a positive supermartingale* (*resp. a positive martingale*) *for all initial states x of the chain, i.e., for all the probabilities $\mathbf{P}_x$ $(x \in E)$.*

PROOF. The definition of a Markov chain implies that for every function $f: E \to \mathbf{R}^+$,

$$E_x^{\mathscr{B}_n}(f(X_{n+1})) = Pf(X_n)$$

because on each event $\{X_0 = x_0, \ldots, X_n = x_n\}$ the equalities

$$E_x^{\mathscr{B}_n}[f(X_{n+1})] = \sum_{y \in E} \mathbf{P}_x(X_{n+1} = y \mid X_0 = x, \ldots, X_n = x_n) f(y)$$

$$= \sum_{y \in E} P(x_n, y) f(y) = Pf(x_n)$$

hold. The sequence $(f(X_n), n \in \mathbf{N})$ is thus a supermartingale (resp. a martingale) whenever $Pf \leqslant f$ (resp. $Pf = f$).

Conversely, if $(f(X_n), n \in \mathbf{N})$ is a supermartingale for the probability $\mathbf{P}_x$, we have

$$f(x) = f(X_0) \geqslant \mathbf{E}_x^{\mathscr{B}_0}(f(X_1)) = Pf(X_0) = Pf(x) \quad P_x\text{-a.s.}$$

and the inequality $f \geqslant Pf$ is thus valid at the point $x \in E$; in the case of a martingale, the equality $f = Pf$ can be established at the point x. ■

REMARKS. (1) For the sequence $(f(X_n), n \in \mathbf{N})$ to be a supermartingale for the single probability $\mathbf{P}_x$ corresponding to a *fixed* state $x \in E$, it is necessary and sufficient that $f \geqslant Pf$ on the set E_x of states which can be attained with positive probability starting at x (the proof is left to the reader). For an *irreducible* Markov matrix P, the quantifier "for all x" in the lemma can thus be replaced by "for an arbitrary but fixed state x in E". On the other hand, this extension of the preceding lemma is not always possible: simply consider the identity Markov matrix!

(2) For a given Markov chain it is natural to look not only at the supermartingales of the form $(f(X_n), n \in \mathbf{N})$, but also at those of the more general form $(f_n(X_n), n \in \mathbf{N})$, where for each $n \in \mathbf{N}$, f_n is a function defined on E. However, the study of these more general supermartingales can be reduced to that of supermartingales of the form $(f(X_n), n \in \mathbf{N})$ by passing to the "space–time" chain defined as follows. If $(X_n, n \in \mathbf{N})$ is a Markov chain with initial state x and transition matrix P and if m is an integer $\geqslant 0$, it is easy to check that the sequence $((m+n, X_n), n \in N)$ of r.v.'s with values in $\mathbf{N} \times E$ (whose first component $m+n$ is "deterministic"!) is a Markov chain on the state space $\mathbf{N} \times E$, with initial state (m, x) and transition matrix $\tilde{P}$ given on $\mathbf{N} \times E$ by

$$\tilde{P}[(p, x), (q, y)] = \begin{cases} P(x, y) & \text{if } q = p+1, \\ 0 & \text{otherwise.} \end{cases}$$

Consequently, given a function $g:\mathbf{N}\times E\to\mathbf{R}_+$, the sequences $(g(m+n, X_n), n\in\mathbf{N})$ will be a positive (super)martingale for all probabilities $\mathbf{P}_x(x\in E)$ and for all $m\in N$ if and only if the function g is P-(super)harmonic on $\mathbf{N}\times E$. [For the same reason as in Remark (1), it is not generally possible to limit ourselves to the case $m=0$ in this condition.] Putting $g_n=g(n, \cdot)$ on $E(n\in\mathbf{N})$, we also note that the condition $Pg\leqslant g$ on $\mathbf{N}\times E$ is equivalent to the conditions $Pg_{n+1}\leqslant g_n$ $(n\in\mathbf{N})$ on the functions g_n from E into $\mathbf{R}_+$. ■

If P is a Markov matrix on E, we will denote by P_n $(n\in E)$ its iterates (P_0 is the identity matrix I) and we denote by U the associated potential matrix defined by

$$U(x, y)=\sum_{\mathbf{N}} P_n(x, y)\leqslant +\infty \quad (x, y\in E).$$

Since the matrix $U=\sum_{\mathbf{N}} P_n$ satisfies the relations $U=I+PU\geqslant PU$, it is clear that for every function $h:E\to\mathbf{R}_+$ the function Uh is superharmonic; this function is called the potential of h. This method of construction of superharmonic functions is made important by the first part of the following proposition.

PROPOSITION III-5-13. *Every positive and finite superharmonic function f can be decomposed on E into the sum $f=Uh+f'$ of the potential of a positive function h and a positive harmonic function f'. The functions h and f' of this decomposition are uniquely determined by the formulae $h=f-Pf$ and $f'=\lim\downarrow_{n\uparrow\infty} P_nf$. The preceding decomposition is called the Riesz decomposition of the superharmonic function f.*

Furthermore, for every probability $\mathbf{P}_x$, the limits $\lim_{n\to\infty}f(X_n)$ and $\lim_{n\to\infty}f'(X_n)$ exist a.s. in $\mathbf{R}_+$ and are equal.

PROOF. Let us put $h=f-Pf\geqslant 0$, which makes sense since f is finite; then the equality $\sum_{m<n}P_mh=f-P_nf$ is true for all $n\in\mathbf{N}$. When $n\uparrow\infty$, the left-hand side of this equality increases to the potential Uh; this potential is finite on E since it is dominated by f, in fact it is equal to $f-f'$ if we put $f'=\lim\downarrow_{n\uparrow\infty}P_nf$. The function f' is harmonic since

$$Pf'=P(\lim_n\downarrow P_n f)=\lim_n\downarrow P_{n+1}f=f'$$

by an application of the dominated convergence theorem; the hypothesis that f is finite on E is essential for the application of this theorem!

To show the uniqueness of the Riesz decomposition, let us just remark that the equality $f=Uh^*+f'^*$, where h^* is a positive function and f'^* a positive harmonic function, implies that $Pf=PUh^*+Pf'^*$ and hence that $f=Pf+h^*$

since $U = I + PU$. We thus see that $h^* = h$ since f is finite and it is then clear that $f'^* = f - Uh^* = f - Uh = f'$.

For every probability $\mathbf{P}_x$ $(x \in E)$ the sequences $(f(X_n),\ n \in \mathbf{N})$, $(Uh(X_n), n \in \mathbf{N})$ and $(f'(X_n), n \in \mathbf{N})$ are finite positive supermartingales by the preceding proposition; they thus converge a.s. to a finite limit by Theorem II-**2**-9. Proving the a.s. equality of the limits $\lim_{n\to\infty} f(X_n)$ and $\lim_{n\to\infty} f'(X_n)$ is the same as proving that $\lim_{n\to\infty} Uh(X_n) = 0$ a.s., since $f = Uh + f'$. But

$$\mathbf{E}_x(Uh(X_n)) = P_n\, Uh(x) = \sum_{m \geqslant n} P_m h(x) \downarrow 0$$

when $n \uparrow \infty$ since the sum $\sum_{\mathbf{N}} P_m h(x) = Uh(x)$ is finite; the sequence

$$(Uh(X_n), n \in \mathbf{N})$$

thus converges in mean to 0 so that its a.s. limit can only be zero as well. ∎

Here is an application of the preceding proposition which already allows us to recover the fundamentals concerning recurrence in a Markov chain.

PROPOSITION III-**5**-14. *The probability of hitting an arbitrary subset F of the state space E of a Markov chain, i.e. the function of the initial state given by*

$$\phi_F(x) = \mathbf{P}_x(\nu_F < \infty) \quad (x \in E),$$

where $\nu_F = \min(n : X_n \in F)$ and $\{\nu_F = \infty\} = \{X_n \notin F, \text{for all } n \in \mathbf{N}\}$, is a superharmonic function with values between 0 and 1. The harmonic component of its Riesz decomposition, say $\phi'_F = \lim\downarrow_n P_n \phi_F$, is equal to the probability of visiting F infinitely often, say

$$\phi'_F(x) = \mathbf{P}_x(N_F = \infty) \quad (x \in E),$$

where $N_F = \sum_{\mathbf{N}} 1_F \circ X_n$.

Furthermore, for all $x \in E$,

$$\lim_n \phi_F(X_n) = \lim_n \phi'_F(X_n) = 1_{\{N_F = \infty\}} \quad \mathbf{P}_x\text{-}a.s.$$

Finally, the function h_F in the Riesz decomposition $\phi_F = Uh_F + \phi'_F$ of the function ϕ_F is equal to the probability of not returning to F, defined by

$$h_F(x) = \begin{cases} \mathbf{P}_x(X_n \notin F \text{ if } n \geqslant 1) & \text{if } x \in F, \\ 0 & \text{if } x \notin F. \end{cases}$$

PROOF. For all $n \in \mathbf{N}$ and all $x \in E$, the Markov property implies that

$$\mathbf{P}_x(\sup_{p \geqslant n} \{X_p \in F\}) = \mathbf{P}_x(\nu_F \circ \theta_n < \infty) = P_n \phi_F(x).$$

As the left-hand side is clearly a decreasing function of n, the function $\phi_F : E \to [0, 1]$ is superharmonic and further its harmonic component ϕ_F' takes values at points in E given by

$$\phi_F'(x) = \lim_n \downarrow P_n \phi_F(x) = \mathbf{P}_x(\limsup_{n \to \infty} \{X_n \in F\}) = \mathbf{P}_x(N_F = \infty).$$

The previous proposition already tells us that the limits $\lim_n \phi_F(X_n)$ and $\lim_n \phi_F'(X_n)$ exist and are a.s. equal; to establish the fact that they coincide with $1_{\{N_F = \infty\}}$ it is enough, by Proposition II-2-11, to prove that the positive martingale $(\phi_F(X_n), n \in \mathbf{N})$ coincides with the sequence $(E_x^{\mathscr{B}_n}(1_{\{N_F = \infty\}}), n \in \mathbf{N})$. Now the event $\{N_F = \infty\}$ is invariant under a translation because N_F and $N_F \circ \theta_n$ never differ by more than n units, so that $\{N_F = \infty\} = \{N_F \circ \theta_n = \infty\}$; but the Markov property then implies that

$$\mathbf{E}_x^{\mathscr{B}_n}(1_{\{N_F = \infty\}}) = \mathbf{E}_x^{\mathscr{B}_n}(1_{\{N_F = \infty\}} \circ \theta_n) = \phi_F'(X_n)$$

$\mathbf{P}_x$-a.s. for all $n \in \mathbf{N}$.

Finally, by the preceding proposition and the formula at the beginning of this proof, the function h_F of the Riesz decomposition of ϕ_F takes values

$$\begin{aligned} h_F(x) &= \phi_F(x) - P\phi_F(x) \\ &= \mathbf{P}_x(\sup_{n \geqslant 0} \{X_n \in F\}) - \mathbf{P}_x(\sup_{n \geqslant 1} \{X_n \in F\}) \\ &= \mathbf{P}_x(X_0 \in F, X_n \notin F \quad \text{if } n \geqslant 1) \qquad (x \in E); \end{aligned}$$

the last formula of the proposition is thus proved. ■

A subset F of the state space is called *transient* if

$$\mathbf{P}_x(N_F = \infty) = 0 \quad \text{for all } x \in E.$$

It is easy to deduce from this definition that every subset of a transient set is itself transient (for $N_F \leqslant N_F'$ if $F' \subset F$) and that any *finite* union of transient subsets is again transient (for $N_{\cup_1^n F_m} \leqslant \sum_1^n N_{F_m}$). A state y is said to be transient whenever the subset $\{y\}$ is. It is clear that a transient set is uniquely formed from transient states, but a Markov chain can admit transient sets which are not finite (a random walk on $\mathbf{Z}$ with strictly positive mean, for example, admits all the intervals $]-\infty, a]$ ($a \in \mathbf{Z}$) as transient subsets by the law of large numbers). Here is a criterion which permits the recognition of a subset F of E as transient; this criterion shows in particular that the state y is transient if $0 < U(y, z) < \infty$ for at least one $z \in E$, and only if $U(y, y) < \infty$.

COROLLARY III-5-15. *For a subset F of the state space E to be transient, it is necessary and sufficient that there exists a positive function h of finite potential Uh on E such that $Uh \geqslant 1$ on F. Further, this function h can always be chosen to be zero outside F.*

PROOF. The condition is sufficient. Indeed, by Proposition III-5-13 the limit $\lim_{n\to\infty} Uh(X_n)$ exists a.s. and is zero for all the probabilities $\mathbf{P}_x$ ($x \in E$) since the function Uh is finite; the inequality $Uh \geqslant 1_F$ then implies that $\limsup_{n\to\infty}\{X_n \in F\} = \emptyset$ $\mathbf{P}_x$-a.s. ($x \in E$), which shows that F is transient.

Conversely, if F is transient, the harmonic component ϕ_F' of the superharmonic function ϕ_F is, by definition, zero on E. Consequently, $\phi_F = Uh_F$; further, h_F is zero outside F by the last part of the preceding proposition. Moreover, we have $\phi_F = 1$ on F by definition of the function ϕ_F. ■

COROLLARY III-5-16. *For every state $x \in E$ the superharmonic function ϕ_x is either a potential or a harmonic function; in the first case, $\mathbf{P}_{\cdot}(N_x = \infty) = 0$ on E and the state x is said to be transient, whilst in the second case, $\mathbf{P}_x(N_x = \infty) = 1$ and the state x is said to be recurrent.*

If x is a recurrent state, only two possibilities exist for any other state y: either $N_y = 0$ $\mathbf{P}_x$-a.s. or $N_y = \infty$ $\mathbf{P}_x$-a.s. If

$$C_x = \{y : \mathbf{P}_x(N_y = \infty) = 1\}$$

denotes the class of states of the second type, every state $y \in C_x$ is recurrent and such that $C_y = C_x$.

PROOF. We will rely on the observation that for any positive superharmonic function f, the a.s. limit of the supermartingale $(f(X_n), n \in \mathbf{N})$ can only be $f(x)$ $\mathbf{P}_x$-a.s. on the set $\{N_x = \infty\}$, since on this event $X_n = x$ and thus $f(X_n) = f(x)$ for infinitely many n.

If the function $\phi_x' = \mathbf{P}_{\cdot}(N_x = \infty)$ is not identically zero on E, i.e. if the state x is not transient, we can choose a state y such that $\mathbf{P}_y(N_x = \infty) \neq 0$. Then by the preceding proposition the positive supermartingale $(\phi_x'(X_n), n \in \mathbf{N})$ converges $\mathbf{P}_y$-a.s. to 1 on the non-null event $\{N_x = \infty\}$; but by the opening remark of this proof, this limit is also a.s. equal to $\phi_x'(x)$. We have thus proved that $\phi_x'(x) = \mathbf{P}_x(N_x = \infty)$ has the value 1. By the last part of the previous proposition, this evidently implies that $h_x(x) = 0$ and thus $h_x = 0$ on E since h_x is always zero outside $\{x\}$ by definition. Hence the function ϕ_x coincides with ϕ_x' and is thus harmonic.

Let x be a recurrent state, i.e. a state such that $N_x = +\infty$ $\mathbf{P}_x$-a.s. and let y be another state of E. For the probability $\mathbf{P}_x$, the supermartingale $(\phi_y(X_n), n \in \mathbf{N})$ converges a.s. to $1_{\{N_y = \infty\}}$ by the preceding proposition; on the other hand, by the opening remark this limit also has the value $\phi_y(x)$ a.s. Thus

$$1_{\{N_y = \infty\}} = \phi_y(x) \quad \mathbf{P}_x \text{ a.s.};$$

this implies that

$$\phi_y(x) = 0, \qquad \mathbf{P}_x(N_y = \infty) = 0,$$

or

$$\phi_y(y) = 1, \qquad \mathbf{P}_x(N_y = \infty) = 1.$$

This dichotomy can be simplified by interpreting the function ϕ_y:

$$\mathbf{P}_x(N_y \neq 0) = 0 \quad \text{or} \quad \mathbf{P}_x(N_y = \infty) = 1.$$

If $y \in C_x$, the function ϕ_y' takes the value 1 at the point x and hence is not identically zero; by the above this implies that y is recurrent. If, moreover, $z \in C_x$, then $\mathbf{P}_x$-a.s. the limit $\lim_n \phi_z'(X_n)$ has the value $1_{\{N_z = \infty\}} = 1$ by the preceding proposition, and equals $\phi_z'(y)$ by the opening remark since $\mathbf{P}_x(N_y = \infty) = 1$; hence $\phi_z'(y) = 1$, which says that $z \in C_y$. We have proved that $C_x \subset C_y$; the reverse inclusion follows by symmetry. ■

CHAPTER IV

CONVERGENCE AND REGULARITY OF MARTINGALES

IV-1. A.s. Convergence of Submartingales

We will begin this section with Doob's proof of the a.s. convergence theorem for martingales (and submartingales). Then, after having studied regular martingales, i.e. martingales converging in L^1, we will introduce the notion of regular stopping time for an arbitrary integrable martingale (ν is said to be regular for the martingale $(X_n, n \in \mathbf{N})$ if the stopped martingale $(X_{\nu \wedge n}, n \in \mathbf{N})$ is regular); this notion is very important for applications, and so we will also be looking for regularity criteria for stopping times.

Definition IV-1-1. An adapted sequence $(X_n, n \in \mathbf{N})$ of integrable real-valued r.v.'s is called an *integrable submartingale* if the a.s. inequality

$$X_n \leqslant E^{\mathscr{B}_n}(X_{n+1})$$

is satisfied for all $n \in \mathbf{N}$. The sequence is called an *integrable martingale* if the inequality is replaced by an equality.

The assumption of integrability of the X_n ($n \in \mathbf{N}$) often only appears in the weaker form $E(X_n^+) < \infty$ ($n \in \mathbf{N}$). (Note that the assumption $E(X^+) < \infty$ allows the conditional expectation $E^{\mathscr{B}}(X)$ to be defined as $E^{\mathscr{B}}(X^+) - E^{\mathscr{B}}(X^-)$ since it implies that $E^{\mathscr{B}}(X^+) < \infty$ a.s.) However, the generalisation which is thus suggested is not very useful beyond the case of negative submartingales, whose study is reduced to that of positive supermartingales by a sign change.

The "submartingale theorem" stated below is in fact a corollary of the convergence theorems for positive supermartingales proved in Chapter II (Theorem II-2-9). Note the difference between positive *super*martingales which always converge a.s., and positive *sub*martingales which do not necessarily converge a.s.; the following theorem due to Doob implies that this convergence only takes place for positive submartingales bounded in L^1 (see also Section IV-5).

THEOREM IV-1-2. *Every integrable submartingale $(X_n, n \in \mathbf{N})$ satisfying the condition $\sup_{\mathbf{N}} E(X_n^+) < \infty$ converges a.s. to a limit which is an integrable r.v.*

In the case of an integrable martingale, the preceding condition is equivalent to the condition $\sup_{\mathbf{N}} E(|X_n|) < \infty$.

PROOF. If $(X_n, n \in \mathbf{N})$ is an integrable submartingale, the sequence $(X_n^+, n \in \mathbf{N})$ is a positive integrable submartingale, for the inequalities

$$E^{\mathscr{B}_p}(X_{p+1}^+) \geqslant E^{\mathscr{B}_p}(X_{p+1}) \geqslant X_p \quad (p \in \mathbf{N})$$

imply that $E^{\mathscr{B}_p}(X_{p+1}^+) \geqslant X_p^+$ for all $p \in \mathbf{N}$. But then for all fixed $n \in \mathbf{N}$, the sequence $(E^{\mathscr{B}_n}(X_p^+), p \geqslant n)$ is an increasing sequence of positive r.v.'s since

$$E^{\mathscr{B}_n}(X_{p+1}^+) = E^{\mathscr{B}_n} E^{\mathscr{B}_p}(X_{p+1}^+) \geqslant E^{\mathscr{B}_n}(X_p^+) \qquad \text{if } p \geqslant n;$$

hence we put

$$M_n = \lim_p \uparrow E^{\mathscr{B}_n}(X_p^+)$$

and show that the sequence $(M_n, n \in \mathbf{N})$ of positive r.r.v.'s thus defined is an integrable martingale, and hence a.s. finite.

It is clear that M_n is $\mathscr{B}_n$-measurable for all $n \in \mathbf{N}$ and the martingale equality then follows from

$$E^{\mathscr{B}_n}(M_{n+1}) = \lim_p \uparrow E^{\mathscr{B}_n} E^{\mathscr{B}_{n+1}}(X_p^+) = \lim_p \uparrow E^{\mathscr{B}_n}(X_p^+) = M_n \qquad (n \in \mathbf{N})$$

due to the increasing continuity of the positive r.v. conditional expectations. The integrability of $(M_n, n \in \mathbf{N})$ follows from the hypothesis

$$\sup_{\mathbf{N}} E(X_n^+) < \infty$$

since we have

$$E(M_n) = \lim_p \uparrow E(E^{\mathscr{B}_n}(X_p^+)) = \lim_p \uparrow E(X_p^+) \qquad (n \in \mathbf{N}).$$

The differences $Y_n = M_n - X_n$ $(n \in \mathbf{N})$ define a positive integrable supermartingale. Indeed it is clear that these r.v.'s are integrable and $\mathscr{B}_n$-measurable; the inequality $M_n \geqslant X_n^+$ surely implies that $Y_n \geqslant 0$; finally, $(Y_n, n \in \mathbf{N})$ is a supermartingale, being the difference between the martingale $(M_n, n \in \mathbf{N})$ and the submartingale $(X_n, n \in \mathbf{N})$. The positive supermartingale theorem then implies that the limits $M_\infty = \lim \text{a.s.}_{n \to \infty} M_n$ and $Y_\infty = \lim \text{a.s.}_{n \to \infty} Y_n$ exist, are positive and integrable, and thus a.s. *finite*. It follows that the submartingale $(X_n, n \in \mathbf{N})$ converges a.s. to the integrable r.r.v. $M_\infty - Y_\infty$, which completes the proof of the convergence theorem for submartingales.

Let us observe that the positive r.r.v.'s M_∞ and Y_∞ introduced above are equal to the positive and negative parts, X_∞^+ and X_∞^- respectively, of the limit

$$X_\infty = \lim_{n\to\infty} \text{a.s. } X_n = M_\infty - Y_\infty.$$

Indeed, the sequence $(M_n - X_n^+, n \in \mathbf{N})$ of positive r.r.v.'s converges in L^1 to zero since by a previous calculation,

$$E(M_n - X_n^+) = E(M_0) - E(X_n^+) \downarrow 0 \qquad (n \uparrow \infty);$$

this implies that the r.v. $M_\infty - X_\infty^+ = \lim \text{a.s.}_{n\to\infty}(M_n - X_n^+)$ must be null. Finally we have

$$Y_\infty = M_\infty - X_\infty = X_\infty^+ - X_\infty = X_\infty^-.$$

To complete the proof of the theorem, let us note that the general relation

$$E(|X|) = 2E(X^+) - E(X) \qquad (X \in L^1)$$

implies that for an integrable martingale $(X_n, n \in \mathbf{N})$ the two conditions $\sup_{\mathbf{N}} E(|X_n|) < \infty$ and $\sup_{\mathbf{N}} E(X_n^+) < \infty$ are equivalent, since the sequence $(E(X_n), n \in \mathbf{N})$ is finite and constant. ∎

The decomposition $X_n = M_n - Y_n$ $(n \in \mathbf{N})$ of the integrable submartingale $(X_n, n \in \mathbf{N})$ as the difference of a positive integrable martingale and a positive integrable supermartingale, which was the basis of the preceding proof, is generally called the *Krickeberg decomposition* of the submartingale $(X_n, n \in \mathbf{N})$. It is not hard to verify that this decomposition is minimal in the sense that any other decomposition $X_n = M_n' - Y_n'$ $(n \in \mathbf{N})$ of the submartingale $(X_n, n \in \mathbf{N})$ as the difference of a positive martingale and a positive supermartingale necessarily satisfies the inequalities $M_n' \geqslant M_n$ and $Y_n' \geqslant Y_n$ for all $n \in \mathbf{N}$. [Let us also recall that the Krickeberg decomposition is based upon the assumption $\sup_{\mathbf{N}} E(X_n^+) < \infty$.]

IV-2. Regularity of integrable martingales

We have already seen in Chapter III in the case of positive martingales that the a.s. convergence of a martingale does not imply its convergence in L^1. The following proposition considers this question more precisely.

PROPOSITION IV-**2**-3. *For every integrable martingale* $(X_n, n \in \mathbf{N})$, *the following conditions are equivalent:*

(a) *The sequence* $(X_n, n \in \mathbf{N})$ *converges in* L^1.

(b) $\sup_{\mathbf{N}} E(|X_n|) < \infty$ *and the a.s. limit* $X_\infty = \lim_n X_n$ *of the martingale which exists by Theorem IV-**1**-2 satisfies the equalities* $X_n = E^{\mathscr{B}_n}(X_\infty)$ *for all* $n \in \mathbf{N}$.

(b′) *There exists an integrable r.r.v.* X *such that* $X_n = E^{\mathscr{B}_n}(X)$ *for all* $n \in \mathbf{N}$.

(c) *The sequence* $(X_n, n \in \mathbf{N})$ *satisfies the uniform integrability condition*

$$\sup_{\mathbf{N}} \int_{\{|X_n|>a\}} |X_n| \mathrm{d}P \downarrow 0 \quad \textit{as } a \uparrow \infty$$

(*this condition is satisfied whenever* $\sup_{\mathbf{N}} |X_n| \in L^1$).

The integrable martingale $(X_n, n \in \mathbf{N})$ will be called *regular* if it satisfies one of these equivalent conditions.

The implication $(b') \Rightarrow (a)$ was the object of Proposition II-**2**-1; nonetheless we will prove it again here via condition (c), which is an important technical condition.

PROOF. $(a) \Rightarrow (b)$. The convergence of the sequence $(X_n, n \in \mathbf{N})$ in L^1 implies that $\sup_{\mathbf{N}} \|X_n\|_1 < \infty$, and the a.s. limit X_∞ of the martingale $(X_n, n \in \mathbf{N})$, which then exists by Theorem IV-**1**-2, coincides with the limit-in-mean of the X_n. The continuity of the conditional expectation $E^{\mathscr{B}_n}$ on L^1 implies that

$$E^{\mathscr{B}_n}(X_p) \xrightarrow[L^1]{} E^{\mathscr{B}_n}(X_\infty) \quad \text{if } p \to \infty;$$

but $E^{\mathscr{B}_n}(X_p) = X_n$ whenever $p \geqslant n$ and hence we have proved that $X_n = E^{\mathscr{B}_n}(X_\infty)$.

$(b) \Rightarrow (b')$. This implication is clear since the r.v. X_∞ of condition (b) is necessarily integrable by Fatou's lemma: the a.s. convergence of $|X_n|$ towards $|X_\infty|$ implies that

$$E|X_\infty| \leqslant \varliminf_n E|X_n| \leqslant \sup_{\mathbf{N}} E|X_n| < \infty.$$

We will prove the implication $(b') \Rightarrow (c)$ by establishing a more general result.

LEMMA IV-**2**-4. *For every r.r.v.* $X \in L^1$, *the family of r.v.'s* $E^{\mathscr{B}}(X)$ *obtained when* $\mathscr{B}$ *varies over all the sub-σ-fields of* $\mathscr{A}$ *is uniformly integrable, i.e., is such that*

$$\sup_{\mathscr{B}} \int_{\{|E^{\mathscr{B}}(X)| \geqslant a\}} |E^{\mathscr{B}}(X)| \,\mathrm{d}P \downarrow 0 \quad \textit{when } a \uparrow \infty.$$

Indeed, we have $|E^{\mathscr{B}}(X)| \leqslant E^{\mathscr{B}}(|X|)$ and consequently

$$\int_{\{|E^{\mathscr{B}}(X)|\geqslant a\}} |E^{\mathscr{B}}(X)|\,\mathrm{d}P \leqslant \int_{\{E^{\mathscr{B}}(|X|)\geqslant a\}} E^{\mathscr{B}}[|X|]\,\mathrm{d}P = \int_{\{E^{\mathscr{B}}(|X|)\geqslant a\}} |X|\,\mathrm{d}P$$

since $\{E^{\mathscr{B}}(|X|) \geqslant a\}$ is an event in $\mathscr{B}$. Let $b \in R_+$; by considering the last integral on $\{|X| \leqslant b\}$ and $\{|X| > b\}$ separately, we obtain the upper bound

$$\int_{\{E^{\mathscr{B}}(|X|)\geqslant a\}} |X|\,\mathrm{d}P \leqslant bP(E^{\mathscr{B}}(|X|) \geqslant a) + \int_{\{|X|>b\}} |X|\,\mathrm{d}P.$$

Collecting the inequalities obtained and taking into account that

$$P(E^{\mathscr{B}}(|X|) > a) \leqslant a^{-1} E(E^{\mathscr{B}}(|X|)) = a^{-1} E|X| \quad (\mathscr{B} \subset \mathscr{A}),$$

we find that

$$\sup_{\mathscr{B}} \int_{\{|E^{\mathscr{B}}(X)|\geqslant a\}} |E^{\mathscr{B}}(X)|\,\mathrm{d}P \leqslant \frac{b}{a} E(|X|) + \int_{\{|X|>b\}} |X|\,\mathrm{d}P,$$

which shows, upon taking $b = \sqrt{a}$ and letting $a \uparrow \infty$, that the family $\{E^{\mathscr{B}}(X), \mathscr{B} \subset \mathscr{A}\}$ is uniformly integrable.

$(c) \Rightarrow (a)$. The proof of this implication rests on the following general lemma, which is applicable to any uniformly integrable submartingale $(X_n, n \in \mathbf{N})$, since this uniform integrability implies that

$$\sup_{\mathbf{N}} E|X_n| \leqslant a + \sup_{\mathbf{N}} \int_{\{|X_n|>a\}} |X_n|\,\mathrm{d}P < \infty$$

and hence (Theorem IV-**1**-2) that the limit $X_\infty = \lim_n X_n$ exists a.s.

LEMMA IV-**2**-5. *Every uniformly integrable sequence $(X_n, n \in \mathbf{N})$ of r.r.v.'s which converges a.s. also converges in L^1.*

For proving this lemma we introduce for all $a \in \mathbf{R}_+$ the bounded continuous function f_a defined on $\mathbf{R}$ by

$$f_a(x) = \begin{cases} x \text{ if } |x| < a \\ \pm\, a \text{ according as } x \geqslant a \text{ or } x \leqslant -a. \end{cases}$$

This function satisfies the inequality $|x - f_a(x)| \leqslant |x|$ for all real x. Next the triangle inequality allows us to write

$$\|X_m - X_n\|_1 \leqslant \|f_a(X_m) - f_a(X_n)\|_1 + \|X_m - f_a(X_m)\|_1 + \|X_n - f_a(X_n)\|_1$$

for every pair m, n of integers in $\mathbf{N}$, and for every $a \in \mathbf{R}_+$.

But if X_∞ is the a.s. limit of the sequence $(X_n, n \in \mathbf{N})$, then $f_a(X_m) \to f_a(X_\infty)$ since f_a is a bounded continuous function, and as the r.v.'s $f_a(X_m)$ are dominated by the constant a, the dominated convergence theorem then shows that $f_a(X_m) \to f_a(X)$ in L^1. On the other hand,

$$\|X_m - f_a(X_m)\|_1 \leqslant \int_{\{|X_m|>a\}} |X_m| \, \mathrm{d}P \qquad (m \in \mathbf{N})$$

by the definition of the f_a. Letting $m, n \to \infty$ and then $a \to \infty$ in the original inequality, the uniform integrability of the sequence $(X_n, n \in \mathbf{N})$ allows us to conclude that $\lim_{m,n\to\infty} \|X_m - X_n\|_1 = 0$. Since L^1 is complete, we have thus proved that the sequence $(X_n, n \in \mathbf{N})$ converges in mean. ■

REMARK. For a systematic study of uniform integrability, the reader is referred to Section II-**5** of [223]. ■

The following corollary is basic to all that follows.

COROLLARY IV-**2**-6. *Let $(X_n, n \in \mathbf{N})$ be a regular martingale. For every stopping time ν, the r.v. X_ν is integrable. For every pair of stopping times ν_1, ν_2 such that $\nu_1 \leqslant \nu_2$ a.s., the "martingale equality"*

$$X_{\nu_1} = E^{\mathscr{B}_{\nu_1}}(X_{\nu_2})$$

is also satisfied.

(For a regular martingale the limit $X_\infty = \lim_n X_n$ exists a.s. and the random variables X_ν by definition equal X_∞ on $\{\nu = \infty\}$.)

PROOF. The a.s. limit X_∞ of a regular martingale $(X_n, n \in \mathbf{N})$ satisfies the equality $X_\nu = E^{\mathscr{B}_\nu}(X_\infty)$ for all stopping times ν. Indeed the two equalities $X_\nu = X_n$ and $E^{\mathscr{B}_\nu}(X_\infty) = E^{\mathscr{B}_n}(X_\infty)$ are valid on $\{\nu = n\}$ $(n \in \bar{\mathbf{N}})$ [Proposition II-**1**-3]; the regularity hypothesis $X_n = E^{\mathscr{B}_n}(X_\infty)$ of the martingale thus implies that $X_\nu = E^{\mathscr{B}_\nu}(X_\infty)$ on each of the events $\{\nu = n\}$ $(n \in \bar{\mathbf{N}})$ and consequently in Ω. The r.v. X_ν is therefore integrable.

The σ-fields $\mathscr{B}_{\nu_1}$ and $\mathscr{B}_{\nu_2}$ associated with a pair of stopping times such that $\nu_1 \leqslant \nu_2$ satisfy $\mathscr{B}_{\nu_1} \subset \mathscr{B}_{\nu_2}$ (Proposition II-**1**-5). For a regular martingale we thus have

$$E^{\mathscr{B}_{\nu_1}}(X_{\nu_2}) = E^{\mathscr{B}_{\nu_1}} E^{\mathscr{B}_{\nu_2}}(X_\infty) = E^{\mathscr{B}_{\nu_1}}(X_\infty) = X_{\nu_1}. \;■$$

The following proposition gives our first example of a regular martingale.

PROPOSITION IV-**2**-7. *Let p be a real number* >1. *Every martingale* $(X_n, n \in \mathbf{N})$ *whose terms belong to the space* L^p *and which is bounded in this space in the sense that*

$$\sup_{\mathbf{N}} \|X_n\|_p < \infty,$$

is regular. Furthermore, the martingale converges to an a.s. limit X_∞ *in the Banach space* L_p.

By the foregoing we know that this proposition is false for $p = 1$. For example, a positive integrable martingale converging a.s. to 0 cannot be regular unless it is identically null (for examples of such martingales see Proposition III-**2**-6).

PROOF. The assumption implies that the martingale $(X_n, n \in \mathbf{N})$ is uniformly integrable and thus regular. Indeed, the elementary inequality

$$a^{p-1} \int_{\{|X_n| > a\}} |X_n| \, dP \leqslant \int_\Omega |X_n|^p \, dP \quad (a \in \mathbf{R}_+, p > 1)$$

implies that

$$\sup_{\mathbf{N}} \int_{\{|X_n| > a\}} |X_n| \, dP \leqslant \frac{C^p}{a^{p-1}} \quad \text{if} \quad C = \sup_{\mathbf{N}} \|X_n\|_p < \infty$$

and since $p > 1$, the last expression decreases to 0 as $a \uparrow \infty$.

Since the martingale $(X_n, n \in \mathbf{N})$ is regular, it can be written in the form $X_n = E^{\mathscr{B}_n}(X_\infty)$, where $X_\infty = \lim \text{a.s.}_{n \to \infty} X_n$ belongs to L^1 (Proposition IV-**2**-3). Fatou's lemma applied to the positive sequence $(|X_n|^p, n \in \mathbf{N})$ which converges a.s. to $|X_\infty|^p$ shows that

$$\int |X_\infty|^p \, dP \leqslant \liminf_{n \to \infty} \int |X_n|^p \, dP < \infty$$

and hence that $X_\infty \in L^p$. But then Proposition II-**2**-11 shows that $(X_n, n \in \mathbf{N})$ converges in L^p to X_∞. ■

The following proposition completes the preceding result.

PROPOSITION IV-**2**-8. *Let p be a real number* >1. *For every martingale* $(X_n, n \in \mathbf{N})$ *bounded in* L^p, *the r.r.v.* $\sup_{\mathbf{N}} |X_n|$ *belongs to* L^p *and, more precisely, satisfies the inequality*

$$\|\sup_{\mathbf{N}} |X_n|\|_p \leqslant \frac{p}{p-1} \sup_{\mathbf{N}} \|X_n\|_p.$$

PROOF. The proof of the proposition rests upon the following lemma which, although its proof is very easy, is interesting in its own right.

LEMMA IV-2-9. *Every positive and integrable submartingale* $(X_n, n \in \mathbf{N})$ *satisfies the inequalities*

$$aP(\sup_{m \leqslant n} X_m > a) \leqslant \int_{\{\sup_{m \leqslant n} X_m > a\}} X_n \, \mathrm{d}P$$

for all $n \in \mathbf{N}$ *and all* $a \in \mathbf{R}_+$.

Indeed, every positive and integrable submartingale $(X_n, n \in \mathbf{N})$ satisfies the inequality

$$X_\nu \leqslant E^{\mathscr{B}_\nu}(X_n) \quad \text{on } \{\nu \leqslant n\} \quad (n \in \mathbf{N})$$

for all stopping times ν, for by Proposition II-1-3 this inequality becomes $X_m \leqslant E^{\mathscr{B}_m}(X_n)$ on $\{\nu = m\}$. Integrating both sides of the inequality applied to the stopping time

$$\nu_a = \inf(n : X_n > a)$$

gives

$$aP(\nu_a \leqslant n) \leqslant \int_{\{\nu_a \leqslant n\}} X_{\nu_a} \mathrm{d}P \leqslant \int_{\{\nu_a \leqslant n\}} E^{\mathscr{B}_{\nu_a}}(X_n) \, \mathrm{d}P = \int_{\{\nu_a \leqslant n\}} X_n \, \mathrm{d}P$$

for all $n \in \mathbf{N}$. The lemma is thus proved since

$$\{\nu_a \leqslant n\} = \{\sup_{m \leqslant n} X_m > a\}.$$

Now let us consider an integrable martingale $(X_n, n \in \mathbf{N})$ and the positive integrable submartingale $(|X_n|, n \in \mathbf{N})$ with which it is associated. Putting $S_n = \sup_{m \leqslant n} |X_m|$, we can write the inequality of the preceding lemma

$$aE(1_{\{S_n > a\}}) \leqslant E(|X_n| 1_{\{S_n > a\}}) \qquad (n \in \mathbf{N}, a \in \mathbf{R}_+).$$

Integrate both sides with respect to the measure $pa^{p-2} \mathrm{d}a$ on R_+ (p real >1); Fubini's theorem then shows that

$$E(S_n^p) = \int_0^\infty pa^{p-1} E(1_{\{S_n > a\}}) \, \mathrm{d}a \leqslant \int_0^\infty pa^{p-2} E(|X_n| 1_{\{S_n > a\}}) \, \mathrm{d}a$$

$$= \frac{p}{p-1} E(|X_n| S_n^{p-1}).$$

On the other hand, Hölder's inequality implies that

$$E(|X_n|S_n^{p-1}) \leqslant \|X_n\|_p \|S_n^{p-1}\|_{p/(p-1)} = \|X_n\|_p \|S_n\|_p^{p-1}.$$

As a consequence, if the r.r.v.'s X_n belong to L^p, in which case the r.v.'s $S_p = \sup_{m \leqslant n} |X_m|$ also belong to L^p, the combination of the two preceding inequalities gives

$$\|S_n\|_p \leqslant \frac{p}{p-1} \|X_n\|_p$$

after division by $\|S_n\|_p^{p-1} < \infty$. As S_n increases to $\sup_{\mathbf{N}} |X_m|$ when $n \uparrow \infty$, it only remains to let $n \uparrow \infty$ to conclude that

$$\|\sup_{\mathbf{N}} |X_n|\|_p \leqslant \frac{p}{p-1} \sup_{\mathbf{N}} \|X_n\|_p. \blacksquare$$

In Proposition IV-2-8 above we showed that a martingale $(X_n, n \in \mathbf{N})$ bounded in L^p for some $p > 1$ (i.e. such that $\sup_{\mathbf{N}} \|X_n\|_p < \infty$) has $\sup_{\mathbf{N}} |X_n| \in L^p$. The same result is not true for $p = 1$ as is seen from the examples of martingales which are bounded in L^1 but not regular (Proposition IV-2-3). On the other hand, relying on the inequality of Lemma IV-2-9 as in the proof of Proposition IV-2-8, one can prove the following result.

PROPOSITION IV-2-10. *For every martingale* $(X_n, n \in \mathbf{N})$ *satisfying the condition*

$$\sup_{\mathbf{N}} E[|X_n| \log^+ |X_n|] < \infty,$$

the r.r.v. $\sup_{\mathbf{N}} |X_m|$ *is integrable (and the martingale* $(X_n, n \in \mathbf{N})$ *is therefore regular).*

The function $\log^+$ is defined on R_+ by $\log^+ x = (\log x)^+$.

PROOF. We will use the elementary inequality

$$a \log^+ b \leqslant a \log^+ a + b/\mathrm{e}$$

valid for all $a, b \in R_+$. This inequality follows from the inequality $\log b \leqslant b/\mathrm{e}$ which is obtained by noting that the concave function $\log b$ is bounded above by its tangent at the point $b = \mathrm{e}$; indeed, this inequality implies that $a \log(b/a) \leqslant a(b/a\mathrm{e}) = b/\mathrm{e}$, and thus that

$$a \log b \leqslant a \log a + b/\mathrm{e} \leqslant a \log^+ a + b/\mathrm{e}.$$

Let us integrate both sides of the inequality in Lemma IV-2-9 applied to the submartingale $(|X_n|, n \in \mathbf{N})$ over $[1, \infty[$, after dividing both sides by a; this gives

$$\int_1^\infty P(\sup_{m \leqslant n} |X_m| > a)\,\mathrm{d}a \leqslant \int_1^{\infty} \left(\int_{\{\sup_{m \leqslant n} |X_m| > a\}} |X_n| \,\mathrm{d}P \right) \frac{da}{a}$$

$$= E(|X_n| \log^+(\sup_{m \leqslant n} |X_m|)) \leqslant E(|X_n| \log^+ |X_n|) + \mathrm{e}^{-1} E(\sup_{m \leqslant n} |X_m|)$$

by the elementary inequality above. But

$$E(\sup_{m \leqslant n} |X_m|) = \int_0^\infty P(\sup_{m \leqslant n} |X_m| > a)\,\mathrm{d}a \leqslant 1 + \int_1^\infty P(\sup_{m \leqslant n} |X_m| > a)\,\mathrm{d}a$$

and consequently the preceding inequality can also be written

$$(1 - e^{-1})\, E(\sup_{m \leqslant n} |X_m|) \leqslant 1 + E(|X_n| \log^+ |X_n|)$$

since the r.v. $\sup_{m \leqslant n} |X_m|$ is integrable. Letting $n \uparrow \infty$, we find that

$$E(\sup_{\mathbf{N}} |X_n|) \leqslant e\,(e-1)^{-1}(1 + \sup_{\mathbf{N}} E(|X_n| \log^+ |X_n|))$$

which proves Proposition IV-**2**-10. ■

R. Gundy has recently shown that for a wide class of martingales the result of Proposition IV-**2**-10 cannot be improved.

PROPOSITION IV-**2**-11. *For every positive martingale* $(X_n, n \in \mathbf{N})$ *satisfying the inequalities*

$$X_{n+1} \leqslant C X_n \quad \text{a.s.} \quad (n \in \mathbf{N})$$

for a constant $C > 0$, *the condition* $E(\sup_{\mathbf{N}} X_n) < \infty$ *implies that*

$$\sup_{\mathbf{N}} E[X_n \log^+ X_n] < \infty$$

whenever $E(X_0 \log^+ X_0) < \infty$.

The latter condition is clearly essential as is already shown by the constant martingales $X_n = X_0$ $(n \in \mathbf{N})$.

PROOF. We begin by proving an inequality reversed to that of Lemma IV-2-9. If $(X_n, n \in \mathbf{N})$ denotes a positive martingale such that $X_{n+1} \leqslant CX_n$ a.s. $(n \in \mathbf{N})$ and if ν_a denotes the stopping time equal to the first $n \in \mathbf{N}$ such that $X_n > a$, we have

$$\int_{\{0<\nu_a \leqslant n\}} X_n \,\mathrm{d}P = \int_{\{0<\nu_a \leqslant n\}} X_{\nu_a} \,\mathrm{d}P \leqslant Ca\, P(0 < \nu_a \leqslant n)$$

for $X_{\nu_a} \leqslant CX_{\nu_a - 1} \leqslant Ca$ on $\{0 < \nu_a \leqslant n\}$. This implies that

$$\int_{\{\sup_{m \leqslant n} X_m > a\}} X_n \,\mathrm{d}P = \int_{\{X_0 > a\}} X_n \,\mathrm{d}P + \int_{\{0<\nu_a \leqslant n\}} X_n \,\mathrm{d}P$$
$$\leqslant \int_{\{X_0 > a\}} X_0 \,\mathrm{d}P + Ca\, P(\sup_{m \leqslant n} X_m > a)$$

and by integrating with respect to the measure $1_{\{a \geqslant 1\}} a^{-1} \mathrm{d}a$, we obtain

$$E(X_n \log^+(\sup_{m \leqslant n} X_m)) \leqslant E(X_0 \log^+ X_0) + CE(\sup_{m \leqslant n} X_m).$$

Since $E(X_n \log^+ X_n)$ is bounded above by the left-hand side, we have shown that

$$\sup_{\mathbf{N}} E(X_n \log^+ X_n) \leqslant E(X_0 \log^+ X_0) + CE(\sup_{\mathbf{N}} X_n) < \infty. \blacksquare$$

REMARKS. (1) Every positive integrable martingale $(X_n, n \in \mathbf{N})$ such that

$$(X_{n+1} - X_n)^2 \leqslant cE^{\mathscr{B}_n}((X_{n+1} - X_n)^2) < \infty \quad \text{a.s.} \quad (n \in \mathbf{N})$$

for a constant c (necessarily $\geqslant 1$) satisfies the preceding hypothesis in the form: $X_{n+1} \leqslant (1 + 2c) X_n$ a.s. $(n \in \mathbf{N})$. Indeed, by this inequality

$$E^{\mathscr{B}_n}((X_{n+1} - X_n)^2) = E^{\mathscr{B}_n}(|X_{n+1} - X_n| \cdot |X_{n+1} - X_n|)$$
$$\leqslant \sqrt{cE^{\mathscr{B}_n}((X_{n+1} - X_n)^2)}\, E^{\mathscr{B}_n}(|X_{n+1} - X_n|),$$

and consequently

$$E^{\mathscr{B}_n}((X_{n+1} - X_n)^2) \leqslant c(E^{\mathscr{B}_n}(|X_{n+1} - X_n|))^2.$$

But since $(X_n, n \in \mathbf{N})$ is a positive martingale,

$$E^{\mathscr{B}_n}(|X_{n+1} - X_n|) = 2E^{\mathscr{B}_n}[(X_{n+1} - X_n)^-] \leqslant 2\, X_n,$$

and consequently,

$$(X_{n+1} - X_n)^2 \leqslant cE^{\mathscr{B}_n}((X_{n+1} - X_n)^2) \leqslant c^2[E^{\mathscr{B}_n}(|X_{n+1} - X_n|)]^2 \leqslant 4\, c^2\, X_n^2.$$

This inequality implies that

$$X_{n+1} \leqslant (2\, c + 1)\, X_n \qquad (n \in \mathbf{N}).$$

(2) (The Vitali–Chow Condition.) If the σ-fields $\mathscr{B}_n$ are atomic (= generated by countable partitions), and if there exists a constant c such that for all atoms B_{n+1} of $\mathscr{B}_{n+1}$ we have $P(B_n) \leqslant cP(B_{n+1})$, where B_n denotes the atom of $\mathscr{B}_n$ containing B_{n+1}, then every positive martingale $(X_n, n \in \mathbf{N})$ defined relative to the sequence $(\mathscr{B}_n, n \in \mathbf{N})$ satisfies the condition $X_{n+1} \leqslant cX_n$ a.s. $(n \in \mathbf{N})$. In fact for almost all points ω of the atom B_{n+1}, we have

$$X_{n+1}(\omega)\, P(B_{n+1}) = \int_{B_{n+1}} X_{n+1}\, \mathrm{d}P \leqslant \int_{B_n} X_{n+1}\, \mathrm{d}P = \int_{B_n} X_n\, \mathrm{d}P = X_n(\omega)\, P(B_n)$$

since X_{n+1} is a.s. constant on B_{n+1} and X_n is a.s. constant on B_n; as $P(B_n) \leqslant cP(B_{n+1})$ this implies that $X_{n+1} \leqslant cX_n$ a.s.

The sequence $(\mathscr{B}_n, n \in \mathbf{N})$ of sub-σ-fields of $[0, 1[$ generated by the dyadic partitions $([2^{-n}k, 2^{-n}(k+1)[, \ 0 \leqslant k < 2^n)$ satisfies the above condition for Lebesgue measure with $c = 2$. ■

IV-3. **Regular stopping times for an integrable martingale**

The third part of this fourth chapter is devoted to the study of "stopping theorems" for martingales.

PROPOSITION IV-3-12. *Let $(X_n, n \in \mathbf{N})$ be an integrable martingale. For every stopping time ν the sequence $(X_{\nu \wedge n}, n \in \mathbf{N})$ obtained by "stopping" the sequence $(X_n, n \in \mathbf{N})$ at the time ν is again an integrable martingale.*

The stopping time ν is said to be regular for the martingale $(X_n, n \in \mathbf{N})$ if the martingale $(X_{\nu \wedge n}, n \in \mathbf{N})$ is regular. For such a stopping time the limit $X_\infty = \lim_n X_n$ exists a.s. on $\{\nu = \infty\}$; the r.v. X_ν, which is defined a.s., is integrable and

$$X_{\nu \wedge n} = E^{\mathscr{B}_n}(X_\nu)$$

for all $n \in \mathbf{N}$. Conversely, a stopping time ν cannot exhibit these three properties without being regular.

Finally, if ν *is regular, the martingale identity remains true for every pair* ν_1, ν_2 *of stopping times such that* $\nu_1 \leqslant \nu_2 \leqslant \nu$; *for such a pair the r.v.'s* X_{ν_1} *and* X_{ν_2} *both exist, are integrable, and satisfy*

$$X_{\nu_1} = E^{\mathscr{B}_{\nu_1}}(X_{\nu_2}) \qquad (\nu_1 \leqslant \nu_2 \leqslant \nu \text{ on } \Omega).$$

In applications, the regularity of a stopping time ν is often used to assert that $E(X_\nu) = E(X_0)$. This equality follows readily from the result $E^{\mathscr{B}_0}(X_\nu) = X_0$ stated above.

PROOF. The r.v.'s $X_{\nu \wedge n}$ are integrable and $\mathscr{B}_n$-measurable respectively, as they are finite sums of such variables:

$$X_{\nu \wedge n} = \sum_{m<n} X_m 1_{\{\nu=m\}} + X_n 1_{\{\nu \geqslant n\}} \qquad (n \in \mathbf{N}).$$

As

$$X_{\nu \wedge (n+1)} - X_{\nu \wedge n} = 1_{\{\nu > n\}}(X_{n+1} - X_n) \qquad (n \in \mathbf{N})$$

we then see that

$$E^{\mathscr{B}_n}(X_{\nu \wedge (n+1)} - X_{\nu \wedge n}) = 1_{\{\nu > n\}} E^{\mathscr{B}_n}(X_{n+1} - X_n) = 0$$

since $\{\nu > n\} \in \mathscr{B}_n$; this proves that $(X_{\nu \wedge n}, n \in \mathbf{N})$ is an integrable martingale.

If the martingale $(Y_n = X_{\nu \wedge n}, n \in \mathbf{N})$ is regular, the limit $Y_\infty = \lim_n Y_n$ exists a.s.; as $Y_n = X_n$ $(n \in \mathbf{N})$ on $\{\nu = \infty\}$, the limit $X_\infty = \lim_n X_n$ thus exists on $\{\nu = \infty\}$ and further the equality $Y_\infty = X_\nu$ is valid a.s. on Ω. By Proposition IV-2-3 the r.v. $Y_\infty = X_\nu$ is integrable, whilst the equality $Y_n = E^{\mathscr{B}_n}(Y_\infty)$ can be written $X_{\nu \wedge n} = E^{\mathscr{B}_n}(X_\nu)$. The converse is immediate from the definition of regular martingales.

Finally, as $Y_{\nu_1} = X_{\nu_1}$ for all stopping times $\nu_1 \leqslant \nu$, the last part of the proposition is an immediate consequence of Corollary IV-2-6 applied to the regular martingale $(Y_n, n \in \mathbf{N})$. ■

COROLLARY IV-3-13. *Let* ν_1 *and* ν_2 *be two stopping times such that* $\nu_1 \leqslant \nu_2$ *a.s. For a given martingale* $(X_n, n \in \mathbf{N})$, *the stopping time* ν_1 *is regular whenever the stopping time* ν_2 *is regular.*

This corollary shows in particular that for a *regular* martingale, every stopping time is regular (take $\nu_2 = +\infty$).

PROOF. Since ν_2 is a regular stopping time and $\nu_1 \leqslant \nu_2$, the last part of the above proposition applied in the case $\nu_2 = \nu$ implies that the r.v. X_{ν_1} exists and is integrable. To show that the stopping time ν_1 is regular it remains to prove that $X_{\nu_1 \wedge n} = E^{\mathscr{B}_n}(X_{\nu_1})$ for all $n \in \mathbf{N}$.

But on the one hand, *on the event* $\{\nu_1 \leqslant n\}$ which belongs to $\mathscr{B}_n$ we have

$$E^{\mathscr{B}_n}(X_{\nu_1}) = E^{\mathscr{B}_n}(X_{\nu_1} 1_{\{\nu_1 \leqslant n\}}) = X_{\nu_1} 1_{\{\nu_1 \leqslant n\}} = X_{\nu_1 \wedge n}$$

since the r.v. $X_{\nu_1} 1_{\{\nu_1 \leqslant n\}} = \sum_{m \leqslant n} X_m 1_{\{\nu_1 = m\}}$ is $\mathscr{B}_n$-measurable. On the other hand, the identity $X_{\nu_1} = E^{\mathscr{B}_{\nu_1}}(X_{\nu_2})$ of the preceding proposition implies that

$$E^{\mathscr{B}_{\nu_1 \wedge n}}(X_{\nu_1}) = E^{\mathscr{B}_{\nu_1 \wedge n}}(X_{\nu_2}) \qquad (n \in \mathbf{N})$$

since $\mathscr{B}_{\nu_1 \wedge n} \subset \mathscr{B}_{\nu_1}$; but in addition the right-hand side is equal to $X_{\nu_1 \wedge n}$ by the same proposition. We have thus shown that

$$X_{\nu_1 \wedge n} = E^{\mathscr{B}_{\nu_1 \wedge n}}[X_{\nu_1}];$$

but by Proposition II-1-3 the conditional expectations $E^{\mathscr{B}_{\nu_1 \wedge n}}(\,.\,)$ and $E^{\mathscr{B}_n}(\,.\,)$ coincide on the event $\{\nu_1 \geqslant n\}$. ■

Now we have two criteria for the regularity of stopping times. The most important of these is given in the following proposition.

Proposition IV-3-14. *For the stopping time ν to be regular relative to the martingale $(X_n, n \in \mathbf{N})$, it is necessary and sufficient that the following two conditions be satisfied:*

(a) $\int_{\{\nu < \infty\}} |X_\nu| \, dP < \infty$,

(b) $(X_n 1_{\{\nu > n\}}, n \in \mathbf{N})$ *is a uniformly integrable sequence.*

Condition (a) is automatically satisfied by every martingale $(X_n, n \in \mathbf{N})$ such that $\sup_{\mathbf{N}} E|X_n| < \infty$, *in particular by every positive integrable martingale.*

Proof. By Proposition IV-2-3 the stopping time ν is regular for the martingale $(X_n, n \in \mathbf{N})$ if and only if the martingale $(X_{\nu \wedge n}, n \in \mathbf{N})$ is uniformly integrable. To this end it suffices that conditions (a), (b) be verified, for we have

$$\begin{aligned}\int_{\{|X_{\nu \wedge n}| > a\}} |X_{\nu \wedge n}| \, dP &= \int_{\{\nu \leqslant n, |X_\nu| > a\}} |X_\nu| \, dP + \int_{\{\nu > n, |X_n| > a\}} |X_n| \, dP \\ &\leqslant \int_{\{\nu < \infty, |X_\nu| > a\}} |X_\nu| \, dP + \int_{\{|X_n 1_{\{\nu > n\}}| > a\}} |X_n 1_{\{\nu > n\}}| \, dP\end{aligned}$$

and the integrability of X_ν on $\{\nu < \infty\}$ implies that the first term in the last line tends to 0 as $a \uparrow \infty$, whilst the uniform integrability of the sequence

$(X_n 1_{\{\nu > n\}}, n \in \mathbf{N})$ ensures that the second term of the last line decreases uniformly in n to 0 as $a \uparrow \infty$.

Conversely, conditions (a), (b) are necessary in order that the sequence $(X_{\nu \wedge n}, n \in \mathbf{N})$ be uniformly integrable. Indeed,

$$\int_{\{\nu < \infty\}} |X_\nu| \, dP = \lim_n \uparrow \int_{\{\nu \leqslant n\}} |X_\nu| \, dP \leqslant \sup_{\mathbf{N}} E|X_{\nu \wedge n}| < \infty$$

whilst the inequality $|X_n 1_{\{\nu > n\}}| \leqslant |X_{\nu \wedge n}|$ $(n \in \mathbf{N})$ implies that the sequence $(X_n 1_{\{\nu > n\}}, n \in \mathbf{N})$ is uniformly integrable whenever the sequence $(X_{\nu \wedge n}, n \in \mathbf{N})$ is uniformly integrable. The necessary and sufficient condition of the proposition is thus proved.

The condition $\sup_{\mathbf{N}} E(|X_n|) < \infty$, which implies the existence of the a.s. limit X_∞ of the sequence $(X_n, n \in \mathbf{N})$, implies that the r.v. X_ν is integrable and thus, a fortiori, that the r.v. $X_\nu 1_{\{\nu < \infty\}}$ is integrable. In fact, Fatou's lemma shows that

$$E(|X_\nu|) \leqslant \liminf_{n \to \infty} E(|X_{\nu \wedge n}|)$$

since $X_{\nu \wedge n} \to X_\nu$ a.s. when $n \uparrow \infty$, whilst the equality

$$X_{\nu \wedge n} = E^{\mathscr{B}_{\nu \wedge n}}(X_n) \qquad (n \in \mathbf{N})$$

(which reduces to $X_m = E^{\mathscr{B}_m}(X_n)$ on $\{\nu \wedge n = m\}$) implies that

$$E(|X_{\nu \wedge n}|) \leqslant E(|X_n|)$$

for all $n \in \mathbf{N}$. Therefore it is clear that

$$E(|X_\nu|) \leqslant \sup_{\mathbf{N}} E(|X_n|) < \infty. \blacksquare$$

The next corollary is an important application of the preceding proposition.

COROLLARY IV-3-15. *For every martingale* $(X_n, n \in \mathbf{N})$ *such that* $\sup_{\mathbf{N}} E|X_n| < \infty$, *in particular for every positive and integrable martingale, the stopping time* ν_a *defined by*

$$\nu_a = \begin{cases} \min(n : |X_n| > a) & \\ +\infty & \text{if } \sup_{\mathbf{N}} |X_n| \leqslant a \end{cases}$$

is regular for all $a \in \mathbf{R}_+$.

PROOF. Condition (a) of Proposition IV-3-14 is satisfied since $\sup_{\mathbf{N}} E|X_n| < \infty$. Condition (b) also follows since the r.v.'s $|X_n| 1_{\{\nu_a > n\}}$ are dominated on Ω by a. ■

PROPOSITION IV-3-16. *Let $(X_n, n \in \mathbf{N})$ be an integrable martingale. In order that the stopping time ν be regular for this martingale and that also $\lim_{n\to\infty} X_n = 0$ a.s. on $\{\nu = \infty\}$, it is necessary and sufficient that the following two conditions be satisfied:*

(1) $\int_{\{\nu<\infty\}} |X_\nu| \, dP < \infty$;

(2) $\lim_{n\to\infty} \int_{\{\nu>n\}} |X_n| \, dP = 0$.

PROOF. The conditions are clearly necessary. Indeed let us suppose that the sequence $(X_{\nu\wedge n}, n \in \mathbf{N})$ converges in L^1 to a limit Y which is zero on $\{\nu = \infty\}$; since this limit Y equals X_ν on $\{\nu < \infty\}$, we have

$$\int_{\{\nu<\infty\}} |X_\nu| \, dP = \int |Y| \, dP < \infty$$

whilst as $n \uparrow \infty$

$$\int_{\{\nu>n\}} |X_n| \, dP = \int_{\{\nu>n\}} |X_{\nu\wedge n}| \, dP \to \int_{\{\nu=\infty\}} |Y| \, dP = 0$$

by the L^1 convergence of the sequence $(X_{\nu\wedge n}, n \in \mathbf{N})$ to Y.

To prove that the conditions are sufficient, let us first remark that they imply that the martingale $(X_{\nu\wedge n}, n \in \mathbf{N})$ is bounded in L^1 since

$$E(|X_{\nu\wedge n}|) = \int_{\{\nu\leqslant n\}} |X_\nu| \, dP + \int_{\{\nu>n\}} |X_n| \, dP$$

$$\to \int_{\{\nu<\infty\}} |X_\nu| \, dP < \infty \quad \text{when } n \uparrow \infty.$$

This martingale therefore converges a.s., which amounts to saying that the limit $\lim_{n\to\infty} X_n$ exists a.s. on $\{\nu = \infty\}$. Moreover, this limit is zero, for Fatou's lemma shows that

$$\int_{\{\nu=\infty\}} |\lim_n X_n| \, dP \leqslant \liminf_{n\to\infty} \int_{\{\nu=\infty\}} |X_n| \, dP \leqslant \lim_n \int_{\{\nu>n\}} |X_n| \, dP = 0.$$

The r.v. X_ν can hence be defined on the entire space by putting $X_\nu = \lim_n X_n = 0$ on $\{\nu = \infty\}$, and the L_1-convergence of the sequence $(X_{\nu\wedge n}, n \in \mathbf{N})$ to X_ν then follows from the relations

$$E(|X_{\nu\wedge n} - X_\nu|) = \int_{\{\nu>n\}} |X_{\nu\wedge n} - X_\nu| \,\mathrm{d}P \leqslant \int_{\{\nu>n\}} |X_n| \,\mathrm{d}P + \int_{\{\nu>n\}} |X_\nu| \,\mathrm{d}P$$

$$\to 0 + \int_{\{\nu=\infty\}} |X_\nu| \,\mathrm{d}P = 0 \quad \text{as } n \uparrow \infty. \blacksquare$$

REMARKS. (1) In the literature the limit in the second condition is often replaced by a lim inf. But this is not a proper generalisation for when the r.v. $X_\nu 1_{\{\nu<\infty\}}$ is integrable, the limit of the integrals $\int_{\{\nu>n\}} X_n \mathrm{d}P$ always exists; indeed,

$$\int_{\{\nu>n\}} |X_n| \,\mathrm{d}P = E(|X_{\nu\wedge n}|) - \int_{\{\nu\leqslant n\}} |X_\nu| \,\mathrm{d}P$$

$$\to \lim_n \uparrow E(|X_{\nu\wedge n}|) - \int_{\{\nu<\infty\}} |X_\nu| \,\mathrm{d}P,$$

when $n \to \infty$ and the last expression is not of the indeterminate form $\infty - \infty$!

(2) A quicker (but less direct) method of proving that the conditions of the preceding proposition imply the regularity of a stopping time ν consists of using Proposition IV-3-14 after remarking that the sequence $(X_n 1_{\{\nu>n\}}, n \in \mathbf{N})$ is uniformly integrable whenever it converges to 0 in L^1. ■

IV-4. Application: An exponential formula and Wald's identity

Let $(Y_n, n \in \mathbf{N}^*)$ be a sequence of independent and identically distributed real-valued r.v.'s defined on the probability space $(\Omega, \mathscr{A}, P)$ and let $\phi: R \to]-\infty, +\infty]$ be the cumulant generating function, defined in terms of the Laplace transform by

$$\phi(u) = \log E(\exp(uY_1)) \qquad (u \in \mathbf{R}).$$

We will suppose that the r.r.v Y_1 is not a.s. constant; Hölder's inequality applied to the r.r.v.'s $\exp(uY_1)$ $(u \in \mathbf{R})$ then implies that the function ϕ is strictly convex on $\mathbf{R}$. The set $\{\phi < +\infty\}$ is thus a real interval, containing the origin since $\phi(0) = 0$. If the interior of this interval is non-empty, we can use Lebesgue's dominated convergence theorem to show that the function ϕ is infinitely differentiable there (in fact, analytic). More precisely the first derivative, for example, of the function ϕ is the strictly increasing function given by

$$\phi'(u) = E(Y_1 \exp[uY_1 - \phi(u)])$$

if u belongs to the interior of the interval $\{\phi < \infty\}$; this formula remains true, moreover, at the endpoints of this interval, provided we take one-sided derivatives and admit infinite values for these derivatives at these points.

In the sequel we denote by $(X_n = \sum_{m=1}^{n} Y_m,\ n \in \mathbf{N})$ the random walk associated with the given sequence $(Y_n,\ n \in \mathbf{N}^*)$, and by $(\mathscr{B}_n,\ n \in \mathbf{N})$ the increasing sequence of sub-σ-fields of $\mathscr{A}$ given by

$$\mathscr{B}_n = \sigma(Y_1, \ldots, Y_n) = \sigma(X_0, \ldots, X_n) \qquad (n \in \mathbf{N});$$

by definition, $X_0 = 0$ and the σ-field $\mathscr{B}_0$ is trivial. The following proposition and its corollaries then give an interesting class of positive martingales.

PROPOSITION IV-4-17. *For every $u \in \mathbf{R}$ such that $\phi(u) < \infty$, the sequence*

$$(\exp[uX_n - n\phi(u)], \quad n \in \mathbf{N})$$

associated with the above random walk is a positive martingale whose initial term equals 1. *If $u \neq 0$, this martingale converges a.s. to* 0 *as $n \uparrow \infty$.*

PROOF. For all $n \in \mathbf{N}$ the r.r.v. $\exp[uX_n - n\phi(u)]$ is finite, strictly positive and $\mathscr{B}_n$-measurable if $\phi(u) < \infty$. Further, the independence of Y_{n+1} and $\mathscr{B}_n$ implies that

$$E^{\mathscr{B}_n}[\exp(uY_{n+1})] = \exp\phi(u);$$

it then follows that

$$\begin{aligned} &E^{\mathscr{B}_n}(\exp[uX_{n+1} - (n+1)\phi(u)]) \\ &\quad = \exp[uX_n - (n+1)\phi(u)]\,E^{\mathscr{B}_n}(\exp[uY_{n+1}]) = \exp[uX_n - n\phi(u)] \end{aligned}$$

and the first part of the proposition is established.

Let $u \neq 0$ be a real number such that $\phi(u) < \infty$, if such a number exists. The strict convexity of the function ϕ then implies that $\phi(\tfrac{1}{2}u) < \tfrac{1}{2}\phi(u)$ since $\phi(0) = 0$. On the other hand, by Theorem II-2-9 there exists a r.v. Z such that

$$\exp[\tfrac{1}{2}uX_n - n\phi(\tfrac{1}{2}u)] \to Z < \infty \quad \text{a.s. when } n \uparrow \infty;$$

thus by squaring

$$\exp[uX_n - 2n\phi(\tfrac{1}{2}u)] \to Z^2 < \infty \quad \text{a.s. when } n \uparrow \infty.$$

Since $2\phi(\tfrac{1}{2}u) < \phi(u)$ this clearly implies that

$$\exp[uX_n - n\phi(u)] \to 0 \quad \text{a.s.} \quad \text{when } n \uparrow \infty. \blacksquare$$

The preceding proposition provides the means of constructing a whole sequence of new martingales from the random walk.

COROLLARY IV-4-18. *If the function ϕ is finite on an open neighbourhood V of 0, for all fixed $x \in \mathbf{R}$ and $n \in \mathbf{N}$ let*

$$\exp [ux - n\phi(u)] = \sum_{k \in \mathbf{N}} \frac{u^k}{k!} f_k(n, x) \qquad (u \in V)$$

be the expansion of the analytic function $u \to \exp [ux - n\phi(u)]$ defined on V. For every integer $k \in \mathbf{N}^*$, the sequence $(f_k(n, X_n), n \in \mathbf{N})$ is then an integrable martingale; in particular,

$$f_0(n, X_n) = 1,$$

$$f_1(n, X_n) = X_n - nE(Y_1) = X_n - E(X_n),$$

$$f_2(n, X_n) = [X_n - nE(Y_1)]^2 - n \operatorname{var}(Y_1) = [X_n - E(X_n)]^2 - \operatorname{var}(X_n).$$

REMARK. Let us observe that $f_k(\cdot)$ is a polynomial of degree k in n and x, whose coefficients depend on the derivatives of order $\leqslant k$ at the origin of the function ϕ, i.e. depend on the first k moments $E(Y_1^j)$ $(j = 1, \ldots, k)$ of the r.v. Y_1. In fact it is not hard to show that f_k exists and $(f_k(n, X_n), n \in \mathbf{N})$ is an integrable martingale whenever $E(|Y_1|^k) < \infty$, without it being necessary to suppose that $\phi < \infty$ on a neighbourhood of 0.■

PROOF. Differentiating the martingale identities

$$E^{\mathscr{B}_m}(\exp [uX_n - n\phi(u)]) = \exp[uX_m - m\phi(u)] \qquad (m, n \in \mathbf{N}, m \leqslant n)$$

k times with respect to u at the origin, formally we obtain

$$E^{\mathscr{B}_m}(f_k(n, X_n)) = f_k(m, X_m),$$

which is exactly what is wanted. To justify the differentiation under the integral sign it suffices to choose for fixed m and n an $\varepsilon > 0$ such that

$$E(\exp [\pm 2\varepsilon Y_1]) < \infty,$$

and thus that

$$E(\exp [\pm 2\varepsilon X_p]) < \infty \quad \text{for all } p \in \mathbf{N}^*,$$

and use bounds of the type

$$|x|^k \exp (ux) \leqslant C_k[\exp (2\varepsilon x) + \exp (-2\varepsilon x)]$$

valid for x on R, for u on the interval $]-\varepsilon, \varepsilon[$ and for all integers $k > 0$ if the constants C_k are suitably chosen. ■

Another example of a positive martingale which one can construct from the exponential martingales of Proposition IV-**4**-17 is given by the formula

$$\int_{\mathbf{R}} du \exp[uX_n - n\phi(u)] \qquad (n \in \mathbf{N}),$$

In the particular case where the r.r.v.'s Y_n are standardised gaussian, we have $\phi(u) = \frac{1}{2}u^2$ on $\mathbf{R}$ and the preceding integral equals

$$\sqrt{\frac{2\pi}{n}} \exp\left(\frac{X_n^2}{2\,n}\right).$$

The sequence of r.r.v.'s thus obtained is a positive martingale, finite from the index 1 onwards, but not integrable since the expected value of the integral above equals $\int_{\mathbf{R}} du = +\infty$ by Fubini's theorem. Theorem II-**2**-9 implies that this martingale converges a.s. to a finite limit (a more detailed analysis would show that this limit is in fact zero) and hence that

$$X_n = O(\sqrt{n \log n}) \quad \text{a.s.} \quad \text{when } n \uparrow \infty.$$

We shall see later (Section VII-**2**) that a much more precise and more general theorem than the preceding result can be proved.

The exponential martingales $(\exp[uX_n - n\phi(u)], n \in \mathbf{N})$ of the above proposition are not regular since they converge to 0 a.s. (if $u \neq 0$), but they admit many interesting regular stopping times.

PROPOSITION IV-**4**-19. (a) *Let $u \neq 0$ be a real number such that $\phi(u) < \infty$ and $\phi'(u) \geq 0$. For every $a \in R_+$, the stopping time*

$$\nu_a^+ = \begin{cases} \min(a : X_n \geq a), \\ +\infty & \textit{if } X_n < a, \quad (n \in \mathbf{N}) \end{cases}$$

is then regular for the martingale $(\exp[uX_n - n\phi(u)], n \in \mathbf{N})$. *Wald's identity*

$$\int_{\{\nu < \infty\}} \exp[uX_\nu - \nu\phi(u)]\, dP = 1$$

is therefore true for the stopping times $\nu = \nu_a^+$ and, more generally, for any stopping time ν bounded above by ν_a^+.

(b) *For every real number $u \neq 0$ such that $\phi(u) < \infty$ and for every pair $a, b > 0$ of real numbers, the stopping time*

$$\nu_{a,b} = \begin{cases} \min(n : X_n \geqslant a \text{ or } X_n \leqslant -b) \\ +\infty & \text{if } -b < X_n < a \text{ for all } n \in \mathbf{N} \end{cases}$$

is regular for the martingale ($\exp[uX_n - n\phi(u)]$, $n \in \mathbf{N}$) *and thus satisfies Wald's identity above.*

PROOF. (a) We apply Proposition IV-**3**-16. The first condition of this proposition is satisfied as the exponential martingale

$$(\exp[uX_n - n\phi(u)], \quad n \in \mathbf{N})$$

is positive; it remains to see that the second condition is satisfied, i.e. that

$$\int_{\{\nu_a^+ > n\}} \exp[uX_n - n\phi(u)]\,\mathrm{d}P \to 0 \quad \text{when } n \uparrow \infty.$$

To this end we use the following lemma taken from the theory of recurrence of random walks.

LEMMA IV-**4**-20. *For every random walk* ($X_n = \sum_{m=1}^{n} Y_m$, $n \in \mathbf{N}$) *such that* $E(Y_1) \geqslant 0$ *and for every real number* $a > 0$, *the stopping time*

$$\nu_a^+ = \begin{cases} \min(n : X_n \geqslant a), \\ +\infty & \text{if } \sup_{\mathbf{N}} X_n < a \end{cases}$$

is a.s. finite.

(The condition $E(Y_1) \geqslant 0$ in this lemma can be interpreted in the wide sense: either $E|Y_1| < \infty$ and $E(Y_1) \geqslant 0$ or $E(Y_1^+) = +\infty$ and $E(Y_1^-) < +\infty$.)

If μ denotes the common probability law of the r.v.'s Y_n ($n \in \mathbf{N}^*$), let us consider an independent sequence (Y_n', $n \in \mathbf{N}^*$) of real-valued r.v.'s on an auxiliary probability space (Ω', $\mathscr{A}'$, P') with the same law

$$\mu^{(u)}(\mathrm{d}y) = \exp[uy - \phi(u)]\,\mu(\mathrm{d}y)$$

and thus also the associated random walk ($X_n' = \sum_{m=1}^{n} Y_m'$, $n \in \mathbf{N}$). As the probability law of the vector (Y_1', ..., Y_n') is given on R^n by

$$\exp[u(y_1' + \dots + y_n') - n\phi(u)]\,\mu(\mathrm{d}y_1)\,\dots\,\mu(\mathrm{d}y_n),$$

it is not hard to check that

$$\int_{\{\nu_a^+>n\}} \exp[uX_n - n\phi(u)]\,\mathrm{d}P = P'(\nu_a'^+ > n)$$

where $\nu_a'^+$ is the first time that the sequence $(X_n', n \in \mathbf{N})$ exceeds the barrier a. As

$$\int_{\mathbf{R}} y'\,\mu^{(u)}(\mathrm{d}y') = \int_{\mathbf{R}} y \exp[uy - \phi(u)]\,\mu(\mathrm{d}y) = \phi'(u),$$

Lemma IV-4-20 above shows that $P'(\nu_a'^+ < \infty) = 1$ if $\phi'(u) \geqslant 0$ and we have thus proved that $P'(\nu_a'^+ > n)\downarrow 0$ when $n\uparrow\infty$.

(b) Since the stopping time $\nu_{a,b}$ is bounded above by the stopping time ν_a^+ which is regular if $\phi'(u) \geqslant 0$, and by the stopping time ν_b^- which is regular if $\phi'(u) \geqslant 0$, $\nu_{a,b}$ is necessarily regular. ■

EXAMPLE. Let us suppose that the r.v. Y_1 can only take integral values $\geqslant 1$ and that $P(Y_1 = 1) \neq 0$. In this case, $\phi < \infty$ on R_+ and there exists a real $u^* \geqslant 0$ such that $\phi(u^*) = \inf_{\mathbf{R}_+} \phi$. If the barrier a is integral (this is the only interesting case), we have $X_{\nu_a^+} = a$ on $\{\nu_a^+ < \infty\}$ and by the preceding proposition we can write

$$\int_{\{\nu_a^+<\infty\}} \exp[-\phi(u)\,\nu_a^+]\,\mathrm{d}P = \exp[-ua] \quad \text{if } u \geqslant u^*.$$

On the interval $[u^*, +\infty[$, the function ϕ increases continuously from $\phi(u^*)$ to $+\infty$; the preceding formula thus gives the Laplace transform of the law of ν_a^+. In particular, if $E(Y_1) < 0$, we have $\phi'(0) < 0$, hence $\phi(u^*) < 0$ and the equation $\phi(u) = 0$ admits exactly one solution $u_0 > u^* > 0$; Wald's identity shows that

$$P(\nu_a^+ < \infty) = \exp[-au_0].$$

(On the other hand, if $E(Y_1) \geqslant 0$, we can deduce from either the law of large numbers or Wald's identity that $P(\nu_a^+ < \infty) = 1$.)

Let us also note that the result in this example can be extended as an approximation to r.v.'s Y_n of arbitrary law concentrated on $]-\infty, c]$ (c a real number > 0) with $P(Y_1 > 0) \neq 0$. For such r.v.'s the function ϕ is also finite on $\mathbf{R}_+$ and the double inequality $a \leqslant X_{\nu_a^+} \leqslant a + c$ which holds on $\{\nu_a^+ < \infty\}$ implies that

$$\exp[-u(a+c)] \leqslant \int_{\{\nu_a^+<\infty\}} \exp[-\nu_a^+\phi(u)]\,\mathrm{d}P \leqslant \exp[-ua] \quad \text{if } u > u^*$$

(where $u^* \in \mathbf{R}_+$ and $\phi(u^*) = \inf_{\mathbf{R}_+} \phi$); this result is mainly of interest when $a \gg c$. ■

The integrable martingales

$$(X_n - nE(Y_1), n \in \mathbf{N}), \qquad ([X_n - nE(Y_1)]^2 - n\operatorname{var}(Y_1), n \in \mathbf{N}),$$

defined whenever $Y_1 \in L^1$ and $Y_1 \in L^2$ respectively, are not regular since they diverge a.s. when $n \uparrow \infty$. The following proposition is a simple result for the regularity of stopping times associated with these martingales.

PROPOSITION IV-4-21. *Every stopping time v such that $E(v) < \infty$ is regular for each of the two martingales*

$$(X_n - nE(Y_1), \quad n \in \mathbf{N}), \qquad ([X_n - nE(Y_1)]^2 - n\operatorname{var}(Y_1), \quad n \in \mathbf{N})$$

provided that $E(|Y_1|) < \infty$ and $E(Y_1^2) < \infty$ respectively. Such a stopping time satisfies the identities

$$E(X_v) = E(v)\,E(Y_1),$$
$$E([X_v - vE(Y_1)]^2) = E(v)\operatorname{var}(Y_1).$$

In particular, if $E(Y_1) > 0$, the stopping time $v_a^+ = \min(n : X_n \geq a)$ is integrable for every real number $a > 0$.

PROOF. In the proof of the first part of the proposition we can suppose that $E(Y_1) = 0$. If $E(v) < \infty$, the stopping time v is a.s. finite and the sequence $(X_{v \wedge n}, n \in \mathbf{N})$ tends to X_v a.s. Let us show that this convergence also takes place in L^1, which will establish the regularity of v. To this end, let us write

$$|X_{v\wedge n} - X_v| = \left| \sum_{m>n} Y_m 1_{\{v \geq m\}} \right| \leq \sum_{m>n} |Y_m| 1_{\{v \geq m\}} \qquad (n \in \mathbf{N})$$

and observe that the assumptions of independence and identical distribution of r.r.v.'s Y_m imply that

$$E\left(\sum_{n \in \mathbf{N}^*} |Y_m| 1_{\{v \geq m\}} \right) = \sum_{m \in \mathbf{N}^*} E(E^{\mathscr{B}_{m-1}}(|Y_m|) 1_{\{v \geq m\}}) = E(|Y_1|)\,E(v) < +\infty;$$

an application of the dominated convergence theorem then completes the proof since this theorem shows that

$$E\left(\sum_{m>n} |Y_m| 1_{\{v > m\}} \right) \downarrow 0 \quad \text{when } n \uparrow \infty.$$

A slight change in this proof shows that the sequence $(X_{\nu\wedge n}, n \in \mathbf{N})$ also converges in L^2 to X_ν if $E(Y_1^2) < \infty$. Indeed, the sequence $(Y_m 1_{\{\nu \geqslant m\}}, n \in \mathbf{N})$ is an orthogonal sequence for

$$E(Y_l 1_{\{\nu \geqslant l\}} Y_m 1_{\{\nu \geqslant m\}}) = E(Y_l 1_{\{\nu \geqslant l\}} E^{\mathscr{B}_{m-1}}(Y_m)) = 0$$

if $l < m$; on the other hand,

$$\sum_{\mathbf{N}^*} E((Y_m 1_{\{\nu \geqslant m\}})^2) = \sum_{\mathbf{N}^*} E(E^{\mathscr{B}_{m-1}}(Y_m^2) 1_{\{\nu \geqslant m\}}) = E(Y_1^2) E(\nu) < \infty.$$

As before, we then have

$$E((X_{\nu\wedge n} - X_\nu)^2) = E\left(\left(\sum_{m>n} Y_m 1_{\{\nu \geqslant m\}}\right)^2\right) = E\left(\sum_{m>n} Y_m^2 1_{\{\nu \geqslant m\}}\right) \downarrow 0 \quad \text{when } n \uparrow \infty$$

by dominated convergence. The convergence of the sequence $(X_{\nu\wedge n}, n \in \mathbf{N})$ in L^2 implies the convergence of the sequence $(X_{\nu\wedge n}^2, n \in \mathbf{N})$ in L^1 to X_ν^2 and thus also that of the martingale $(X_{\nu\wedge n}^2 - (\nu \wedge n)\operatorname{var}(Y_1), \ n \in \mathbf{N})$ to $X_\nu^2 - \nu \operatorname{var}(Y_1)$. Thus the first part of the proposition is established.

Now drop the assumption that $E(Y_1) = 0$. The martingale property of the stopped sequence $(X_{\nu_a^+\wedge n} - (\nu_a^+ \wedge n)E(Y_1), n \in \mathbf{N})$ implies that

$$E(\nu_a^+ \wedge n) E(Y_1) = E(X_{\nu_a^+\wedge n}) \qquad (n \in \mathbf{N}).$$

Let us suppose then that $Y_1 \leqslant c$ a.s. for a constant c; the definition of ν_a^+ implies that

$$X_{\nu_a^+} = X_{(\nu_a^+ - 1)} + Y_{\nu_a^+} \leqslant a + c \quad \text{on } \{\nu_a^+ < \infty\}$$

thus also that $X_{\nu_a^+\wedge n} \leqslant a + c$ a.s. $(n \in \mathbf{N})$ since $X_n \leqslant a$ if $n < \nu_a^+$. Under the stated assumption we thus see that

$$E(\nu_a^+) E(Y_1) = \lim_{n\uparrow\infty}\uparrow E(\nu_a^+ \wedge n) E(Y_1) \leqslant a + c$$

which shows that $E(\nu_a^+) < \infty$ if $E(Y_1) > 0$.

If there does not exist a constant bounding Y_1 above a.s., choose c sufficiently large that $E(Y_1 \wedge c) > 0$, which is certainly possible when $E(Y_1) > 0$. The random walk $(\sum_{m=1}^n (Y_m \wedge c), n \in \mathbf{N})$ is then dominated by the sequence $(X_n, n \in \mathbf{N})$ and can only reach the barrier at $a > 0$ at a time ν_a^* subsequent to ν_a^+; but this new random walk satisfies the boundedness assumption of the previous paragraph and so $E(\nu_a^*) < \infty$; a fortiori, $E(\nu_a^+) < \infty$. ■

REMARK. The argument of the first part of the preceding proof can be extended to cover arbitrary integrable martingales in the following way: *For every integrable martingale $(X_n, n \in \mathbf{N})$, the stopping time ν is regular whenever it satisfies the integrability condition*

$$E\left(\sum_{n<\nu} E^{\mathscr{B}_n}(|X_{n+1} - X_n|)\right) < \infty$$

(for an integrable random walk, this condition reduces to $E(\nu) < \infty$).

Indeed, this assumption can also be written $E(\sum_{n<\nu} |X_{n+1} - X_n|) < \infty$ because $\{\nu > n\} \in \mathscr{B}_n$ for all $n \in \mathbf{N}$ and, by the dominated convergence theorem, this implies the a.s. and L^1 convergence of the r.v.'s

$$X_{\nu \wedge p} = X_0 + \textstyle\sum_{n<p} (X_{n+1} - X_n) 1_{\{n<\nu\}}.$$

This regularity condition is not very refined! ■

IV-5. Supplements

IV-5-A. *Doob's inequalities.*

We have deduced the submartingale convergence theorem (Theorem IV-**1**-2) from the convergence theorem for positive supermartingales (Theorem II-**2**-9), whose proof depends on Dubins' inequalities concerning the "upcrossings" of positive supermartingales. The classical proof [106] of Theorem IV-**1**-2 is slightly different: it is based on Doob's inequalities below concerning the upcrossings and downcrossing of submartingales.

PROPOSITION IV-**5**-22. *Let $(X_m, n \in \mathbf{N})$ be an integrable submartingale. For every pair a, b of real numbers such that $a < b$ and for every ω, let us denote by $\beta_{a,b}$ (resp. $\beta'_{a,b}$) the number of times the sequence $(X_n(\omega), n \in \mathbf{N})$ upcrosses (resp. downcrosses) the interval $[a, b]$. Then $\beta_{a,b}$ and $\beta'_{a,b}$ are random variables satisfying the inequalities*

$$(b-a)\, E^{\mathscr{B}_0}(\beta_{a,b}) \leqslant \lim_n \uparrow E^{\mathscr{B}_0}((X_n - a)^+) - (X_0 - a)^+,$$

$$(b-a)\, E^{\mathscr{B}_0}(\beta'_{a,b}) \leqslant \lim_n \uparrow E^{\mathscr{B}_0}((X_n - b)^+) - (X_0 - b)^+.$$

For an integrable submartingale $(X_n, n \in \mathbf{N})$ satisfying the condition $\sup_{\mathbf{N}} E(X_n^+) < \infty$, the preceding proposition implies that the r.v. $\beta_{a,b}$ is integrable and thus a.s. finite, for all $a < b$ in $\mathbf{R}$. By Lemma II-**2**-10 this implies that the sequence $(X_n, n \in \mathbf{N})$ converges a.s. Thus Theorem IV-**1**-2 will be proved a second time when the above inequalities are established.

PROOF. (1) We will be using the following switching principle for submartingales, whose proof is entirely analogous to that of Lemma II-**2**-8:

Given two integrable submartingales $(X_n^{(i)}, n \in \mathbf{N})$ $(i = 1, 2)$ *and a stopping time* ν *such that* $X_\nu^{(1)} \leqslant X_\nu^{(2)}$ *on* $\{\nu < \infty\}$, *the formula*

$$X_n^* = X_n^{(1)} 1_{\{n<\nu\}} + X_n^{(2)} 1_{\{\nu \leqslant n\}} \qquad (n \in \mathbf{N})$$

defines a new integrable submartingale $(X_n^*, n \in \mathbf{N})$.

Let us associate with our given submartingale $(X_n, n \in \mathbf{N})$ the stopping times ν_k $(k \in \mathbf{N})$ defined as in Chapter II by the formulae

$$\nu_1 = \min(n : n \geqslant 0, X_n \leqslant a),$$
$$\nu_2 = \min(n : n \geqslant \nu_1, X_n \geqslant b),$$
$$\nu_3 = \min(n : n \geqslant \nu_2, X_n \leqslant a),$$

and so on indefinitely (by convention, $\nu_k = +\infty$ when ν_k is not defined by the above formulae). Using the switching principle given above, it is easy to see that the sequence $(Z_n, n \in \mathbf{N})$ of r.r.v.'s defined by

$$Z_n = \begin{cases} X_n - a & \text{if } 0 \leqslant n < \nu_1, \\ 0 & \text{if } \nu_1 \leqslant n < \nu_2, \\ (X_n - a) - (b - a) & \text{if } \nu_2 \leqslant n < \nu_3, \\ -(b - a) & \text{if } \nu_3 \leqslant n < \nu_4, \\ (X_n - a) - 2(b - a) & \text{if } \nu_4 \leqslant n < \nu_5, \\ \vdots & \vdots \end{cases}$$

and so on, is an integrable submartingale; this follows essentially from the ν_k being stopping times increasing to $+\infty$, with $X_n - a \geqslant b - a$ at times ν_{2k} $(k \in \mathbf{N}^*)$ and from $X_n - a \leqslant 0$ at times ν_{2k+1} $(k \in \mathbf{N})$. If $\beta_{a,b}^n$ denotes the number of upcrossings that the finite sequence $(X_0, \ldots, X_n)$ makes over $[a, b]$, the submartingale $(Z_n, n \in \mathbf{N})$ can also be written

$$Z_n = \begin{cases} (X_n - a) - (b - a)\,\beta_{a,b}^n & \text{on } \sum_{k \in \mathbf{N}} \{\nu_{2k} \leqslant n < \nu_{2k+1}\}, \\ -(b - a)\,\beta_{a,b}^n & \text{on } \sum_{k \in \mathbf{N}^*} \{\nu_{2k-1} \leqslant n < \nu_{2k}\}, \end{cases}$$

putting $\nu_0 \equiv 0$ by convention. It is therefore clear that

$$Z_n \leqslant (X_n - a)^+ - (b-a)\,\beta^n_{a,b} \qquad (n \in \mathbf{N}).$$

Since on the other hand

$$Z_0 = (X_0 - a)\,1_{\{\nu_1 > 0\}} = (X_0 - a)^+,$$

here the submartingale inequality $Z_0 \leqslant E^{\mathscr{B}_0}(Z_n)$ implies that

$$(b-a)\,E^{\mathscr{B}_0}(\beta^n_{a,b}) \leqslant E^{\mathscr{B}_0}((X_n - a)^+) - (X_0 - a)^+.$$

The first inequality of the proposition is then obtained by letting $n \uparrow \infty$. Let us also note that the above inequality is quite "sharp"; indeed as $X_n - a > 0$ on each of the events $\{\nu_{2k} \leqslant n < \nu_{2k+1}\}$ $(k \in \mathbf{N})$ and $X_n - a < b - a$ on each of the events $\{\nu_{2k-1} \leqslant n < \nu_{2k}\}$ $(k \in \mathbf{N}^*)$, the definition of Z_n implies that

$$(X_n - a)^+ - (b-a)\,\beta^n_{a,b} \leqslant Z_n + (b-a) \qquad (n \in \mathbf{N}).$$

The sequence $((X_n - a)^+ - (b-a)\beta^n_{a,b},\ n \in \mathbf{N})$ hence does not differ from the submartingale $(Z_n,\ n \in \mathbf{N})$ by more than $b - a$.

(2) The second of Doob's inequalities is proved in a similar way to the first. After having defined the stopping times ν'_k $(k \in \mathbf{N}^*)$ by the formulae

$$\begin{aligned}
\nu'_1 &= \min(n : n \geqslant 0,\ X_n \geqslant b),\\
\nu'_2 &= \min(n : n \geqslant \nu'_1,\ X_n \leqslant a),\\
\nu'_3 &= \min(n : n \geqslant \nu'_2,\ X_n \geqslant b),\\
&\ \vdots
\end{aligned}$$

and so on indefinitely, we construct a new integrable submartingale $(Z'_n,\ n \in \mathbf{N})$ by putting

$$Z'_n = \begin{cases} 0 & \text{if } 0 \leqslant n < \nu'_1,\\ X_n - b & \text{if } \nu'_1 \leqslant n < \nu'_2,\\ -(b-a) & \text{if } \nu'_2 \leqslant n < \nu'_3,\\ (X_n - b) - (b-a) & \text{if } \nu'_3 \leqslant n < \nu'_4,\\ -2(b-a) & \text{if } \nu'_4 \leqslant n < \nu'_5,\\ \ \vdots & \qquad \vdots \end{cases}$$

and so on, for all $n \in \mathbf{N}$; the submartingale property of the sequence $(Z'_n,\ n \in \mathbf{N})$ follows from the switching principle, since $X_n - b \geqslant 0$ at the times ν'_{2k-1}

($k \in \mathbf{N}^*$) and $X_n - b \leqslant a - b$ at times ν'_{2k} ($k \in \mathbf{N}^*$). If $\beta'^n_{a,b}$ then denotes the number of downcrossings of the finite sequence $(X_0, \ldots, X_n)$ over the interval $[a, b]$, the submartingale $(Z'_n, n \in \mathbf{N})$ satisfies the inequalities

$$Z'_n \leqslant (X_n - b)^+ - (b - a)\,\beta'^n_{a,b} \qquad (n \in \mathbf{N})$$

as one can easily check. But

$$Z'_0 = (X_0 - b)\,1_{\{\nu'_1 = 0\}} = (X_0 - b)^+$$

so that the submartingale inequality $Z'_0 \leqslant E^{\mathscr{B}_0}(Z'_n)$ implies the inequality

$$(b - a)\,E^{\mathscr{B}_0}(\beta'^n_{a,b}) \leqslant E^{\mathscr{B}_0}((X_n - b)^+) - (X_0 - b)^+$$

for all $n \in \mathbf{N}$. To complete the proof, it only remains to let $n \uparrow \infty$. ■

REMARK. The random variables $\beta_{a,b}$ and $\beta'_{a,b}$ differ by at most unity when they are finite, since a downcrossing (resp. upcrossing) is necessarily located between two upcrossings (resp. downcrossings) of the sequence $(X_n, n \in \mathbf{N})$. [More formally, the inequalities $|\beta^n_{a,b} - \beta'^n_{a,b}| \leqslant 1$ result from the fact that for all $\omega \in \Omega$, either $\nu_k(\omega) = \nu'_{k+1}(\omega)$ for all $k \in \mathbf{N}^*$, or $\nu'_k(\omega) = \nu_{k+1}(\omega)$ for all $k \in \mathbf{N}^*$.] Thus the left-hand sides of Doob's inequalities differ by at most $(b - a)$. We then observe that the right-hand sides of Doob's inequalities also differ by at most $(b - a)$; this is easy to prove using the elementary inequalities

$$(X_n - b)^+ \leqslant (X_n - a)^+ \leqslant (X_n - b)^+ + (b - a) \qquad (n \in \mathbf{N}).$$ ■

We now go on to show that it is possible, using the same method, to deduce Doob's inequalities as well as those of Dubins from the same family of inequalities.

PROPOSITION IV-5-23. *For every integrable submartingale $(X_n, n \in \mathbf{N})$, the numbers of upcrossings $\beta^n_{a,b}$ and downcrossings $\beta'^n_{a,b}$ up to time n ($n \in \mathbf{N}$) satisfy the inequalities*

(1) $$(b - a)\,P^{\mathscr{B}_0}(\beta^n_{a,b} \geqslant k) \leqslant E^{\mathscr{B}_0}((X_n - a)^+\,1_{\{\beta^n_{a,b} = k\}}) \quad \text{if } k \geqslant 1,$$

$$(X_0 - a)^+ \leqslant E^{\mathscr{B}_0}((X_n - a)^+\,1_{\{\beta^n_{a,b} = 0\}}).$$

(2) $$(b - a)\,P^{\mathscr{B}_0}(\beta'^n_{a,b} \geqslant k) \leqslant E^{\mathscr{B}_0}((X_n - b)^+\,1_{\{\beta'^n_{a,b} = k-1\}}) \quad \text{if } k > 1,$$

$$(X_0 - b)^+ + (b - a)\,P^{\mathscr{B}_0}(\beta'^n_{a,b} \geqslant 1) \leqslant E^{\mathscr{B}_0}((X_n - b)^+\,1_{\{\beta'^n_{a,b} = 0\}}).$$

PROOF. The first of these inequalities is proved for a fixed value of $k \geqslant 1$ by constructing the submartingale

$$Z_n^{(k)} = \begin{cases} 0 & \text{if } \quad n < \nu_{2k}, \\ X_n - b & \text{if } \quad \nu_{2k} \leqslant n < \nu_{2k+1}, \\ a - b & \text{if } \nu_{2k+1} \leqslant n, \end{cases}$$

which can also be written

$$Z_n^{(k)} = (X_n - a)\, 1_{\{\nu_{2k} \leqslant n < \nu_{2k+1}\}} - (b - a)\, 1_{\{\nu_{2k} \leqslant n\}} \qquad (n \in \mathbf{N});$$

the notation is as in the proof of Doob's inequalities. Since $Z_0^{(k)} = 0$ and $\{\nu_{2k} \leqslant n\} = \{\beta_{a,b}^n \geqslant k\}$, here the submartingale inequality $Z_0^{(k)} \leqslant E^{\mathscr{B}_0}(Z_n^{(k)})$ is

$$(b - a)\, P^{\mathscr{B}_0}(\beta_{a,b}^n \geqslant k) \leqslant E^{\mathscr{B}_0}((X_n - a)\, 1_{\{\nu_{2k} \leqslant n < \nu_{2k+1}\}}),$$

but the r.v. $(X_n - a)\, 1_{\{\nu_{2k} \leqslant n < \nu_{2k+1}\}}$ is clearly dominated by

$$(X_n - a)^+\, 1_{\{\beta_{a,b}^n = k\}},$$

proving the first line of (1).

The definition of $Z_n^{(k)}$ makes sense for $k = 0$ (if we agree to put $\nu_0 \equiv 0$) and moreover

$$Z_n^{(0)} = (X_n - a)\, 1_{\{n < \nu_1\}} \qquad (n \in \mathbf{N})$$

if we add $b - a$ to all terms. The submartingale inequality $Z_0^{(0)} \leqslant E^{\mathscr{B}_0}(Z_n^{(0)})$ in this case shows that

$$(X_0 - a)^+ \leqslant E^{\mathscr{B}_0}((X_n - a)^+\, 1_{\{n < \nu_1\}}) \leqslant E^{\mathscr{B}_0}((X_n - a)^+\, 1_{\{\beta_{a,b}^n = 0\}}).$$

The proof of the second inequality uses the submartingales $(Z_n'^{(k)}, n \in \mathbf{N})$ defined for all integers $k \geqslant 1$ by

$$Z_n'^{(k)} = \begin{cases} 0 & \text{if } n < \nu'_{2k-1}, \\ X_n - b & \text{if } \nu'_{2k-1} \leqslant n < \nu'_{2k}, \\ a - b & \text{if } \nu'_{2k} \leqslant n \end{cases}$$

or, equivalently, by

$$Z_n'^{(k)} = (X_n - b)\, 1_{\{\nu'_{2k-1} \leqslant n < \nu'_{2k}\}} - (b - a)\, 1_{\{\beta'^n_{a,b} \geqslant k\}} \qquad (n \in \mathbf{N}).$$

As $Z_0'^{(k)} = (X_0 - b)1_{\{v_1'=0\}} = (X_0 - b)^+$ if $k = 1$ and is zero if $k > 1$, the submartingale inequality $Z_0'^{(k)} \leqslant E^{\mathscr{B}_0}(Z_n^{(k)})$ becomes

$$(b-a)P^{\mathscr{B}_0}(\beta_{a,b}'^n \geqslant k) \leqslant E^{\mathscr{B}_0}((X_n - b)1_{\{v_{2k-1}' \leqslant n < v_{2k}'\}})$$

at least when $k > 1$; if $k = 1$, it is necessary to subtract $(X_0 - b)^+$ from the right-hand side. As the r.v. in the right-hand side is dominated by $(X_n - b)^+ 1_{\{\beta_{a,b}'^n = k-1\}}$, the second inequality is proved. ∎

Summing both sides of the preceding inequalities over k ($k \in \mathbf{N}^*$), we obtain Doob's inequalities. On the other hand, for a *negative* submartingale (which it is not necessary to suppose integrable) and for negative barriers $a < b$, the two inequalities of the previous proposition imply that

$$(b-a)P^{\mathscr{B}_0}(\beta_{a,b}'^n \geqslant k) \leqslant -bP^{\mathscr{B}_0}(\beta_{a,b}'^n = k-1) \quad \text{if } k \geqslant 2$$

and that

$$(X_0 - b)^+ + (b-a) \leqslant (-a)P^{\mathscr{B}_0}(\beta_{a,b}'^n = 0).$$

This can also be written

$$(-a)P^{\mathscr{B}_0}(\beta_{a,b}'^n \geqslant k+1) \leqslant (-b)P^{\mathscr{B}_0}(\beta_{a,b}'^n \geqslant k) \qquad (k \in \mathbf{N}^*)$$

and

$$(-a)P^{\mathscr{B}_0}(\beta_{a,b}'^n > 0) \leqslant (-b) - (X_0 - b)^+ = \min(-X_0, -b);$$

this implies Dubin's inequalities for the negative submartingale

$$P^{\mathscr{B}_0}(\beta_{a,b}'^n \geqslant k) \leqslant \left(\frac{-b}{-a}\right)^k \min\left(\frac{X_0}{b}, 1\right).$$

IV-5-B. *Regularity of integrable submartingales.*

The study of regularity of integrable martingales can be very easily extended to integrable submartingales by using the Krickeberg decomposition of such submartingales.

PROPOSITION IV-5-24. *For every integrable submartingale $(X_n, n \in \mathbf{N})$, the following conditions are equivalent:*

(a) *The sequence $(X_n^+, n \in \mathbf{N})$ converges in L^1.*

(b) $\sup_{\mathbf{N}} E(X_n^+) < \infty$, *and the a.s. limit* $X_\infty = \lim_n X_n$ *of the submartingale* $(X_n, n \in \mathbf{N})$, *which exists and is integrable by Theorem IV-1-2, satisfies the inequalities* $X_n \leqslant E^{\mathscr{B}_n}(X_\infty)$ *for all* $n \in \mathbf{N}$.

(b′) *There exists an integrable r.r.v. Y such that* $X_n \leqslant E^{\mathscr{B}_n}(Y)$ *for all* $n \in \mathbf{N}$.

(c) *The sequence* $(X_n^+, n \in \mathbf{N})$ *satisfies the uniform integrability condition*

$$\sup_{\mathbf{N}} \int_{\{X_n > a\}} X_n \, \mathrm{d}P \downarrow 0 \text{ when } a \uparrow \infty$$

(*in particular this condition holds if* $E(\sup_{\mathbf{N}} X_n^+) < \infty$).

The integrable submartingale $(X_n, n \in \mathbf{N})$ is said to be *regular* if it satisfies the preceding equivalent conditions.

For a *negative* integrable submartingale (i.e. for a positive integrable supermartingale with its sign changed), the conditions of the proposition hold trivially (the second part of condition (b) holds by Fatou's lemma). Observe that such a submartingale does not necessarily converge in mean, although it always converges a.s., and hence condition (a) of the preceding proposition is strictly less restrictive than the convergence of the submartingale $(X_n, n \in \mathbf{N})$ in L^1. On the other hand it is clear that for a *positive* submartingale condition (a) gives L^1-convergence of the submartingale!

Proof. The implication (b) ⇒ (b′) is obvious. The proof of implications (b′) ⇒ (c) and (c) ⇒ (a) can be obtained from the proofs of the corresponding implications in Proposition IV-2-3 by taking absolute values in these expressions. It only remains to prove the implication (a) ⇒ (b).

If the sequence $(X_n^+, n \in \mathbf{N})$ of positive r.v.'s converges in L^1, it is clear that $\sup_{\mathbf{N}} E(X_n^+) < \infty$. Theorem IV-1-2 then implies the existence of the a.s. limit $X_\infty = \lim_n X_n$ on Ω; the L^1 limit of the sequence $(X_n^+, n \in \mathbf{N})$ can only be the positive part X_∞^+ of X_∞. Now let us consider the Krickeberg decomposition $X_n = M_n - Y_n$ $(n \in \mathbf{N})$ of the submartingale $(X_n, n \in \mathbf{N})$; the r.v.'s of the martingale $(M_n, n \in \mathbf{N})$ in this decomposition are defined by $M_n = \lim\uparrow_p E^{\mathscr{B}_n}(X_p^+)$ and hence in this case are $M_n = E^{\mathscr{B}_n}(X_\infty^+)$ by the continuity of $E^{\mathscr{B}_n}$ on L^1. On the other hand, Theorem II-2-9 applied to the positive supermartingale $(Y_n, n \in \mathbf{N})$ shows that $Y_n \geqslant E^{\mathscr{B}_n}(Y_\infty)$ for all $n \in \mathbf{N}$; but the r.v. $Y_\infty = \lim \text{a.s.}_{n\to\infty} Y_n$ is equal to X_∞^-, because $Y_n = M_n - X_n \to X_\infty^+ - X_\infty = X_\infty^-$ a.s. by Proposition II-2-11. Finally we deduce that for all $n \in \mathbf{N}$,

$$X_n = M_n - Y_n \leqslant E^{\mathscr{B}_n}(X_\infty^+) - E^{\mathscr{B}_n}(X_\infty^-) = E^{\mathscr{B}_n}(X_\infty),$$

which establishes condition (b). ■

The preceding proof shows that the martingale (M_n, $n \in \mathbf{N}$) of the Krickeberg decomposition of the submartingale (X_n, $n \in \mathbf{N}$) is regular if the submartingale itself is regular. The converse is true: the equalities

$$M_n = E^{\mathscr{B}_n}(M_\infty) \qquad (n \in \mathbf{N}),$$

which express the regularity of the martingale (M_n, $n \in \mathbf{N}$), and the inequalities $Y_n \geqslant E^{\mathscr{B}_n}(Y_\infty)$ ($n \in \mathbf{N}$) of Theorem II-**2**-9 indeed imply that

$$X_n = M_n - Y_n \leqslant E^{\mathscr{B}_n}(M_\infty - Y_\infty) = E^{\mathscr{B}_n}(X_\infty) \qquad (n \in \mathbf{N}).$$

COROLLARY IV-**2**-25. *For every regular submartingale* (X_n, $n \in \mathbf{N}$) *and for every stopping time* ν, *the r.v.* X_ν *is integrable; for every pair* ν_1, ν_2 *of stopping times such that* $\nu_1 \leqslant \nu_2$ *a.s., the submartingale inequality* $X_{\nu_1} \leqslant E^{\mathscr{B}_{\nu_1}}(X_{\nu_2})$ *remains true a.s.*

PROOF. Again we use the Krickeberg decomposition of the regular submartingale (X_n, $n \in \mathbf{N}$). Since the martingale (M_n, $n \in \mathbf{N}$) of this decomposition is regular, Corollary IV-**2**-6 shows that every r.v. M_ν is integrable and that $M_{\nu_1} = E^{\mathscr{B}_{\nu_1}}(M_{\nu_2})$. On the other hand, every r.v. Y_ν of the positive supermartingale (Y_n, $n \in \mathbf{N}$) is integrable since Y_0 is, and the stopping time theorem for positive supermartingales (Proposition II-**2**-13) implies that $Y_{\nu_1} \geqslant E^{\mathscr{B}_{\nu_1}}(Y_{\nu_2})$ if $\nu_1 \leqslant \nu_2$. Taking differences, the results of the corollary are then clear. ■

Finally the Krickeberg decomposition also permits the easy extension to integrable submartingales of Proposition IV-**3**-12 on the regularity of stopping times. The only changes required in the statement of this proposition consist of replacing the word "martingales" by "submartingales" and writing the inequalities $X_{\nu \wedge n} \leqslant E^{\mathscr{B}_n}(X_\nu)$ and $X_{\nu_1} \leqslant E^{\mathscr{B}_{\nu_1}}(X_{\nu_2})$ instead of the corresponding equalities. By the Krickeberg decomposition, the proof is immediate from the results of Proposition IV-**3**-12 and the stopping theorem (Proposition II-**2**-13) for positive supermartingales.

IV-6. Exercises

IV-1. Let ($X_n = \sum_{m=1}^n Y_m$, $n \in \mathbf{N}$) be the sequence of partial sums of a sequence of independent integrable r.v.'s centred at their means. Show that if this martingale converges a.s., and if its limit is integrable, it is regular. Thus for this particular type of martingale, Doob's condition $\sup_{\mathbf{N}} E(|X_n|) < \infty$ already implies regularity. [First show that $E^{\mathscr{B}_n}(X_\infty - X_n)$ is constant if $\mathscr{B}_n$ denotes the σ-field generated by $Y_1, \ldots, Y_n$ and $X_\infty = \lim_{n \to \infty}$a.s. X_n.]

IV-2. If $(X_n, n \in \mathbf{N})$ is a martingale defined on the space $[\Omega, \mathscr{A}, P; (\mathscr{B}_n, n \in \mathbf{N})]$ such that $\sup_{\mathbf{N}} E(|X_n|) < \infty$, show that if there exists an integrable r.r.v. U such that $X_n \leqslant E^{\mathscr{B}_n}(U)$ for all $n \in \mathbf{N}$, then we already have $X_n \leqslant E^{\mathscr{B}_n}(X_\infty)$ for all $n \in \mathbf{N}$, where X_∞ denotes the a.s. limit of the martingale $(X_n, n \in \mathbf{N})$.

IV-3. Let h be a positive finite-valued harmonic function for a Markov chain $(X_n, n \in \mathbf{N})$ defined on a state space E with transition probability P; in other words, h is a mapping of E into $\mathbf{R}_+$ such that $Ph = h$. Show that if F is a subset of E such that $\sup_F h < \infty$, the martingale $(Z_n = h(X_n), n \in \mathbf{N})$ satisfies the equality

$$\mathbf{E}_x(Z_{\nu_F}) = h(x) \quad \text{for all } x \in E,$$

where ν_F denotes the time of the first entry of the chain into F; how should Z_{ν_F} be defined on the event $\bigcap_{\mathbf{N}} \{X_n \in F^c\}$? (We denote by $\mathbf{E}_x$ the expectation associated with the initial state x of the chain.)

IV-4. Let $(Y_n, n \in \mathbf{N})$ be a sequence of independent and identically distributed real-valued r.v.'s, and let $g: \mathbf{R} \to \mathbf{R}_+$ be a positive Borel-measurable function such that $E(g(Y_0)) = 1$. Show that under these conditions the sequence $(X_n = \Pi_{m \leqslant n}\, g(Y_m), n \in \mathbf{N})$ is a positive martingale converging a.s. to zero, at least if $P(g(Y_0) = 1) \neq 1$.

IV-5. An integrable martingale $(X_n, n \in \mathbf{N})$ cannot converge in L^1 without also converging a.s.; on the other hand, an integrable martingale may converge in probability while being a.s. divergent. Here is an example.

Let $(Y_n, n \in \mathbf{N}^*)$ be a sequence of independent r.v.'s each taking the values ± 1 with probabilities $\frac{1}{2}, \frac{1}{2}$; let $\mathscr{B}_n = \sigma(Y_1, \ldots, Y_n)$ $(n \in \mathbf{N})$ and let $(B_n, n \in \mathbf{N})$ be a sequence of events adapted to $(\mathscr{B}_n, n \in \mathbf{N})$ such that $\lim_n P(B_n) = 0$ and $P(\limsup_{n \to \infty} B_n) = 1$. Then the formulae

$$X_0 = 0, \qquad X_{n+1} = X_n(1 + Y_{n+1}) + 1_{B_n} Y_{n+1} \qquad (n \in \mathbf{N})$$

define an integrable martingale such that

$$\lim_{n \to \infty} P(X_n = 0) = 1, \qquad P(\{X_n \to\}) = 0.$$

[Note that $P(X_{n+1} \neq 0) \leqslant \frac{1}{2} P(X_n \neq 0) + P(B_n)$, and that on $\{X_n \to\}$ the limit $\lim_n 1_{B_n}$ exists.]

IV-6. On the discrete space $\Omega = \mathbf{N}^*$ equipped with the probability P defined by

$$P(\{n\}) = \frac{1}{n} - \frac{1}{n+1} \qquad (n \in \mathbf{N}^*)$$

and the increasing sequence $\mathscr{B}_n$ of σ-fields generated respectively by the partitions $\{\{1\}, \{2\}, \ldots, \{n\}, [n+1, \infty[\}$ $(n \in \mathbf{N})$, show that the sequence

$$(X_n = (n+1)\, 1_{[n+1, \infty[}, \; n \in \mathbf{N})$$

of r.r.v.'s is a positive martingale, that $E(X_n) = 1$ for all $n \in \mathbf{N}$, but that $\sup_{\mathbf{N}} X_n(\omega) = \omega$ is not integrable.

CHAPTER V

EXTENSIONS OF THE NOTION OF MARTINGALE

The sections of this chapter are essentially independent of one another.

V-1. Martingales with a directed index set

Let $(\Omega, \mathscr{A}, P)$ be a probability space. Instead of an increasing sequence $(\mathscr{B}_n, n \in \mathbf{N})$ of the sub-σ-fields of $\mathscr{A}$, in this section we will take an increasing directed family, say $(\mathscr{B}_t, t \in T)$, of sub-σ-fields of $\mathscr{A}$; this family will be fixed once and for all. Recall that a family $(\mathscr{B}_t, t \in T)$ of sub-σ-fields of $\mathscr{A}$ indexed by an ordered set T is said to be increasing directed (or directed upwards) if the ordered set T is directed in the sense that for all pairs $t_1, t_2 \in T$ there exists at least one $t_3 \in T$ such that $t_1 \leqslant t_3$ and $t_2 \leqslant t_3$, and if the mapping $t \to \mathscr{B}_t$ is increasing for inclusion, i.e. if $\mathscr{B}_s \subset \mathscr{B}_t$ whenever $s \leqslant t$ in T. For example, if T denotes the set of all measurable partitions of Ω which are finite (resp. countable), if this set is given the usual ordering, and if for every partition $t \in T$, $\mathscr{B}_t$ denotes the σ-field generated by this partition, the family $(\mathscr{B}_t, t \in T)$ is an increasing directed family of sub-σ-fields of $\mathscr{A}$.

Of course any increasing sequence $(\mathscr{B}_n, n \in \mathbf{N})$ of sub-σ-fields of $\mathscr{A}$ is a particular case of an increasing directed family. On the other hand, the definitions of martingale, sub- and supermartingale given earlier for an increasing sequence $(\mathscr{B}_n, n \in \mathbf{N})$ easily extend to the present more general case of an increasing directed family $(\mathscr{B}_t, t \in T)$; for example, a family $(X_t, t \in T)$ of positive (resp. integrable) r.r.v.'s will be called a positive (resp. integrable) supermartingale if it is adapted to the family $(\mathscr{B}_t, t \in T)$, i.e. if X_t is $\mathscr{B}_t$-measurable for all $t \in T$, and if it satisfies the inequality $X_s \geqslant E^{\mathscr{B}_s}(X_t)$ for every pair $s \leqslant t$ in T. On the contrary, however, the generalisation of the notion of stopping time has little interest except in very special cases, and this is undoubtedly the origin of later difficulties!

We will be showing that the convergence in mean and convergence in probability theorems for the usual martingales extend easily to martingales with a directed index set; we will see that the possibility of this extension is essentially due to the elementary lemma below, from topology.

Recall that a family $(x_t, t \in T)$ of elements from a metric space indexed by a directed set T converges to an element x of the space if for a suitably chosen t_ε in T ($\varepsilon > 0$ arbitrary) the distance $d(x_t, x)$ between x_t and x is less than ε whenever $t \geqslant t_\varepsilon$. Similarly such a family $(x_t, t \in T)$ is said to be Cauchy if for every $\varepsilon > 0$ we have $d(x_s, x_t) < \varepsilon$ whenever $s, t \geqslant t_\varepsilon$ for a suitably chosen t_ε in T; if every Cauchy *sequence* $(x_n, n \in \mathbf{N})$ in the space under consideration is convergent, i.e. in the usual terminology if this space is complete, then every Cauchy family $(x_t, t \in T)$ is also convergent. [To see this, recursively define an increasing sequence $(t_n, n \in \mathbf{N}^*)$ in T such that

$$d(x_s, x_t) \leqslant \frac{1}{n} \quad \text{if } s, t \geqslant t_n,$$

observe that the sequence $(x_{t_n}, n \in \mathbf{N}^*)$ converges to a limit x since it is Cauchy, and conclude by writing

$$d(x_s, x) \leqslant d(x_s, x_{t_n}) + d(x_{t_n}, x) \leqslant \frac{2}{n} \quad \text{if } s \geqslant t_n.$$

Here is a sufficient criterion for convergence which we will be using in what follows.

LEMMA V-1-1. *In order that the family $(x_t, t \in T)$ indexed by the directed set T converge in the complete metric space on which it is defined, it suffices that $(x_{t_n}, n \in \mathbf{N})$ be convergent for all increasing sequences $(t_n, n \in \mathbf{N})$ in T.*

PROOF. By the definitions above, if the family $(x_t, t \in T)$ fails to converge, it cannot be Cauchy; hence there exists an $\varepsilon > 0$ such that for all $s \in T$ there exists $t \in T$ with $t \geqslant s$ and $d(x_s, x_t) > \varepsilon$. Thus we can construct a sequence $(t_n, n \in \mathbf{N})$, beginning at an arbitrary $t_0 \in T$, such that $t_{n+1} > t_n$ and $d(x_{t_n}, x_{t_{n+1}}) > \varepsilon$; the sequence $(x_{t_n}, n \in \mathbf{N})$ clearly does not converge. The lemma is thus proved. ∎

The following proposition collects the extensions of the convergence theorems already proved to martingales with a directed index set.

PROPOSITION V-1-2. *For all real numbers $p \in [1, \infty[$ and for all r.r.v.'s $X \in L^p$, the martingale $(E^{\mathscr{B}_t}(X), t \in T)$ converges in L^p to $E^{\mathscr{B}_\infty}(X)$, where $\mathscr{B}_\infty$ denotes the σ-field $\bigvee_T \mathscr{B}_t$ generated by the $\mathscr{B}_t (t \in T)$.*

For an integrable martingale $(X_t, t \in T)$ to be of the form $X_t = E^{\mathscr{B}_t}(X)$ $(t \in T)$ for a r.v. $X \in L^1$, it is necessary and sufficient that it be uniformly integrable, i.e., that

$$\sup_T \int_{\{|X_t| \geqslant a\}} |X_t| \mathrm{d}P \downarrow 0 \quad \textit{when } a \uparrow \infty.$$

Similarly, if $p \in]1, \infty[$, *a martingale* $(X_t, t \in T)$ *in* L^p *can be written in the form* $X_t = E^{\mathscr{B}_t}(X)$ $(t \in T)$ *for a r.v.* $X \in L^p$ *if and only if it is bounded in* L^p, *i.e., if* $\sup_T E(|X_t|^p) < \infty$.

Every integrable submartingale $(X_t, t \in T)$ *such that* $\sup_T E(X_t^+) < \infty$ *is convergent in probability.*

PROOF. (1) Since every sequence $(E^{\mathscr{B}_{t_n}}(X), n \in \mathbf{N})$ associated with a r.v. $X \in L^p$ and an increasing sequence $(t_n, n \in \mathbf{N})$ in T converges in L^p by Proposition II-**2**-11, it follows from the preceding lemma that $(E^{\mathscr{B}_t}(X), t \in T)$ also converges in L^p. It only remains to identify the limit $X_\infty = \lim_T E^{\mathscr{B}_t}(X)$. Since $L^p(\mathscr{B}_\infty)$ is closed in L^p, it is clear that X_∞ is $\mathscr{B}_\infty$-measurable. On the other hand, for all $B \in \bigcup_T \mathscr{B}_t$, we have

$$\int_B X_\infty \, \mathrm{d}P = \lim_T \int_B X_t \, \mathrm{d}P = \int_B X \, \mathrm{d}P;$$

since the class $\bigcup_T \mathscr{B}_t$ is a Boolean algebra because the family $(\mathscr{B}_t, t \in T)$ is increasing, the preceding identity remains valid on the σ-field $\mathscr{B}_\infty$ generated by this algebra. We have thus proved that $X_\infty = E^{\mathscr{B}_\infty}(X)$ and the first part of the proposition is proved.

(2) If the martingale $(X_t, t \in T)$ is uniformly integrable, the same is true of every martingale $(X_{t_n}, n \in \mathbf{N})$ associated with an increasing sequence $(t_n, n \in \mathbf{N})$ in T; by Proposition IV-**2**-3, such a martingale $(X_{t_n}, n \in \mathbf{N})$ converges in L^1. Lemma V-**1**-1 above thus shows that the martingale $(X_t, t \in T)$ converges in L^1; the limit X_∞ then satisfies the desired equality $X_t = E^{\mathscr{B}_t}(X_\infty)$ $(t \in T)$ by the continuity of conditional expectations on L^1. Conversely, every martingale of the form $(E^{\mathscr{B}_t}(X), t \in T)$ is uniformly integrable by Lemma IV-**2**-4. The first part of the second paragraph of the proposition is thus established; the second part is proved similarly, this time using Proposition IV-**2**-7 instead of Proposition IV-**2**-3.

(3) If $(X_t, t \in T)$ is an integrable submartingale such that $\sup_T E(X_t^+) < \infty$, every submartingale $(X_{t_n}, n \in \mathbf{N})$ associated with an increasing sequence $(t_n, n \in \mathbf{N})$ taken from T satisfies Doob's condition $\sup_N E(X_{t_n}^+) < \infty$ and thus converges a.s. to an integrable limit. But lemma V-**1**-1 is inapplicable since a.s. convergence cannot be defined by a metric; on the other hand, it can be applied to convergence in probability (which is weaker than a.s. convergence) as this is the convergence associated with the metric

$$d(X, Y) = \int \min(|X - Y|, 1) \, \mathrm{d}P$$

defined on the space of equivalence classes of a.s. finite r.r.v.'s. [Cauchy's criterion for this metric is proved as for L^1. On the other hand, let us note the equivalence

$$\lim_T d(X_t, X) = 0 \Leftrightarrow \lim_T P(|X_t - X| > \varepsilon) = 0 \text{ for all } \varepsilon > 0$$

which follows immediately from the elementary inequalities

$$(\min(\varepsilon, 1))^{-1} P(|X - Y| > \varepsilon) \leqslant d(X, Y) \leqslant \varepsilon + P(|X - Y| > \varepsilon).]$$

In conclusion, Lemma V-**1**-1 shows that the submartingale $(X_t, t \in T)$ converges in probability. ■

REMARK. Another way of proving the first part of the proposition consists of extending Corollary I-**1**-3 to increasing directed families, which is immediate, then using the reasoning in the latter part of Proposition V-**2**-6. In this way it is also easy to see that the first part of the preceding proposition extends to the vector r.v.'s studied in the second section. ■

A directed family $(X_t, t \in T)$ of r.r.v.'s is said to *essentially converge* if the essential upper and lower bounds

$$\operatorname{ess} \limsup_T X_t = \operatorname{ess} \inf_s (\operatorname{ess} \sup_{t \geqslant s} X_t),$$
$$\operatorname{ess} \liminf_T X_t = \operatorname{ess} \sup_s (\operatorname{ess} \inf_{t \geqslant s} X_t)$$

coincide a.s. (The definitions of essential infimum and essential supremum are given on p. 44 of [222] or in Proposition VI-**1**-1 below.) For sequences $(X_n, n \in \mathbf{N})$ of r.r.v.'s this convergence notion clearly coincides with a.s. convergence; essential convergence is the natural generalisation to directed families of r.r.v.'s of a.s. convergence for sequences of r.r.v.'s since we are only interested in the random variables through their equivalence classes.

A directed family $(X_t, t \in T)$ satisfying Doob's condition

$$\sup_T E(|X_t|) < \infty$$

is not necessarily essentially convergent; a counterexample was given in 1950 by J. Dieudonné [102]. On the other hand, if the directed family of sub-σ-fields $\mathscr{B}_t$ $(t \in T)$ is totally ordered or, more generally, if it satisfies the Vitali condition of the following proposition, the essential convergence of every martingale bounded in L^1 is assured.

PROPOSITION V-1-3. *Let* $(\mathscr{B}_t, t \in T)$ *be an increasing directed family of sub-σ-fields of* $\mathscr{A}$ *in the probability space* $(\Omega, \mathscr{A}, P)$; *let us suppose that this family satisfies the following "Vitali condition":*

For every A belonging to the σ-field $\bigvee_T \mathscr{B}_t$ *generated by the* $\mathscr{B}_t (t \in T)$, *for every family of* $A_t \in \mathscr{B}_t$ $(t \in T)$ *such that* $A \subset \operatorname{ess\,lim\,sup}_T A_t$, *and for every* $\varepsilon > 0$, *there exists finitely many indices* $t_1, \ldots, t_n \in T$ *and* $B_i \in \mathscr{B}_{t_i}$ $(i = 1, \ldots, n)$ *such that*

$$B_i B_j = \emptyset \quad \text{if } i, j \in [1, n],\ i \neq j,$$

$$B_i \subset A_{t_i}, \qquad P\left(A \cap \left(\sum_1^n B_i\right)^c\right) \leqslant \varepsilon.$$

Then every integrable martingale $(X_t, t \in T)$ *defined on the space* $[\Omega, \mathscr{A}, P; (\mathscr{B}_t, t \in T)]$ *and satisfying Doob's condition* $\sup_T E(|X_t|) < \infty$ *essentially converges.*

PROOF. We can restrict ourselves to the case where the martingale $(X_t, t \in T)$ is positive, the general case being reduced to this one using the Krickeberg decomposition (cf. IV-1-1) duly extended to martingales with a directed index set.

Let s be an element of T, B an element of $\mathscr{B}_s$ and a a positive real number. Let us put

$$A_t = \begin{cases} B\{X_t \geqslant a\} & \text{if } t \geqslant s, \\ \emptyset & \text{otherwise,} \end{cases}$$

$$A = B \cap \operatorname{ess\,lim\,sup}_T \{X_t \geqslant a\}.$$

These sets satisfy the Vitali condition above so that by hypothesis for all $\varepsilon > 0$ there exists finitely many elements $t_1, \ldots, t_n$ of T and $B_i \in \mathscr{B}_{t_i}$ $(i = 1, \ldots, n)$ pairwise disjoint such that $B_i \subset A_{t_i}$ $(i = 1, \ldots, n)$ and $P(A \cap (\sum_{i=1}^n B_i)^c) < \varepsilon$. Then using the positive martingale property of the family $(X_t, t \in T)$ and the fact that $\sum_{i=1}^n B_i \subset B$, we can write for $t \geqslant s$,

$$\int_B X_s \,\mathrm{d}P = \int_B X_t \,\mathrm{d}P \geqslant \int_{\sum_{i=1}^n B_i} X_t \,\mathrm{d}P = \sum_{i=1}^n \int_{B_i} X_{t_i} \,\mathrm{d}P$$

$$\geqslant \sum_{i=1}^n aP(B_i) \geqslant a[P(A) - \varepsilon].$$

Letting $\varepsilon \downarrow 0$ we have proved that $\int_B X_s \,\mathrm{d}P \geqslant aP(A)$; but this inequality is only possible for all $B \in \mathscr{B}_s$ if

$$X_s \geqslant aP^{\mathscr{B}_s}(\operatorname{ess\,lim\,sup}_T \{X_t \geqslant a\}).$$

By Proposition V-1-2, both sides of this inequality have a limit in probability as s increases; denoting the limit for the family $(X_t, t \in T)$ by X_∞, we find that

$$X_\infty \geqslant a 1_{\text{ess lim sup}_T \{X_t \geqslant a\}}$$

(for convergence in probability preserves the inequalities). This inequality can only hold for all $a \in R_+$ if $X_\infty \geqslant \text{ess lim sup}_T X_t$. In the same way we show that $X_\infty \leqslant \text{ess lim inf}_T X_t$, which completely proves that the family $(X_t, t \in T)$ essentially converges to X_∞. ■

REMARKS. (1) Every directed family of sub-σ-fields which is *totally ordered* satisfies the Vitali condition. Indeed, since $A \subset \text{ess sup}_T A_t$, there exists a sequence $(t_n, n \in \mathbf{N})$ in T such that $A \subset \bigcup_N A_{t_n}$ a.s.; for every given $\varepsilon > 0$, we can then choose $n \in \mathbf{N}$ such that $P(A \cap (\bigcup_{m \leqslant n} A_{t_m})^c) \leqslant \varepsilon$. Since T is totally ordered, by permuting the elements of the finite sequence $(t_0, t_1, \ldots, t_n)$ we can suppose that this sequence is increasing. It then remains to put

$$B_i = A_{t_i} \cap (\bigcup_{j<i} A_{t_j})^c \qquad (i = 1, \ldots, n)$$

in order to see that the Vitali condition is satisfied with the B_i.

(2) The preceding proposition can only be extended to submartingales by strengthening the Vitali condition. ■

V-2. Vector-valued martingales

The notion of martingale extends easily to vector-valued random variables; the treatment which follows is limited to the case of random variables with values in a separable Banach space and begins with several basic notions concerning the integration of such random variables.

Let E be a real Banach space which we will suppose to be separable and we will denote the norm by $|\cdot|$. We denote by $\mathscr{B}$ the σ-fields of Borel subsets of E, i.e. the σ-field generated by the topology on E.

Given a probability space $(\Omega, \mathscr{A}, P)$, a mapping $X : \Omega \to E$ is called a random variable (r.v.) with values in E if it is measurable as a map of $(\Omega, \mathscr{A})$ into $(E, \mathscr{B})$. A r.v. $X : \Omega \to E$ is said to be a step r.v. if it only takes a finite number of values on E; then, if $x_1, \ldots, x_n$ is an enumeration of the finite set $X(\Omega)$, the sets $\{X = x_m\}$ $(1 \leqslant m \leqslant n)$ form a finite measurable partition of $(\Omega, \mathscr{A})$ and on Ω we can write $X = \sum_{m=1}^n x_m 1_{\{X = x_m\}}$ in E. The following lemma is basic; it is simpler than those usually found in the literature.

LEMMA V-2-4. *For every r.v. $X : \Omega \to E$, there exists at least one sequence $(X_n, n \in \mathbf{N})$ of step r.v.'s with values in E such that $X(\omega) = \lim_{n\to\infty} X_n(\omega)$ in E for all $\omega \in \Omega$ and $|X_n(\omega)| \leqslant |X(\omega)|$ for all $n \in \mathbf{N}$ and all $\omega \in \Omega$. Conversely, the limit $X(\omega) = \lim_{n\to\infty} X_n(\omega)$ of every sequence $(X_n, n \in \mathbf{N})$ of r.v.'s with values in E (step or not) which converges in E at each point of Ω is again a r.v.*

PROOF. Let $(y_n, n \in \mathbf{N})$ be a sequence which is dense in E and such that $y_0 = 0$. For every $n \in \mathbf{N}$, let us denote by $f_n : E \to \{y_0, y_1, \ldots, y_n\}$ the function defined at each point $x \in E$ whose value is the first y_l $(0 \leqslant l \leqslant n)$ for which the minimum $\min_{0 \leqslant m \leqslant n} |x - y_n|$ is achieved; equivalently, for all $l \leqslant n$

$$f_n(\,.\,) = y_l \quad \text{on } \{x : |x - y_l| < |x - y_m| \text{ for all } m \in [0, l[,\; |x - y_l| \leqslant |x - y_m| \text{ for all } m \in [l, n]\}.$$

This formula shows explicitly that the functions $f_n : E \to E$ are Borel measurable, for the functions $|\cdot - y_m|$ are continuous and thus Borel measurable mappings from E into R_+. On the other hand, $\lim_{n\to\infty} f_n(x) = x$ for all $x \in E$ for

$$|x - f_n(x)| = \min_{0 \leqslant m \leqslant n} |x - y_m| \downarrow 0 \qquad (n \uparrow \infty, x \in E)$$

by the denseness of the sequence $(y_m, m \in \mathbf{N})$ in E; further $|x - f_n(x)| \leqslant |x|$ for all $n \in \mathbf{N}$ and all $x \in E$ since $y_0 = 0$, hence $|f_n(x)| \leqslant 2|x|$.

If $X : \Omega \to E$ is a random variable, it is clear that the $f_n \circ X$ are step r.v.'s with values in E such that $\lim_{n\to\infty} f_n \circ X = X$ in E at every point of Ω and that $|f_n(X(\omega))| \leqslant |X(\omega)|$ for every $\omega \in \Omega$ and $n \in \mathbf{N}$; the first part of the lemma is thus proved.

If the $X_n : \Omega \to E$ are r.v.'s converging at every point of Ω to a mapping $X : \Omega \to E$, this map is necessarily a random variable. By definition of the Borel σ-field of E, it is enough to show that $\{X \in G\} \in \mathscr{A}$ for every open subset G of E. But if G_k denotes the open subset of E consisting of all points x whose distance from G^c is strictly larger than $1/k$ $(k \in \mathbf{N}^*)$, we have $\lim\uparrow_k G_k = G$ and $\bar{G}_k \subset G$ for all $k \in \mathbf{N}^*$; it is easy to deduce that in Ω,

$$\{X \in G\} = \bigcup_{k \in \mathbf{N}^*} \liminf_{n\to\infty} \{X_n \in G_k\}.$$

The measurability of the X_n $(n \in \mathbf{N})$ implies that the right-hand side belongs to $\mathscr{A}$, and it then follows that X is measurable. ■

(A common error consists of replacing the preceding equality by the false one $\{X \in G\} = \liminf_{n\to\infty} \{X_n \in G\}$!)

After having observed that the step r.v.'s $X : \Omega \to E$ form a vector space, it follows easily from the preceding lemma that the set of all r.v.'s $X : \Omega \to E$ is also a *vector space*.

For every r.v. $X : \Omega \to E$, the mapping $|X| : \Omega \to R_+$ is measurable, for it is obtained by composing the measurable mapping $X : (\Omega, \mathscr{A}) \to (E, \mathscr{B})$ with the continuous (and thus Borel-measurable) mapping $x \to |x|$ of E into R_+. The r.v. X is said to be null, resp. integrable, if the positive r.r.v. $|X|$ is null, resp. integrable, on the space $(\Omega, \mathscr{A}, P)$. Two r.v.'s $X_i : \Omega \to E\,(i = 1, 2)$ are said to be equivalent if they are equal a.s. on Ω or, equivalently, if the r.r.v. $|X_1 - X_2|$ is null.

We will assemble in a single proposition the more basic results concerning the space $L^1_E(\Omega, \mathscr{A}, P)$ of equivalence classes of integrable r.v.'s with values in E, the integral (= expectation) and the conditional expectations of such random variables.

PROPOSITION V-2-5. *The set $L^1_E = L^1_E(\Omega, \mathscr{A}, P)$ of equivalence classes of integrable r.v.'s $X : \Omega \to E$ is a Banach space for the norm $\|X\|_1 = \int_\Omega |X| \,\mathrm{d}P$. The equivalence classes of step r.v.'s form a vector subspace of L^1_E which is dense in this Banach space.*

There exists a unique continuous linear mapping of L^1_E onto E, called the integral or expectation and written $X \to \int_\Omega X \mathrm{d}P$, such that

$$\int_\Omega X \,\mathrm{d}P = \sum_{X(\Omega)} xP(X = x) \quad \textit{if } X \textit{ is a step r.v.}$$

This integral decreases the norm, i.e.,

$$\left| \int_\Omega X \,\mathrm{d}P \right| \leqslant \|X\|_1 \quad \textit{for all } X \in L^1_E.$$

On the other hand, for every continuous linear functional x' on E, the real-valued r.v. (x', X) is integrable on $(\Omega, \mathscr{A}, P)$ if $X \in L^1_E$, and its integral satisfies

$$\int_\Omega (x', X) \,\mathrm{d}P = \left(x', \int_\Omega X \,\mathrm{d}P \right).$$

For every sub-σ-field $\mathscr{B}$ of $\mathscr{A}$ there exists a unique continuous linear mapping of L^1_E into itself, denoted by $E^{\mathscr{B}}$ and called conditional expectation with respect to the sub-σ-field $\mathscr{B}$, such that

$$E^{\mathscr{B}}(X) = \sum_{X(\Omega)} xP^{\mathscr{B}}(X = x) \quad \textit{if } X \textit{ is a step r.v.}$$

The image of L_E^1 under this operator $E^{\mathscr{B}}$ is the closed vector subspace $L_E^1(\mathscr{B})$ of equivalence classes of $\mathscr{B}$-measurable r.v.'s with values in E; further, the operator $E^{\mathscr{B}}$ is idempotent. Every $X \in L_E^1$ satisfies the a.s. inequality

$$|E^{\mathscr{B}}(X)| \leqslant E^{\mathscr{B}}(|X|),$$

which implies that $\|E^{\mathscr{B}}(X)\|_1 \leqslant \|X\|_1$. Finally, for every continuous linear functional x' on E, the integrable r.r.v. (x', X) is such that

$$E^{\mathscr{B}}(x', X) = (x', E^{\mathscr{B}}(X)).$$

PROOF (succinct). The Banach space property of L_E^1 can be proved exactly as in the scalar case ($E = R$) (Riesz–Fischer theorem); we will not repeat this proof. The step r.v.'s are dense in L^1 because the step r.v.'s X_n ($n \in \mathbf{N}$) associated with a r.v. $X : \Omega \to E$ in Lemma V-**1**-1 converge in L_E^1 to X by the dominated convergence theorem if $X \in L_E^1$: since

$$\lim_{n\to\infty} |X_n - X| = 0 \qquad |X_n - X| \leqslant 2|X|$$

on Ω, we have

$$\lim_{n\to\infty} \int_\Omega |X_n - X| \, \mathrm{d}P = 0.$$

It is immediate that the integral of step r.v.'s $X : \Omega \to E$ defined as in the statement of the proposition by

$$\int_\Omega X \, \mathrm{d}P = \sum_{X(\Omega)} xP(X = x)$$

is a linear mapping of the vector space of step r.v.'s in E which decreases the norm:

$$\left| \int_\Omega X \, \mathrm{d}P \right| \leqslant \int_\Omega |X| \, \mathrm{d}P.$$

This integral can hence be extended by continuity to the space L_E^1 since E is complete; the extension is a linear mapping of L_E^1 into E which again decreases the norm. For every continuous linear functional x' on E, the mapping (x', X) of Ω into R is the composition of the mappings $X : \Omega \to E$ and $x' : E \to R$ and so is measurable; since $|(x', X)| \leqslant |x'| |X|$ on Ω if $|x'| = \sup_{|x|=1} (x', x)$ denotes the norm of x' in the dual E' of E, the r.r.v. (x', X) is integrable whenever $X \in L_E^1$. If X is a step r.v., it is immediately checked that

$$\int_\Omega (x', X) \, \mathrm{d}P = \left(x', \int_\Omega X \, \mathrm{d}P \right);$$

this formula easily extends by continuity to all r.v.'s $X \in L_E$.

The proof of the last part of the proposition—concerning conditional expectations—is essentially the same as that of the second part; we will leave the details to the reader. ■

For every real number $p \in [1, \infty[$, we will further define the space L_E^p as the space of equivalence classes of r.v.'s $X : \Omega \to E$ whose norm $|X|$ is pth power integrable. This space is a Banach space for the norm

$$\|X\|_p = \left(\int_\Omega |X|^p \, dP \right)^{1/p} \quad (X \in L_E^p)$$

and the step r.v.'s (after passing to equivalence classes) form a dense linear subspace there. Further $L_E^p \subset L_E^q$ when $1 \leqslant q \leqslant p < \infty$ and $\|X\|_q \leqslant \|X\|_p$ if $X \in L_E^p$. Finally, for every sub-σ-field $\mathscr{B}$ of $\mathscr{A}$, the conditional expectation $E^{\mathscr{B}}$ is an idempotent norm-decreasing linear operator when restricted to L_E^p; the norm-decreasing property on L_E^p follows from the analogous property on L_R^p and the a.s. inequality $|E^{\mathscr{B}}(X)| \leqslant E^{\mathscr{B}}(|X|)$, which allows us to write

$$\|E^{\mathscr{B}}(X)\|_p^p = \int_\Omega |E^{\mathscr{B}}(X)|^p \, dP \leqslant \int_\Omega [E^{\mathscr{B}}(|X|)]^p \, dP \leqslant \int |X|^p \, dP = \|X\|_p^p.$$

After these simple notions of vector integration, we begin the study of vector-valued martingales. The following proposition generalises Proposition II-**2**-11, to which it reduces when $E = \mathbf{R}$. On the other hand, the proof of this proposition will give a new proof of Proposition II-**2**-11 in the case $E = \mathbf{R}$.

PROPOSITION V-**2**-6. *Let $(\mathscr{B}_n, n \in \mathbf{N})$ be an increasing sequence of sub-σ-fields of $\mathscr{A}$ in the probability space $(\Omega, \mathscr{A}, P)$, and let E be a separable Banach space. Then for every integrable r.v. $X : \Omega \to E$, the convergence*

$$E^{\mathscr{B}_n}(X) \to E^{\mathscr{B}_\infty}(X),$$

where $\mathscr{B}_\infty$ denotes the σ-field generated by the $\mathscr{B}_n$ $(n \in \mathbf{N})$, takes place in E for almost all ω. Furthermore, the sequence $(E^{\mathscr{B}_n}(X), n \in \mathbf{N})$ converges to $E^{\mathscr{B}_\infty}(X)$ in L_E^1, and similarly in L_E^p if $X \in L_E^p$ (p real $\in [1, \infty[$).

PROOF. (1) The proof which follows is based upon a generalisation of Corollary I-**1**-3 to the vector case, which gives the following lemma, and then on the maximal inequality of Proposition II-**2**-7.

LEMMA V-2-7. *Under the assumptions of the preceding proposition we have for all real numbers $p \in [1, \infty[$,*

$$\overline{\bigcup_N L_E^p(\mathscr{B}_n)} = L_E^p(\mathscr{B}_\infty) \text{ in } L_E^p.$$

PROOF. Since $\mathscr{B}_\infty$-measurable step r.v.'s are dense in $L_E^p(\mathscr{B}_\infty)$, it is enough to show that every $\mathscr{B}_\infty$-measurable step r.v. is a limit in L_E^p of r.v.'s in $\bigcup_N L_E^p(\mathscr{B}_n)$. But if X is a $\mathscr{B}_\infty$-measurable step r.v., Corollary I-1-3 implies that each of the real-valued r.v.'s $1_{X\{=x\}}$ corresponding to a possible value $x \in X(\Omega)$ of X is the limit of a sequence $f_{n,x}$ ($n \in \mathbf{N}$) of r.r.v.'s belonging respectively to the spaces $L_R^p(\mathscr{B}_n)$. Then the finite sums

$$X_n = \sum_{X(\Omega)} x f_{n,x}$$

define r.v.'s in $L_E^p(\mathscr{B}_n)$ respectively for $n \in \mathbf{N}$, and the sequence $(X_n, n \in \mathbf{N})$ converges in L_E^p to X since

$$|X_n - X| = \left| \sum_{X(\Omega)} x(f_{n,x} - 1_{\{X=x\}}) \right| \leqslant \sum_{X(\Omega)} |x|\, |f_{n,x} - 1_{\{X=x\}}|$$

on Ω; consequently,

$$\|X_n - X\|_p \leqslant \sum_{X(\Omega)} |x|\, \|f_{n,x} - 1_{\{X=x\}}\|_{L_R^p} \to 0$$

when $n \to \infty$. ■

The lemma thus being proved, we pass on to the proof of the proposition. (2) Let us denote by Λ the subset of those $X \in L_E^1$ for which

$$\lim_{n\to\infty} E^{\mathscr{B}_n}(X) = E^{\mathscr{B}_\infty}(X) \quad \text{a.s. on } \Omega.$$

This set is clearly a vector subspace of L_E^1 and, since $E^{\mathscr{B}_n}(X) = X$ for all $n \in [p, \infty]$ if $X \in L_E^1(\mathscr{B}_p)$, it contains the vector subspace $\bigcup_N L_E^1(\mathscr{B}_p)$. We will now show that Λ is closed in L_E^1; the preceding lemma will then imply that $\Lambda \supset L_E^1(\mathscr{B}_\infty)$. But every $X \in L_E^1$ satisfies $E^{\mathscr{B}_n} E^{\mathscr{B}_\infty}(X) = E^{\mathscr{B}_n}(X)$ ($n \in \mathbf{N}$ or $n = +\infty$) and is such that $E^{\mathscr{B}_\infty}(X) \in \Lambda$ by the preceding inclusion; we will thus have shown that every $X \in L_E^1$ belongs to Λ.

To see that Λ is closed in L_E^1 consider a sequence $(X_n, n \in \mathbf{N})$ in Λ converging in L_E^1 to an element X of L_E^1 and, for all $p \in \mathbf{N}$, let us write

$$\begin{aligned}|E^{\mathscr{B}_n}(X) - E^{\mathscr{B}_\infty}(X)| &\leqslant |E^{\mathscr{B}_n}(X_p) - E^{\mathscr{B}_\infty}(X_p)| + |E^{\mathscr{B}_n}(X_p - X)| + |E^{\mathscr{B}_\infty}(X_p - X)| \\ &\leqslant |E^{\mathscr{B}_n}(X_p) - E^{\mathscr{B}_\infty}(X_p)| + E^{\mathscr{B}_n}(|X_p - X|) + E^{\mathscr{B}_\infty}(|X_p - X|)\end{aligned}$$

on Ω. When $n \to \infty$, the first term in the last expression tends to zero a.s. so that for every $p \in \mathbf{N}$ fixed,

$$\limsup_{n\to\infty} |E^{\mathscr{B}_n}(X) - E^{\mathscr{B}_\infty}(X)| \leqslant \sup_{n\in\mathbf{N}} E^{\mathscr{B}_n}(|X_p - X|) + E^{\mathscr{B}_\infty}(|X_p - X|).$$

To check the first term on the right-hand side, we shall apply the maximal inequality of Proposition II-**2**-7 to the positive martingale $(E^{\mathscr{B}_n}(|X - X_p|), n \in \mathbf{N})$; for every $\varepsilon > 0$ this inequality can be written

$$P(\sup_{n\in\mathbf{N}} E^{\mathscr{B}_n}(|X - X_p|) \geqslant \varepsilon) \leqslant \frac{1}{\varepsilon}\int_\Omega |X - X_p| \, dP.$$

Also, taking into account that

$$P(E^{\mathscr{B}_\infty}(|X - X_p|) \geqslant \varepsilon) \leqslant \frac{1}{\varepsilon}\int_\Omega E^{\mathscr{B}_\infty}(|X - X_p|) \, dP = \frac{1}{\varepsilon}\int_\Omega |X - X_p| \, dP,$$

we deduce from the above that

$$\begin{aligned}&P(\limsup_{n\to\infty} |E^{\mathscr{B}_n}(X) - E^{\mathscr{B}_\infty}(X)| \geqslant 2\varepsilon) \\ &\leqslant P(\sup_{n\in\mathbf{N}} E^{\mathscr{B}_n}(|X - X_p|) \geqslant \varepsilon) + P(E^{\mathscr{B}_\infty}(|X - X_p|) \geqslant \varepsilon) \\ &\leqslant \frac{2}{\varepsilon}\int_\Omega |X - X_p| \, dP\end{aligned}$$

for all $p \in \mathbf{N}$ and for all real $\varepsilon > 0$. It remains to let $p \uparrow \infty$ and $\varepsilon \downarrow 0$ to see that $\limsup_{n\to\infty} |E^{\mathscr{B}_n}(X) - E^{\mathscr{B}_\infty}(X)| = 0$ a.s., and hence that $X \in \Lambda$.

(3) We now have to show that the sequence $E^{\mathscr{B}_n}(X)$ converges in L_E^p to $E^{\mathscr{B}_\infty}(X)$ for all $X \in L_E^p$, whatever fixed real number $p \in [1, \infty]$ we choose. As above we will consider the set Λ_p consisting of all $X \varepsilon L_E^p$ for which $E^{\mathscr{B}_n}(X) \to E^{\mathscr{B}_\infty}(X)$ in L_E^p when $n \to \infty$; this set is a vector subspace of L_E^p containing $\bigcup_{\mathbf{N}} L_E^p(\mathscr{B}_n)$. By Lemma V-**2**-7 it again suffices to show that Λ_p is closed in L_E^p for it to follow that $\Lambda_p \supset L_E^p(\mathscr{B}_\infty)$, and finally, taking into account the identities $E^{\mathscr{B}_n} E^{\mathscr{B}_\infty} = E^{\mathscr{B}_n}$, that $\Lambda_p = L_E^p$.

Now it is easy to show that Λ_p is closed in L_E^p. Indeed, if $(X_j, j \in \mathbf{N})$ is a sequence in Λ_p converging in L_E^p to an element X of L_E^p, the same triangle inequalities as above imply that for all $j \in \mathbf{N}$ and $n \in \mathbf{N}$,

$$\|E^{\mathscr{B}_n}(X) - E^{\mathscr{B}_\infty}(X)\|_p \leqslant \|E^{\mathscr{B}_n}(X_j) - E^{\mathscr{B}_\infty}(X_j)\|_p + 2\|X_j - X\|_p.$$

It then suffices to let $n \uparrow \infty$ and then $j \uparrow \infty$, giving

$$\|E^{\mathscr{B}_n}(X) - E^{\mathscr{B}_\infty}(X)\|_p \to 0$$

when $n \to \infty$. ■

Next we will study the extension of Doob's theorem to vector-valued martingales; we find that the natural generalisation of this theorem is only possible for particular Banach spaces. Given a probability space $(\Omega, \mathscr{A}, P)$ equipped with an increasing sequence $(\mathscr{B}_n, n \in \mathbf{N})$ of sub-σ-fields of $\mathscr{A}$ and given a separable Banach space E, we will obviously call every sequence $(X_n, n \in \mathbf{N})$ of random variables belonging respectively to the spaces $L_E^1(\mathscr{B}_n)$ $(n \in \mathbf{N})$ and satisfying $E^{\mathscr{B}_n}(X_{n+1}) = X_n$ for all $n \in \mathbf{N}$ an "integrable martingale with values in E".

PROPOSITION V-2-8. *Let E be a separable Banach space which is the dual of a separable Banach space F (every separable reflexive Banach space, in particular every separable Hilbert space, satisfies this hypothesis). For such a space E, every integrable martingale* $(X_n, n \in \mathbf{N})$ *with values in E which satisfies "Doob's condition"*

$$\sup_{\mathbf{N}} E(|X_n|) < \infty$$

converges a.s. to an integrable r.v. X_∞ *with values in E.*

PROOF. (1) We begin by establishing two results valid without further hypotheses on a separable Banach space E. If $(X_n, n \in \mathbf{N})$ is an integrable martingale with values in E and satisfying Doob's condition, then for every element $x' \in E'$ the sequence $((x', X_n), n \in \mathbf{N})$ of real-valued r.v.'s is a martingale since Proposition V-2-5 shows that

$$E^{\mathscr{B}_n}((x', X_{n+1})) = (x', E^{\mathscr{B}_n}(X_{n+1})) = (x', X_n)$$

for all $n \in \mathbf{N}$, and further, by hypothesis,

$$\sup_{\mathbf{N}} E(|(x', X_n)|) \leqslant |x'| \sup_{\mathbf{N}} E(|X_n|) < \infty;$$

Theorem IV-**1**-2 therefore implies that the limit $\lim_{n\to\infty}(x', X_n)$ exists except on an event $\Omega_{x'}$ of probability zero. On the other hand, the sequence $(|X_n|, n \in \mathbf{N})$ of norms is a real-valued positive submartingale since by Proposition V-**2**-5 we have

$$E^{\mathscr{B}_n}(|X_{n+1}|) \geqslant |E^{\mathscr{B}_n}(X_{n+1})| = |X_n|$$

for all $n \in \mathbf{N}$. The hypothesis $\lim\uparrow_n E(|X_n|) < \infty$ and Theorem IV-**1**-2 hence imply that this submartingale converges a.s. to an integrable limit when $n \to \infty$; this allows us to write $Z \stackrel{\text{def}}{=} \sup_{\mathbf{N}} |X_n| < \infty$ except on an event Ω_0 of probability zero.

(2) Now suppose that the separable Banach space E is the dual of a (necessarily) separable Banach space F and let us identify this space F with a subspace of E', the dual of E. This assumption implies the following property: for every *bounded* sequence $(x_n, n \in \mathbf{N})$ such that the limit $\lim_{n\to\infty}(x', x_n)$ exists for all x' belonging to a set D dense in F, there exists an element $x_\infty \in E$ such that $\lim_{n\to\infty}(x', x_n) = (x', x_\infty)$ for all $x' \in F$.

Indeed it is not hard to check that if $(x_n, n \in \mathbf{N})$ is a bounded sequence in the Banach space E, the subset $\{x' : x' \in F, (x', x_n) \to\}$ of elements $x' \in F$ for which the sequence $((x', x_n), n \in \mathbf{N})$ converges is closed in F. It thus suffices to assume that this subset contains a dense subset D of F in order that $\lim_{n\to\infty}(x', x_n)$ exist for all $x' \in F$. But this limit is then a linear functional on F which is bounded as

$$|\lim_{n\to\infty} (x', x_n)| \leqslant \sup_{\mathbf{N}} |x_n| < \infty \quad \text{if } |x'| \leqslant 1;$$

since E is the dual of F, this limit can be written (x', x_∞) for an element $x_\infty \in E$, and the above property is proved.

If the dense subset D of F which we have introduced above is taken to be countable, and this is possible since F is separable, then the event

$$\Omega^* = \Omega_0 \cup \bigcup_D \Omega_{x'}$$

is again null, and by the first part of the proof for all $\omega \notin \Omega^*$ the sequence $(X_n(\omega), n \in \mathbf{N})$ is bounded in norm, whereas for all $x' \in D$ the limit $\lim_{n\to\infty}(x', X_n(\omega))$ exists. By what we have just shown, this implies the existence of a mapping $X_\infty : \Omega \setminus \Omega^* \to E$ such that $\lim_{n\to\infty}(x', X_n(\omega)) = (x', X_\infty(\omega))$ for all $x' \in F$ and all $\omega \notin \Omega^*$.

(3) We will now go on to show that $\lim_{n\to\infty}|X_n(\omega) - X_\infty(\omega)| = 0$ a.s.; to this end we avail ourselves of the following lemma.

LEMMA V-2-9. *Let* $\{(X_n^i, n \in \mathbf{N}), i \in I\}$ *be a countable family of real-valued integrable submartingales such that* $\sup_{n\in\mathbf{N}} E[\sup_{i\in I}(X_n^i)^+] < \infty$. *Each of the submartingales then converges a.s. to an integrable limit* X_∞^i *as* $n\uparrow\infty$ $(i \in I)$ *and*

$$\sup_{i\in I} X_n^i \to \sup_{i\in I} X_\infty^i \quad a.s. \quad as\ n\uparrow\infty.$$

PROOF. The assumption clearly implies that $\sup_{\mathbf{N}} E[(X_n^i)^+] < \infty$ for all $i \in I$, so that by Theorem IV-1-2 the limit $X_\infty^i = \lim_{n\to\infty}^{\mathbf{N}} X_n^i$ exists a.s. for all $i \in I$. Further it is easy to check that the sequence $(\sup_I X_n^i, n \in \mathbf{N})$ is also an integrable submartingale satisfying the condition of Theorem IV-1-2; the limit $X_\infty = \lim_{n\to\infty}(\sup_I X_n^i)$ thus exists a.s. and is integrable; this limit clearly dominates each r.v. X_∞^i (i ∈ I) and thus also their supremum $\sup_I X_\infty^i$, i.e. $X_\infty \geqslant \sup_I X_\infty^i$. To prove that this inequality is in fact an equality and thus to obtain the lemma, it will suffice to show that $E(X_\infty) = E(\sup_I X_\infty^i)$.

Let $(I_p, p \in \mathbf{N})$ be a sequence of finite subsets of I increasing to I as $p\uparrow\infty$; the expectation $E(\sup_{I_p} X_n^i)$ then clearly increases with p $(p \in \mathbf{N})$ and it also increases with n $(n \in \mathbf{N})$ since $(\sup_{I_p} X_n^i, n \in \mathbf{N})$ is a submartingale for every $p \in \mathbf{N}$. As the upper bound S defined by

$$S = \sup_{p,\, n\in\mathbf{N}} E(\sup_{I_p} X_n^i) = \sup_{n\in\mathbf{N}} E(\sup_{i\in I} X_n^i)$$

is finite—since it is dominated by the expression $\sup_{\mathbf{N}} E(\sup_I (X_n^i)^+)$ which is finite by assumption—for every $\varepsilon > 0$ there exists at least one pair $p_\varepsilon, n_\varepsilon \in \mathbf{N}$ of integers such that

$$E(\sup_{I_p} X_n^i) \geqslant S - \varepsilon$$

if $p = p_\varepsilon$ and $n = n_\varepsilon$, and thus if $p \geqslant p_\varepsilon$ and $n \geqslant n_\varepsilon$. But the r.v. $X_\infty - \sup_{I_p} X_\infty^i$ is the a.s. limit of the sequence $(\sup_I X_n^i - \sup_{I_p} X_n^i, n \in \mathbf{N})$ of positive r.r.v.'s, so that Fatou's lemma implies that

$$E(X_\infty - \sup_{I_p} X_\infty^i) \leqslant \liminf_{n\to\infty} E(\sup_I X_n^i - \sup_{I_p} X_n^i)$$

$$\leqslant S - (S - \varepsilon) = \varepsilon \quad \text{whenever } p \geqslant p_\varepsilon.$$

It then follows that $E(X - \sup_I X_\infty^i) \leqslant \varepsilon$ for all $\varepsilon > 0$, and the lemma is proved. ■

Now let us finish the proof of Proposition V-**2**-8. By what we have shown above, if D' denotes a countable dense subset of the unit ball of F, the preceding lemma implies that for all fixed $a \in E$ the sequence

$$|X_n(\omega) - a| = \sup_{\substack{x' \in F \\ |x'| \leqslant 1}} (x', X_n(\omega) - a) = \sup_{x' \in D'} (x', X_n(\omega) - a)$$

converges in **R** to the limit

$$\sup_{x' \in D'} (x', X_\infty(\omega) - a) = |X_\infty(\omega) - a|$$

for all ω not belonging to an event $\Omega_a^* \supset \Omega^*$ of zero probability. Then if A denotes a countable dense subset of E, let us put $\Omega^{**} = \bigcup_A \Omega_a^*$; this event is null and for all $\omega \notin \Omega^{**}$ we have

$$\lim_{n \to \infty} |X_n(\omega) - a| = |X_\infty(\omega) - a|$$

when $a \in A$. But such a limiting relation can only be valid for all $a \in A$ if it is also valid for all $a \in \bar{A} = E$; thus we have shown that

$$\lim_{n \to \infty} |X_n(\omega) - a| = |X_\infty(\omega) - a| \quad \text{if } a \in E \text{ and if } \omega \notin \Omega^{**}.$$

Taking $a = X_\infty(\omega)$ in this relation, we finally obtain

$$\lim_{n \to \infty} |X_n(\omega) - X_\infty(\omega)| = 0 \quad \text{when } \omega \notin \Omega^{**} \; [P(\Omega^{**}) = 0].$$

We have not yet proved that the mapping $\omega \to X_\infty(\omega)$ is measurable; this measurability now follows, at least on the set $\Omega \backslash \Omega^{**}$, from the convergence in norm of the sequence $X_n(\omega)$ to $X_\infty(\omega)$ just as in Lemma V-**2**-4. The integrability of X_∞ follows from Fatou's lemma: since $\lim_{n \to \infty} |X_n| = |X_\infty|$ outside Ω^{**}, we have $E(|X_\infty|) \leqslant \sup_{\mathbf{N}} E(|X_n|) < \infty$. ■

Remark and Example. If E is a separable Banach space whose dual E' is also separable, the argument prior to Lemma V-**2**-9 in the previous proof applies to every martingale $(X_n, n \in \mathbf{N})$ with values in E which satisfies Doob's condition, except that now the mapping X_∞ takes its values in the dual space E'' of E'; of course the Banach space E is canonically isomorphic to a closed subspace of E'' (proper if E is not reflexive), and the result states that $X_n(\omega) \to X_\infty(\omega)$ in E'' for the topology $\sigma(E'', E')$ for almost all ω. If the space E'' is separable, we are led to the hypotheses of the proposition in passing from the space E'' to the space E; on the other hand, if the space E'' is not separable, or at least if the mapping X_∞ does not take its values in a separable subspace of E'' a.s., it is not possible to extend the second part of the proof to show convergence in norm. On this point an example will enlarge our understanding of the situation.

The space $E = C_0(\mathbf{N})$ of all real-valued sequences $(y_n, n \in \mathbf{N})$ converging to zero is a separable Banach space for the supremum norm. Its dual E' is the separable Banach space $L^1(N)$ of summable real-valued sequences, whereas its bidual E'' is the Banach space $L^\infty(\mathbf{N})$ of all bounded real-valued sequences; the latter space is not separable for it contains the uncountable family $(1_J, J \in \mathscr{P}(\mathbf{N}))$ of indicator functions of subsets of $\mathbf{N}$, which is such that $\|1_J - 1_{J'}\|_{L^\infty(\mathbf{N})} = 1$ if $J \neq J'$. On the other hand, if $(Y_n, n \in \mathbf{N})$ denotes an independent sequence of r.v.'s taking the values ± 1 with probabilities $\frac{1}{2}, \frac{1}{2}$, it is not hard to show that the formula

$$X_n = (Y_0, Y_1, \ldots, Y_n, 0, 0, \ldots) \qquad (n \in \mathbf{N})$$

defines a martingale with values in E such that $|X_n| = 1$ everywhere $(n \in \mathbf{N})$; this martingale evidently satisfies Doob's condition. Also, the formula

$$X_\infty = (Y_0, Y_1, \ldots, Y_n, Y_{n+1}, \ldots)$$

defines a mapping with values in $E'' = L^\infty(\mathbf{N})$ and it is immediately checked that here we have

$$(x', X_n) \to (x', X_\infty) \qquad \text{everywhere, for all } x' \in E'.$$

Despite this convergence of X_n to X in E'' for the topology $\sigma(E'', E')$, let us observe that in this example the sequence $(X_n, n \in \mathbf{N})$ cannot converge in any reasonable sense in the original space E (since none of the values of the mapping X_∞ are taken in E) and that the sequence $(X_n, n \in \mathbf{N})$ cannot converge in norm in any subspace of E either (since $|X_{n+1} - X_n|_E = 1$ everywhere for all $n \in \mathbf{N}$)! ■

Remark. The difficulty of the problem of the convergence of vector-valued martingales is largely due to the difficulty of identifying the limit; when this identification is easy, as for example in the study of the convergence of the martingale $(E^{\mathscr{B}_n}(X), n \in \mathbf{N})$ of Proposition V-2-6 $(X \in L_E^1)$, the situation is simpler. Here is a second proof of the a.s. convergence of the martingale $(E^{\mathscr{B}_n}(X), n \in \mathbf{N})$ towards $E^{\mathscr{B}_\infty}(X)$ when $X \in L_E^1$.

Firstly it is clear that for all $x' \in E'$, when $n \to \infty$

$$(x', E^{\mathscr{B}_n}(X)) = E^{\mathscr{B}_n}[(x', X)] \to E^{\mathscr{B}_\infty}[(x', X)] = (x', E^{\mathscr{B}_\infty}(X))$$

outside a null event $\Omega_{x'}$. Then let us choose a countable subset D of the unit ball of E' such that $\sup_{x' \in D}(x', x) = |x|$ for all $x \in E$ (as this identity is satisfied for all $x \in E$ whenever it is satisfied for a dense subset of E which we can choose to be countable, it is easy to establish the existence of such subsets D of E'). For every fixed $a \in E$, we next apply Lemma V-**2**-9 to the countable family of martingales

$$\{((x', E^{\mathscr{B}_n}(X) - a), n \in \mathbf{N}), x' \in D\};$$

as

$$|(x', E^{\mathscr{B}_n}(X))| \leqslant |E^{\mathscr{B}_n}(X)| \leqslant E^{\mathscr{B}_n}(|X|)$$

for all $x' \in D$, the assumptions of the lemma are satisfied and we find that

$$|E^{\mathscr{B}_n}(X) - a| \to |E^{\mathscr{B}_\infty}(X) - a| \quad \text{a.s.}$$

when $n \uparrow \infty$, for all $a \in E$. It is then easy to deduce that

$$P(\lim_{n \to \infty} |E^{\mathscr{B}_n}(X) - a| = |E^{\mathscr{B}_\infty}(X) - a| \text{ for all } a \in E) = 1$$

Since E is separable, and finally, taking $a = E^{\mathscr{B}_n}(X)$ at every point ω, we find that $E^{\mathscr{B}_n}(X) \to E^{\mathscr{B}_\infty}(X)$ a.s. when $n \uparrow \infty$. ■

We end this section with the following proposition which, however, we only partly prove.

PROPOSITION V-**2**-10. *Let $(\Omega, \mathscr{A}, P)$ be a probability space. For every separable Banach space E, the following properties are equivalent:*

(a) *Every integrable martingale $(X_n, n \in \mathbf{N})$ with values in E which satisfies Doob's condition $\sup_{\mathbf{N}} E(|X_n|) < \infty$ converges a.s. and in norm to an integrable r.v. $X_\infty : \Omega \to E$.*

(b) *Every integrable martingale $(X_n, n \in \mathbf{N})$ with values in E such that $(|X_n|, n \in \mathbf{N})$ is uniformly integrable converges in L^1_E.*

(c) *The following vector-valued Radon–Nikodym theorem is true: For every set function $\mu : \mathscr{A} \to E$ having the following three properties:*

(1) $\sup\{\sum_{k \in \mathbf{N}} |\mu(A_k)|\} < \infty$, *the supremum being taken over all sequences $(A_k, k \in \mathbf{N})$ of pairwise disjoint events,*

(2) $\sum_{k \in \mathbf{N}} \mu(A_k) = \mu(A)$ *for every* $A \in \mathscr{A}$ *and every countable partition* $(A_k, k \in \mathbf{N})$ *of* A *in* $\mathscr{A}$, *the series converging in* E *in norm,*

(3) $\mu(A) = 0$ *if* $P(A) = 0$ $(A \in \mathscr{A})$,

there exists a r.v. $X \in L_E^1$ *such that* $\mu(A) = \int_A X \mathrm{d}P$ *for all* $A \in \mathscr{A}$.

REMARK. By Proposition V-**2**-6, property (b) is also equivalent to the following property:

(b′) Every integrable martingale $(X_n, n \in \mathbf{N})$ with values in E such that $(|X_n|, n \in \mathbf{N})$ is uniformly integrable is of the form $X_n = E^{\mathscr{B}_n}(X)$ for a r.v. $X \in L_E^1$.

It is no more difficult to show that property (b) is equivalent to each of the following properties (b_p) or (b'_p), where p denotes a finite real number > 1.

(b_p) (resp. (b'_p)) Every martingale $(X_n, n \in \mathbf{N})$ which is bounded in L_E^p converges in L_E^p (resp. is of the form $X_n = E^{\mathscr{B}_n}(X)$ for a r.v. $X \in L_E^p$). ■

PROOF. We will content ourselves with proving the implications $(a) \Rightarrow (b) \Rightarrow (c)$; for a proof of the implication $(c) \Rightarrow (a)$—which is more a study of the decomposition of measures than of martingales—we refer the reader to [53]. Nevertheless, this incomplete proof will already show that the vector-valued Radon–Nikodym is valid in every separable dual of a separable Banach space.

$(a) \Rightarrow (b)$. If $(X_n, n \in \mathbf{N})$ is an integrable martingale and if the sequence $(|X_n|, n \in \mathbf{N})$ is uniformly integrable, Doob's condition is satisfied, and thus by (a) there exists a r.v. $X_\infty \in L_E^1$ such that $|X_n - X_\infty| \to 0$ a.s. As the sequence $(|X_n - X_\infty|, n \in \mathbf{N})$ is also uniformly integrable, Lemma IV-**2**-5 shows that $\lim_{n \to \infty} E(|X_n - X_\infty|) = 0$.

$(b) \Rightarrow (c)$. Let $\mu : \mathscr{A} \to E$ be a mapping satisfying the properties (1)–(3) of condition (c). Then for every $A \in \mathscr{A}$ we will define

$$|\mu|(A) = \sup \sum_{\mathbf{N}} |\mu(A_k)|$$

taking the supremum on the right-hand side over all countable partitions $(A_k, k \in \mathbf{N})$ of A in $\mathscr{A}$. By property (1) it is clear that $0 \leqslant |\mu|(A) < \infty$; using property (2) it is easy to check that $|\mu|$ is σ-additive; since $|\mu|(A) = 0$ whenever $P(A) = 0$ by property (3), the scalar Radon–Nikodym theorem (Corollary III-**2**-5) then establishes the existence of a positive integrable function g such that

$$|\mu|(A) = \int_A g \, \mathrm{d}P \quad \text{for all } A \in \mathscr{A}.$$

First suppose that the σ-field $\mathscr{A}$ is separable; hence there exists an increasing sequence $(\mathscr{B}_n, n \in \mathbf{N})$ of σ-fields generated by finite partitions $(B_n^p, 1 \leqslant p \leqslant p_n)$ $(n \in \mathbf{N})$ and generating the σ-field $\mathscr{A}$ (i.e. $\mathscr{A} = \bigvee_{\mathbf{N}} \mathscr{B}_n$). The formula

$$X_n = \sum_{1 \leqslant p \leqslant p_n} \frac{\mu(B_n^p)}{P(B_n^p)} 1_{B_n^p} \quad (n \in \mathbf{N}),$$

where by convention $\mu(B_n^p)/P(B_n^p) = 0$ if $P(B_n^p) = 0$, then defines r.v.'s $X_n : \Omega \to E$ which are integrable and respectively $\mathscr{B}_n$-measurable $(n \in \mathbf{N})$ and which furthermore satisfy the equalities $\mu(A) = \int_A X_n \mathrm{d}P$ if $A \in \mathscr{B}_n$ $(n \in \mathbf{N})$. It is then clear that the sequence $(X_n, n \in \mathbf{N})$ is an integrable martingale with values in E, relative to the sequence of σ-algebras $(\mathscr{B}_n, n \in \mathbf{N})$. This martingale is uniformly integrable because the definition of the X_n implies that

$$\int_A |X_n| \,\mathrm{d}P = \sum_{1 \leqslant p \leqslant p_n} \mu(AB_n^p) \leqslant |\mu|(A) = \int_A g \,\mathrm{d}P$$

if $A \in \mathscr{B}_n$ and $n \in \mathbf{N}$; therefore $|X_n| \leqslant E^{\mathscr{B}_n}(g)$ for all $n \in \mathbf{N}$ and the uniform integrability of the sequence $(|X_n|, n \in \mathbf{N})$ follows from that of the positive martingale $(E^{\mathscr{B}_n}(g), n \in \mathbf{N})$ by which it is dominated.

Condition (b) then implies that the vector-valued martingale $(X_n, n \in \mathbf{N})$ converges in L_E^1 to an integrable r.v. X_∞. The identity

$$\mu(A) = \int_A X_n \,\mathrm{d}P \quad (A \in \mathscr{B}_n, n \in \mathbf{N})$$

implies, upon passing to the limit, that

$$\mu(A) = \int_A X_\infty \,\mathrm{d}P \text{ if } A \in \bigcup_{\mathbf{N}} \mathscr{B}_n.$$

Finally this identity extends by continuity to all $A \in \bigvee_{\mathbf{N}} \mathscr{B}_n = \mathscr{A}$; indeed, since $\bigcup_{\mathbf{N}} L^1(\mathscr{B}_n)$ is dense in $L^1(\mathscr{A})$, for all $A \in \mathscr{A}$ it is possible to find a sequence $(A_p, p \in \mathbf{N})$ in $\bigcup_{\mathbf{N}} \mathscr{B}_n$ such that

$$P(A_m \triangle A) = \|1_{A_m} - 1_A\|_1 \to 0 \text{ when } m \to \infty;$$

on the other hand,

$$|\mu(A_m) - \mu(A)| \leqslant |\mu|(A_m \triangle A) = \int_{A_m \triangle A} g \,\mathrm{d}P \to 0,$$

$$\left| \int_{A_m} X_\infty \mathrm{d}P - \int_A X_\infty \mathrm{d}P \right| \leqslant \int_{A_m \triangle A} |X_\infty| \,\mathrm{d}P \to 0$$

when $m \to \infty$, because g and $|X_\infty|$ are both integrable.

It remains to extend the preceding proof to the case where the σ-field $\mathscr{A}$ is not separable. To this end we will consider the increasing directed family of finite sub-σ-fields of $\mathscr{A}$, and with each of these σ-fields $\mathscr{B}$ we will associate a r.v. $X_{\mathscr{B}} : \Omega \to E$ exactly as we associated the r.v.'s X_n with the σ-fields $\mathscr{B}_n$ above; thus we obtain a martingale $(X_{\mathscr{B}}, \mathscr{B} \text{ finite} \subset \mathscr{A})$ in L_E^1 with a directed index set such that also $|X_{\mathscr{B}}| \leqslant E^{\mathscr{B}}(g)$ for every finite sub-σ-field $\mathscr{B}$ of $\mathscr{A}$. By Lemma V-**1**-1, this martingale with a directed index set converges in L_E^1 since by property (*b*) every subsequence $(X_{\mathscr{B}_n}, n \in \mathbf{N})$ corresponding to an increasing sequence $(\mathscr{B}_n, n \in \mathbf{N})$ converges in L_E^1. The proof is completed even more simply than as above: for all $A \in \mathscr{A}$, the identity $\mu(A) = \int_A X_{\mathscr{B}} \mathrm{d}P$ holds whenever $A \in \mathscr{B}$ and, as $X_{\mathscr{B}} \to X_\infty$ in L_E^1 when $\mathscr{B} \uparrow \mathscr{A}$, we have $\mu(A) = \int_A X_\infty \mathrm{d}P$ for all $A \in \mathscr{A}$ (note that $\mathscr{A} = \bigcup_{\mathscr{B} \text{ finite}} \mathscr{B}$!). ∎

V-3. Reversed martingales

Let $(\mathscr{B}_n, n \in \mathbf{N})$ be a *decreasing* sequence of sub-σ-fields of $\mathscr{A}$ in the probability space $(\Omega, \mathscr{A}, P)$. A sequence $(X_n, n \in \mathbf{N})$ of positive or integrable r.r.v.'s is called a *reversed supermartingale* (relative to the sequence $(\mathscr{B}_n, n \in \mathbf{N})$) if

(a) the r.r.v. X_n is $\mathscr{B}_n$-measurable for all $n \in \mathbf{N}$,
(b) $E^{\mathscr{B}_{n+1}}(X_n) \leqslant X_{n+1}$ for all $n \in \mathbf{N}$.

A reversed martingale or submartingale is defined analogously.

The aim of this section is to establish analogues of the theorems in the preceding chapters for these objects; we will see that the results are simpler in the case of "reversed sequences".

The following remark, whose verification is immediate, is the basis of all that follows: a sequence $(X_n, n \in \mathbf{N})$ is a positive reversed supermartingale relative to the decreasing sequence $(\mathscr{B}_n, n \in \mathbf{N})$ of σ-fields if and only if for all n, the sequence

$$(X_n, X_{n-1}, \ldots, X_1, X_0, X_0, X_0, \ldots)$$

is a positive supermartingale relative to the increasing sequence $(\mathscr{B}_n, \mathscr{B}_{n-1}, \ldots, \mathscr{B}_1, \mathscr{B}_0, \mathscr{B}_0, \mathscr{B}_0, \ldots)$ of σ-fields. Let us also note, especially to justify the terminology, that a reversed supermartingale $(X_n, n \in \mathbf{N})$ can be considered as an ordinary supermartingale indexed by the negative integers; indeed, putting $X'_n = X_{-n}$ and $\mathscr{B}'_n = \mathscr{B}_{-n}$ if $n \leqslant 0$, the reversed supermartingale inequality can also be written $X'_n \geqslant E^{\mathscr{B}'_n}(X'_{n+1})\,(n \leqslant -1)$ and the sequence $(\mathscr{B}'_n, n \leqslant 0)$ is increasing.

PROPOSITION V-3-11. *Let $(\mathscr{B}_n, n \in \mathbf{N})$ be a decreasing sequence of sub-σ-fields of $\mathscr{A}$ in a probability space $(\Omega, \mathscr{A}, P)$. Every positive reversed supermartingale $(X_n, n \in \mathbf{N})$ converges a.s. to a limit X_∞ which is a positive r.r.v. measurable with respect to the σ-field $\mathscr{B}_\infty = \bigcap_{\mathbf{N}} \mathscr{B}_n$. Further, $E^{\mathscr{B}_\infty}(X_n) \uparrow X_\infty$ a.s. when $n \uparrow \infty$.*

In order that the sequence $(X_n, n \in \mathbf{N})$ be uniformly integrable, it is necessary and sufficient that $\lim\uparrow_n E(X_n) < \infty$, and in this case the reversed supermartingale $(X_n, n \in \mathbf{N})$ also converges in L^1 to the r.r.v. X_∞.

PROOF. (1) When $a < b$ are given in R_+, the random number $\delta^n_{a,b}(\omega)$ of times that the sequence $(X_0(\omega), \ldots, X_n(\omega))$ downcrosses the interval $[a, b]$ coincides with the number $\gamma^n_{a,b}(\omega)$ of times that the sequence $(X_n(\omega), \ldots, X_0(\omega), X_0(\omega), \ldots)$ upcrosses the interval $[a, b]$. The remark at the beginning and Dubins' inequalities of Chapter II thus imply that

$$P^{\mathscr{B}_n}(\delta^n_{a,b} \geqslant k) \leqslant \left(\frac{a}{b}\right)^k \min\left(\frac{X_n}{a}, 1\right) \qquad (n \in \mathbf{N}, a < b \text{ in } R_+);$$

taking the conditional expectation $E^{\mathscr{B}_\infty}$ of both sides, we have

$$P^{\mathscr{B}_\infty}(\delta^n_{a,b} \geqslant k) \leqslant \left(\frac{a}{b}\right)^k E^{\mathscr{B}_\infty}\left[\min\left(\frac{X_n}{a}, 1\right)\right].$$

Let $n \uparrow \infty$. The r.v.'s $\delta^n_{a,b}$ then increase to the number $\delta_{a,b}$ of times that the sequence $(X_n, n \in \mathbf{N})$ downcrosses the interval $[a, b]$, and the preceding inequality gives us

$$P^{\mathscr{B}_\infty}(\delta_{a,b} \geqslant k) \leqslant \left(\frac{a}{b}\right)^k \sup_{\mathbf{N}} E^{\mathscr{B}_\infty}\left[\min\left(\frac{X_n}{a}, 1\right)\right].$$

This implies that $\delta_{a,b} < \infty$ a.s. for every pair $a < b$ in R_+; by section **II-2**, the sequence $(X_n, n \in \mathbf{N})$ of positive r.r.v.'s will then converge a.s. to a positive limit, finite or infinite, X_∞; in order that X_∞ be everywhere defined, we put $X_\infty = \limsup_{n \to \infty} X_p$. Since the r.v.'s X_n are $\mathscr{B}_p$-measurable whenever $n \geqslant p$, the r.v. X_∞ must be $\mathscr{B}_p$-measurable; this being so for all $p \in \mathbf{N}$, the r.r.v. X_∞ is $\mathscr{B}_\infty$-measurable.

The reversed supermartingale inequality $E^{\mathscr{B}_{n+1}}(X_n) \leqslant X_{n+1}$ $(n \in \mathbf{N})$ then implies that $E^{\mathscr{B}_{n+1}}(\min(X_n, a)) \leqslant \min(X_{n+1}, a)$ and hence also that

$$E^{\mathscr{B}_\infty}(\min(X_n, a)) \leqslant E^{\mathscr{B}_\infty}(\min(X_{n+1}, a))$$

for all $n \in \mathbf{N}$ and all $a \in [0, \infty]$. Now the r.v.'s $(\min(X_n, a), n \in \mathbf{N})$ converge a.s. to the r.r.v. $\min(X_\infty, a)$ when $n \to \infty$ and, as these r.r.v.'s are bounded between 0 and a, the sequence $(E^{\mathscr{B}_\infty}(\min(X_n, a)), n \in \mathbf{N})$ of r.r.v.'s tends a.s. to the r.v.

$$E^{\mathscr{B}_\infty}(\min(X_\infty, a)) = \min(X_\infty, a) \quad \text{for all } a \in R_+$$

by the dominated convergence theorem; since this sequence is increasing, we have shown that

$$E^{\mathscr{B}_\infty}(\min(X_n, a)) \uparrow \min(X_\infty, a) \quad \text{a.s.} \quad \text{when } n \uparrow \infty \ (a \in R_+).$$

It follows that

$$\begin{aligned} \lim_n \uparrow E^{\mathscr{B}_\infty}(X_n) &= \lim_n \uparrow \lim_a \uparrow E^{\mathscr{B}_\infty}(\min(X_n, a)) \\ &= \lim_a \uparrow \lim_n \uparrow E^{\mathscr{B}_\infty}(\min(X_n, a)) \\ &= \lim_a \uparrow \min(X_\infty, a) \\ &= X_\infty \quad \text{a.s.} \end{aligned}$$

The first part of the proposition is thus proved. Let us observe that the convergence that we have just shown allows Dubin's inequalities to be rewritten in the form

$$P^{\mathscr{B}_\infty}(\delta_{a,b} \geqslant k) \leqslant \left(\frac{a}{b}\right)^k \min\left(\frac{X_\infty}{a}, 1\right) \qquad (k \in \mathbf{N};\ a < b)$$

for every positive reversed supermartingale.

(2) Every uniformly integrable sequence $(X_n, n \in \mathbf{N})$ of r.r.v.'s is necessarily bounded in L^1 and hence such that $\sup_{\mathbf{N}} E(X_n) < \infty$. Conversely, let $(X_n, n \in \mathbf{N})$ be a positive reversed supermartingale such that $s = \lim \uparrow_n E(X_n)$ is finite; we show that this sequence is uniformly integrable. For any given $\varepsilon > 0$, choose $p \in \mathbf{N}$ such that $E(X_p) > s - \varepsilon$. The reversed supermartingale inequality $E^{\mathscr{B}_n}(X_p) \leqslant X_n$, which is valid if $n \geqslant p$, implies that

$$\int_{\{X_n \leqslant a\}} X_n \, dP \geqslant \int_{\{X_n \leqslant a\}} X_p \, dP \qquad (n \geqslant p,\ a \in R_+)$$

and hence that

$$\int_{\{X_n>a\}} X_n \mathrm{d}P = E(X_n) - \int_{\{X_n \leqslant a\}} X_n \mathrm{d}P \leqslant E(X_n) - \int_{\{X_n \leqslant a\}} X_p \mathrm{d}P$$

$$= [E(X_n) - E(X_p)] + \int_{\{X_n>a\}} X_p \mathrm{d}P \leqslant \varepsilon + \int_{\{X_n>a\}} X_p \mathrm{d}P$$

if $n \geqslant p$ and $a \in R_+$.

On the other hand,

$$P(X_n > a) \leqslant \frac{1}{a} E(X_n) \leqslant \frac{1}{a} \sup_{\mathbf{N}} E(X_m) \qquad (n \in \mathbf{N})$$

and consequently

$$\sup_{n \in \mathbf{N}} \int_{\{X_n>a\}} X_p \mathrm{d}P \downarrow 0$$

when $a \uparrow \infty$, since the r.v. X_p is integrable. Choose a real number a_ε such that

$$\int_{\{X_n>a_\varepsilon\}} X_p \mathrm{d}P \leqslant \varepsilon \qquad \text{for all } n \in \mathbf{N},$$

and also such that

$$\int_{\{X_n>a_\varepsilon\}} X_n \mathrm{d}P \leqslant 2\varepsilon \qquad \text{for all } n < p.$$

The inequality proved in the preceding paragraph then implies that

$$\int_{\{X_n>a_\varepsilon\}} X_n \mathrm{d}P \leqslant 2\varepsilon \qquad \text{for all } n \in \mathbf{N},$$

which clearly suffices to establish the uniform integrability of the sequence $(X_n, n \in \mathbf{N})$. The last part of the proposition is a consequence of Lemma IV-**2**-5. ■

Corollary V-**3**-12. *Let $(\mathscr{B}_n, n \in \mathbf{N})$ be a decreasing sequence of sub-σ-fields of $\mathscr{A}$ in the probability space $(\Omega, \mathscr{A}, P)$. For every positive r.r.v. X, finite or not, the sequence $(E^{\mathscr{B}_n}(X), n \in \mathbf{N})$ converges a.s. to $E^{\mathscr{B}_\infty}(X)$, whilst for every integrable r.r.v. X the same sequence converges both a.s. and in L^1 to the integrable r.v. $E^{\mathscr{B}_\infty}(X)$.*

Proof. If X is a positive r.r.v., the r.r.v.'s $X_n = E^{\mathscr{B}_n}(X)$ $(n \in \mathbf{N})$ form a positive reversed martingale; moreover, let us observe that we obtain the most general positive reversed martingale in this way. By the preceding proposition, the sequence $(X_n,\ n \in \mathbf{N})$ converges a.s. to an r.r.v. X_∞ and further $X_\infty = \lim\uparrow_n E^{\mathscr{B}_\infty}(X_n)$; but here

$$E^{\mathscr{B}_\infty}(X_n) = E^{\mathscr{B}_\infty}(X) \qquad \text{for all } n \in \mathbf{N}$$

and consequently $X_\infty = E^{\mathscr{B}_\infty}(X)$. If the positive r.r.v. is integrable as well, the second part of the preceding proposition shows that $X_n \to X_\infty$ in L^1 since $E(X_n) = E(X) < \infty$ for all $n \in \mathbf{N}$. Finally, if the r.r.v. X is integrable without being positive, the decomposition $X = X^+ - X^-$ allows a reduction to the previous case, and thus the corollary is completely proved. ■

COROLLARY V-3-13. *Let $(\mathscr{B}_n, n \in \mathbf{N})$ be a decreasing sequence of sub-σ-fields of $\mathscr{A}$ in the probability space $(\Omega, \mathscr{A}, P)$. Every integrable reversed submartingale $(X_n, n \in \mathbf{N})$ then converges a.s. to a limit X_∞ which is a r.r.v. measurable with respect to the σ-field $\mathscr{B}_\infty = \bigcap_{\mathbf{N}} \mathscr{B}_n$; further,*

$$X_\infty \leqslant E^{\mathscr{B}_\infty}(X_n) \quad \textit{for all } n \in \mathbf{N}.$$

Finally, if $\lim\downarrow_n E(X_n) > -\infty$, the r.r.v. X_∞ is integrable and the reversed submartingale also converges in L^1 to the r.r.v. X_∞.

When we do not suppose that $\lim\downarrow_n E(X_n) > -\infty$, the r.r.v. X_∞ can take the value $-\infty$.

PROOF. It is easy to check that the sequence formed from the r.r.v.'s $Z_n = E^{\mathscr{B}_n}(X_0^+) - X_n$ $(n \in \mathbf{N})$ is a positive reversed supermartingale: the sequence converges therefore a.s. to a positive limit, say Z_∞, which is such that $E^{\mathscr{B}_\infty}(Z_n) \leqslant Z_\infty$ for all $n \in \mathbf{N}$. On the other hand, the positive reversed martingale $(E^{\mathscr{B}_n}(X_0^+), n \in \mathbf{N})$ converges a.s. to the r.r.v. $E^{\mathscr{B}_n}(X_0^+)$, which is a.s. finite; it is therefore clear that

$$X_n = E^{\mathscr{B}_n}(X_0^+) - Z_n \to E^{\mathscr{B}_\infty}(X_0^+) - Z_\infty \text{ a.s.} \qquad (n\uparrow\infty)$$

and that

$$E^{\mathscr{B}_\infty}(X_n) = E^{\mathscr{B}_\infty}(X_0^+) - E^{\mathscr{B}_\infty}(Z_n) \geqslant E^{\mathscr{B}_\infty}(X_0^+) - Z_\infty \text{ a.s.} \qquad (n \in \mathbf{N}).$$

Finally, if $\lim\downarrow_n E(X_n) > -\infty$, we have $\lim\uparrow_n E(Z_n) < \infty$ and by the proposition the sequence $(Z_n, n \in \mathbf{N})$ converges in L^1 to Z_∞. As the sequence $(E^{\mathscr{B}_n}(X_0^+), n \in \mathbf{N})$ always converges in L^1 to the r.r.v. $E^{\mathscr{B}_\infty}(X_0^+)$, it is clear that the condition $\lim\downarrow_n E(X_n) > -\infty$ suffices to ensure the L^1-convergence of the reversed submartingale $(X_n, n \in \mathbf{N})$. ■

CHAPTER VI

OPTIMISATION PROBLEMS

VI-1. Snell's problem

Let $(Z_n, n \in \mathbf{N})$ be a sequence of integrable r.r.v.'s representing, for example, the random sequence of winnings of a gambler at successive times n $(n \in \mathbf{N})$; we must suppose that the random variable $\sup_{\mathbf{N}} Z_n^+$ is integrable.

For every finite stopping time ν, the expectation $E(Z_\nu)$ represents the expected winnings of a gambler who decides to leave the game at the random time ν; let us remark that by supposing ν a stopping time we are obliging the gambler to be honest, i.e. not to leave the game at time n (event $\{\nu = n\}$) taking into account information other than that available at that moment (σ-field $\mathscr{B}_n$). The aim of this section is to study the upper bound $\sup_\nu E(Z_\nu)$ and the existence of stopping times which attain this bound.

In the sequel we will only consider those a.s. finite stopping times such that $E(Z_\nu^-) < \infty$; on the other hand, as the assumption $E(\sup_{\mathbf{N}} Z_n^+) < \infty$ implies that $E(Z_\nu^+) < \infty$, this condition assures the integrability of the r.r.v. Z_ν. We will also be making the quite natural assumption that $(Z_n, n \in \mathbf{N})$ is adapted.

In order to render intuitive the solution of Snell's problem that we shall be giving, let us first consider the case where the r.r.v.'s Z_n are equal to constants z_n $(n \in \mathbf{N})$ and where all the σ-fields $\mathscr{B}_n$ reduce to the smallest σ-field $\{\emptyset, \Omega\}$, so that every stopping time is necessarily constant; in this case Snell's problem becomes the "study" of the least upper bound $\sup_{\mathbf{N}} z_n$ and the indices p for which $z_p = \sup_{\mathbf{N}} z_n$! A sophisticated method of attacking this "problem" could consist, if $\sup_{\mathbf{N}} z_n < \infty$, in introducing the sequence $(x_p = \sup_{n \geqslant p} z_n, p \in \mathbf{N})$ and showing (exercise!) that the first index p for which $x_p = z_p$, p_0 say if such exists, satisfies the equalities $x_0 = x_{p_0} = z_{p_0} = \sup_{\mathbf{N}} z_n$. By analogy with this "solution", we will, in the general case, introduce r.v.'s X_p generalising the x_p $(p \in \mathbf{N})$; among other things, we will then show that these r.v.'s form not a decreasing sequence, but a supermartingale and that the stopping time $\nu_0 = \inf(p : X_p = Z_p)$ is optimal whenever it is a.s. finite.

The analysis and then the solution of the posed problem rests on the introduction of the random variables

$$X_n = \operatorname*{ess\,sup}_{\Lambda_n} E^{\mathscr{B}_n}(Z_\nu) \qquad (n \in \mathbf{N}),$$

where for each n the upper bound is taken over the set Λ_n of all stopping times ν such that $n \leqslant \nu < \infty$ a.s. and $E(Z_\nu^-) < +\infty$. These sets are not in general countable and the preceding upper bounds thus have to be taken in the sense of a.s. inequality; the proposition below gives the definition and existence of these essential upper bounds. First let us note that the r.v.'s X_n denote the supremum of the conditional expectations of winnings at time n of a gambler who decides to play at least until time n.

PROPOSITION VI-1-1. *For every family F of real-valued measurable functions $f : \Omega \to \bar{R}$ defined on a probability space $(\Omega, \mathscr{A}, P)$, there exists one and up to equivalence only one measurable function $g : \Omega \to \bar{R}$ such that*

(a) *$g \geqslant f$ a.s. for all $f \in F$,*
(b) *if h is a measurable function such that $h \geqslant f$ a.s. for all $f \in F$, then $h \geqslant g$ a.s.*

This function g, which is the least upper bound of the family F in the sense of a.s. inequality, is denoted by ess sup(F).

Further there exists at least one sequence $(f_n, n \in \mathbf{N})$ taken from F such that $\operatorname{ess\,sup}(F) = \sup_{\mathbf{N}} f_n$ *a.s. If the family F is directed upwards, the sequence $(f_n, n \in \mathbf{N})$ can be chosen to be increasing a.s. and then* $\operatorname{ess\,sup}(F) = \lim\uparrow_n f_n$ *a.s.*

PROOF. Since only the order structure of $\bar{R}$ is involved in this proposition, we can restrict ourselves to the case where the functions f take their values in $[0, 1]$ by mapping $\bar{R}$ onto $[0, 1]$ by an increasing bijection.

Let $\mathscr{J}$ be the class of all countable sub-families of F. For every $G \in \mathscr{J}$ introduce the measurable function f_G defined as the countable supremum $f_G = \sup_{f \in G} f$ on Ω. Next let us consider the supremum $\alpha = \sup_{\mathscr{J}} E(f_G)$. This supremum is attained because if $(G_n, n \in \mathbf{N})$ is a sequence in $\mathscr{J}$ such that $E(f_{G_n}) \to \alpha$, then $G* = \bigcup_{\mathbf{N}} G_n \in \mathscr{J}$ and $E(f_{G*}) = \alpha$ since $E(f_{G_n}) \leqslant E(f_{G*}) \leqslant \alpha$ for all $n \in \mathbf{N}$. We show that the function $g = f_{G*}$ satisfies the properties (a) and (b) of the proposition.

For every $f \in F$, the function f_G corresponding to the countable subfamily $G = G* \cup \{f\}$ is equal to $g \vee f$; we thus have $\alpha = E(g) \leqslant E(g \vee f) \leqslant \alpha$ which, since α is finite, is only possible if $g \vee f = g$ a.s., i.e. if $f \leqslant g$ a.s. Property (a) is thus established. Property (b) is clear by the definition $g = f_{G*}$ of g. The uniqueness of the function g up to equivalence is an immediate consequence of property (b).

Arrange G_* in a sequence (f_n, $n \in \mathbf{N}$); then $\operatorname{ess\,sup}(F) = f_{G_*} = \sup_{\mathbf{N}} f_n$ a.s. Now if the family F is directed upwards (i.e., if for every pair f_1, f_2 of functions in F there exists a third function f_3 of F a.s. larger than the other two), then it is possible to construct an a.s. *increasing* sequence (f'_n, $n \in \mathbf{N}$) in F such that $\lim\uparrow_n f'_n = \operatorname{ess\,sup}(F)$. To this end, it suffices to put $f'_0 = f_0$ and to take for f'_{n+1} a function in F dominating f'_n and f_{n+1}; we then have $\sup_{\mathbf{N}} f_n \leqslant \lim\uparrow_{\mathbf{N}} f'_n \leqslant \operatorname{ess\,sup}(F)$ a.s. which implies the desired equality. ■

REMARK. Although the essential supremum coincides with the supremum *modulo* null sets for countable families of functions, it is not the same for uncountable families. Thus on the real interval [0, 1] equipped with Lebesgue measure the family ($1_{\{a\}}$, $a \in A$) of functions, where A is an arbitrary subset of [0, 1], admits 0 as an essential supremum since $1_{\{a\}} = 0$ a.s. for every $a \in A$. On the other hand, the supremum $\sup_A 1_{\{a\}} = 1_A$ is not measurable if A is not a Borel set or is measurable but a.s. different from 0 if A is a Borel set of positive measure.

After these preliminaries we will establish the following crucial proposition. Recall that Λ_n denotes the set of all stopping times such that $n \leqslant \nu < \infty$ a.s. and $E(Z_\nu^-) < \infty$.

PROPOSITION VI-1-2. *For every adapted sequence (Z_n, $n \in \mathbf{N}$) of integrable r.r.v.'s such that* $\sup_{\mathbf{N}} Z_n^+ \in L^1$, *the r.v.'s* $X_n = \operatorname{ess\,sup}_{\Lambda_n} E^{\mathscr{B}_n}(Z_\nu)$ *form an adapted sequence of integrable r.r.v.'s satisfying the equalities*

$$X_n = \max(Z_n, E^{\mathscr{B}_n}(X_{n+1})) \qquad (n \in \mathbf{N}).$$

In particular, if the r.v.'s Z_n are positive, the sequence (X_n, $n \in \mathbf{N}$) is the smallest positive supermartingale dominating the sequence (Z_n, $n \in \mathbf{N}$). Finally we have

$$E(X_n) = \sup_{\nu \in \Lambda_n} E(Z_\nu) \quad \textit{for all } n \in \mathbf{N}.$$

PROOF. From their definition, the r.v.'s X_n are $\mathscr{B}_n$-measurable (or can always be chosen to be so) and satisfy the inequalities $Z_n \leqslant X_n$ and $X_n \leqslant E^{\mathscr{B}_n}(\sup_{p \geqslant n} Z_p^+)$, which imply that they are integrable.

On the other hand, for every fixed $n \in \mathbf{N}$, the family ($E^{\mathscr{B}_n}(Z_\nu)$, $\nu \in \Lambda_n$) of r.r.v.'s of which X_n is the essential supremum is closed under the operation sup; indeed it is easy to check that if ν_1 and $\nu_2 \in \Lambda_n$, then the same is true of the mapping ν defined by

$$\nu = 1_{B^c}\nu_1 + 1_B \nu_2 \quad \text{with } B = \{E^{\mathscr{B}_n}(Z_{\nu_1}) < E^{\mathscr{B}_n}(Z_{\nu_2})\}$$

and that

$$E^{\mathscr{B}_n}(Z_\nu) = 1_{B^c} E^{\mathscr{B}_n}(Z_{\nu_1}) + 1_B E^{\mathscr{B}_n}(Z_{\nu_2}) = \sup(E^{\mathscr{B}_n}(Z_{\nu_1}), E^{\mathscr{B}_n}(Z_{\nu_2})).$$

By Proposition VI-1-1 there exists a sequence $(\nu_k, k \in \mathbf{N})$ in Λ_n such that

$$\lim_k \uparrow E^{\mathscr{B}_n}(Z_{\nu_k}) = X_n \text{ a.s.}$$

As the sequence $(E^{\mathscr{B}_n}(Z_{\nu_k}), k \in \mathbf{N})$ of r.r.v.'s dominates the integrable r.r.v. $E^{\mathscr{B}_n}(Z_{\nu_0})$, by taking conditional expectations $E^{\mathscr{B}_{n-1}}$ of both sides we conclude that

$$E^{\mathscr{B}_{n-1}}(X_n) = \lim_k \uparrow E^{\mathscr{B}_{n-1}}(Z_{\nu_k}) \leqslant X_{n-1} \text{ a.s.},$$

at least if $n \geqslant 1$.

After having thus established that $X_n \geqslant \max(Z_n, E^{\mathscr{B}_n}(X_{n+1}))$ for all $n \in \mathbf{N}$, we establish the reverse inequality. To this end we write

$$Z_\nu = Z_n 1_{\{\nu=n\}} + Z_{\nu \vee (n+1)} 1_{\{\nu>n\}}$$

for all stopping times $\nu \in \Lambda_n$ and remark that $\nu \vee (n+1) \in \Lambda_{n+1}$; from the inequality

$$E^{\mathscr{B}_{n+1}}(Z_{\nu \vee (n+1)}) \leqslant X_{n+1}$$

we then deduce that

$$\begin{aligned} E^{\mathscr{B}_n}(Z_\nu) &= Z_n 1_{\{\nu=n\}} + E^{\mathscr{B}_n}(Z_{\nu \vee (n+1)}) 1_{\{\nu>n\}} \\ &\leqslant Z_n 1_{\{\nu=n\}} + E^{\mathscr{B}_n}(X_{n+1}) 1_{\{\nu>n\}} \\ &\leqslant \max(Z_n, E^{\mathscr{B}_n}(X_{n+1})) \quad \text{for all } \nu \in \Lambda_n. \end{aligned}$$

The r.v. X_n thus satisfies the desired inequality.

Since the family $(E^{\mathscr{B}_n}(Z_\nu), \nu \in \Lambda_n)$ is directed and increasing, it is easy to check that

$$E(\operatorname*{ess\,sup}_{\Lambda_n} E^{\mathscr{B}_n}(Z_\nu)) = \sup_{\Lambda_n} E(Z_\nu)$$

Indeed, choosing the sequence $(\nu_k, k \in \mathbf{N})$ as above, we have

$$E(\operatorname*{ess\,sup}_{\Lambda_n} E^{\mathscr{B}_n}(Z_\nu)) = \lim_k \uparrow E(E^{\mathscr{B}_n}(Z_{\nu_k})) = \lim_k \uparrow E(Z_{\nu_k}) \leqslant \sup_{\Lambda_n} E(Z_\nu);$$

but clearly $E(Z_\nu) \leqslant E(\operatorname{ess\,sup}_{\Lambda_n} E^{\mathscr{B}_n}(Z_\nu))$ for all $\nu \in \Lambda_n$.

If the r.v.'s Z_n are positive, the same is true of the X_n and it is clear that $(X_n, n \in \mathbf{N})$ is a positive supermartingale dominating the sequence $(Z_n, n \in \mathbf{N})$.

If $(X'_n, n \in \mathbf{N})$ is another positive supermartingale dominating the sequence $(Z_n, n \in \mathbf{N})$, Proposition II-**2**-13 shows that

$$X'_n \geqslant E^{\mathscr{B}_n}(X'_\nu) \geqslant E^{\mathscr{B}_n}(Z_\nu) \quad \text{if } \nu \in \Lambda_n$$

and it then follows that $X'_n \geqslant X_n$ a.s. $(n \in \mathbf{N})$. ■

REMARK. In general, if the sequence $(Z_n, n \in \mathbf{N})$ is not positive (or bounded below by an integrable r.v.), it is not true that $(X_n, n \in \mathbf{N})$ is the smallest integrable supermartingale dominating the sequence $(Z_n, n \in \mathbf{N})$; nonetheless Lemma VI-**1**-5 below allows it to be easily proved that $(X_n, n \in \mathbf{N})$ is the smallest integrable martingale dominating the sequence $(Z_n, n \in \mathbf{N})$ and such that $X_n \geqslant E^{\mathscr{B}_n}(X_\nu)$ if $\nu \in \Lambda_n$ $(n \in \mathbf{N})$. ■

The following proposition (due to Snell) then resolves the problem posed at the beginning of this section.

PROPOSITION VI-**1**-3. *Let $(Z_n, n \in \mathbf{N})$ be a sequence of integrable r.r.v.'s such that $E(\sup_{\mathbf{N}} Z_n^+) < \infty$. In order that the supremum $\sup_\nu E(Z_\nu)$ taken over the set of all finite stopping times (such that $E(Z_\nu^-) < \infty$) be attained, it is necessary and sufficient that the stopping time ν_0 defined in terms of the supermartingale $(X_n, n \in \mathbf{N})$ by*

$$\nu_0 = \begin{cases} \inf(n : X_n = Z_n), \\ +\infty & \text{if } X_n > Z_n \text{ for all } n \in \mathbf{N} \end{cases}$$

be a.s. finite. When this condition is satisfied, $E(Z_{\nu_0}) = \sup_\nu E(Z_\nu)$, and ν_0 is the smallest finite stopping time satisfying this equality.

Further, for all real $\varepsilon > 0$, the formula

$$\nu_\varepsilon = \inf(n : X_n < Z_n + \varepsilon)$$

always defines an a.s. finite stopping time such that $E(Z_{\nu_\varepsilon}) + \varepsilon \geqslant \sup_\nu E(Z_\nu)$.

In view of this result the stopping time ν_0 is said to be optimal when it is a.s. finite, and the finite stopping times ν_ε will be called ε-optimal. (Let us remark that the integrability hypothesis on the r.v. $\sup_{\mathbf{N}} Z_n^+$ implies that the supremum $\sup_\nu E(Z_\nu)$ is between $E(Z_0)$ and $E(\sup_{\mathbf{N}} Z_n^+)$ and hence finite.)

PROOF. (1) The essential points of this rather long proof are taken in the form of lemmas.

LEMMA VI-1-4. *For every stopping time ν dominated by ν_0, the stopped sequence $(X_{\nu\wedge n}, n \in \mathbf{N})$ is an integrable martingale. When ν is a.s. finite as well, it follows that $E(X_\nu) \geqslant E(X_0)$.*

PROOF. In fact, on the event $\{\nu_0 > n\}$ and thus *a fortiori* on the smaller event $\{\nu > n\}$, the inequality $X_n > Z_n$ is satisfied a.s.; consequently Proposition VI-1-2 implies that $X_n = E^{\mathscr{B}_n}(X_{n+1})$ on these events. As a result,

$$E^{\mathscr{B}_n}(X_{\nu\wedge(n+1)}) = X_\nu 1_{\{\nu\leqslant n\}} + E^{\mathscr{B}_n}(X_{n+1}) 1_{\{\nu>n\}}$$
$$= X_\nu 1_{\{\nu\leqslant n\}} + X_n 1_{\{\nu>n\}} = X_{\nu\wedge n}.$$

The sequence $(X_{\nu\wedge n}, n \in \mathbf{N})$ is therefore an integrable martingale and we then have $E(X_{\nu\wedge n}) = E(X_0)$ for all $n \in \mathbf{N}$. Since the sequence $(X_n, n \in \mathbf{N})$ is clearly dominated by the martingale $(U_n = E^{\mathscr{B}_n}(U), n \in \mathbf{N})$, where $U = \sup_{\mathbf{N}} Z_p^+ \in L^1$, Fatou's lemma applies to the positive sequence $(U_{\nu\wedge n} - X_{\nu\wedge n}, n \in \mathbf{N})$ and shows that

$$\varliminf_n E(U_{\nu\wedge n} - X_{\nu\wedge n}) \geqslant E(U_\nu - X_\nu)$$

if ν is a.s. finite; but $E(U_{\nu\wedge n}) = E(U_0) = E(U_\nu)$, so that the preceding inequality can be rewritten

$$E(X_\nu) \geqslant \varlimsup_n E(X_{\nu\wedge n}).$$

This completes the proof of the lemma. ■

It follows from this lemma that if the stopping time ν_0 is a.s. finite, we have

$$E(Z_{\nu_0}) = E(X_{\nu_0}) \geqslant E(X_0)$$

since $Z_{\nu_0} = X_{\nu_0}$ a.s. But $E(X_0) = \sup_{\Lambda_0} E(Z_\nu)$, where Λ_0 denotes the set of all a.s. finite stopping times, and consequently ν_0 is optimal in the sense that

$$E(Z_{\nu_0}) = \sup_{\Lambda_0} E(Z_\nu).$$

(2) To prove that ν_0 is a.s. finite (and thus optimal) whenever there exists a finite optimal stopping time ν^*, we will use the following lemma which is an extension of the definition of the X_n.

LEMMA VI-1-5. *For every finite stopping time ν_1 such that $E(Z_{\nu_1}^-) < \infty$, we have*

$$X_{\nu_1} = \operatorname*{ess\,sup}_{\nu \in \Lambda_{\nu_1}} E^{\mathscr{B}_{\nu_1}}(Z_\nu)$$

if Λ_{ν_1} *denotes the non-empty set of stopping times* ν *such that* $\nu_1 \leqslant \nu < \infty$ *a.s. and* $E(Z_\nu^-) < \infty$. *Consequently,*

$$E(X_{\nu_1}) = \sup_{\Lambda_{\nu_1}} E(Z_\nu).$$

PROOF. The set Λ_{ν_1} contains ν_1 and hence is non-empty. Let us denote by $X_{(\nu_1)}$ the essential supremum above. For every stopping time $\nu \in \Lambda_{\nu_1}$ the r.v. $\nu \vee n$ is a stopping time in Λ_n and $\nu = \nu \vee n$ on $\{\nu_1 = n\}$, so that on this event

$$E^{\mathscr{B}_{\nu_1}}(Z_\nu) = E^{\mathscr{B}_n}(Z_\nu) = E^{\mathscr{B}_n}(Z_{\nu \wedge n}) \leqslant X_n = X_{\nu_1};$$

it already follows that $X_{(\nu_1)} \leqslant X_{\nu_1}$ a.s. Conversely, if ν' is a stopping time belonging to Λ_n, the stopping time $\nu' \vee \nu_1$ belongs to Λ_{ν_1} and $\nu' = \nu' \vee \nu_1$ on $\{\nu_1 = n\}$, so that on this event

$$E^{\mathscr{B}_n}(Z_{\nu'}) = E^{\mathscr{B}_n}(Z_{\nu' \vee \nu_1}) = E^{\mathscr{B}_{\nu_1}}(Z_{\nu' \vee \nu_1}) \leqslant X_{(\nu_1)};$$

we have thus shown that $X_n \leqslant X_{(\nu_1)}$ a.s. on $\{\nu_1 = n\}$ and hence that $X_{\nu_1} \leqslant X_{(\nu_1)}$ a.s. The first part of the lemma is thus established.

The equality $E(X_{\nu_1}) = \sup_{\Lambda_{\nu_1}} E(Z_\nu)$ is proved as for a constant stopping time by showing to begin with that the family $(E^{\mathscr{B}_{\nu_1}}(Z_\nu), \nu \in \Lambda_{\nu_1})$ is directed upwards, which is done as in the case when ν_1 is constant.

If ν^* is a finite stopping time such that $E(Z_{\nu^*}) = \sup_{\Lambda_0} E(Z_\nu)$, the r.v. Z_{ν^*} is integrable, and the preceding lemma shows that

$$E(X_{\nu^*}) = \sup_{\Lambda_{\nu^*}} E(Z_\nu) \leqslant \sup_{\Lambda_0} E(Z_\nu) = E(Z_{\nu^*}).$$

Now this equality is only compatible with the inequality $Z_{\nu^*} \leqslant X_{\nu^*}$ if $Z_{\nu^*} = X_{\nu^*}$ a.s. But then the definition of the stopping time ν_0 implies that $\nu_0 \leqslant \nu^*$ and hence that ν_0 is finite a.s. ■

(3) In order to show the finiteness and ε-optimality of the stopping times ν_ε, we will use the following third and last lemma.

LEMMA VI-1-6. *The following relations are a.s. valid:*

$$\limsup_{n \to \infty} X_n = \limsup_{n \to \infty} Z_n \in [-\infty, \infty[.$$

PROOF. Indeed, the inequality $Z_\nu \leqslant \sup_{p \geqslant m} Z_p$, valid for every stopping time in Λ_m and thus in Λ_n if $n \geqslant m$, implies that

$$X_n = \operatorname{ess\,sup}_{\Lambda_n} E^{\mathscr{B}_n}(Z_\nu) \leqslant E^{\mathscr{B}_n}(\sup_{p \geqslant m} Z_p) \quad \text{if } n \geqslant m.$$

Since the r.v. $\sup_{p \geq m} Z_p$, which is bounded between Z_m and $\sup_{p \geq m} Z_p^+$, is integrable and $\mathscr{B}_\infty$-measurable, the preceding inequalities give

$$\limsup_{n \to \infty} X_n \leqslant \sup_{p \geqslant m} Z_p < + \infty \qquad (m \in \mathbf{N})$$

by the convergence Theorem II-**2**-11. Letting $m \uparrow \infty$, we obtain the inequality

$$\limsup_{n \to \infty} X_n \leqslant \limsup_{p \to \infty} Z_p < + \infty,$$

and the lemma is then an immediate consequence of the fact that $Z_n \leqslant X_n$ for all $n \in \mathbf{N}$. ■

Now let us consider the stopping time ν_ε defined in the statement of the proposition for any fixed $\varepsilon > 0$. Since $\nu_\varepsilon \leqslant \nu_0$, Lemma VI-**1**-4 shows that the sequence $(X_{\nu_\varepsilon \wedge n}, n \in \mathbf{N})$ is an integrable martingale. But the bound $X_n^+ \leqslant E^{\mathscr{B}_n}(U)$, where $U = \sup_p Z_p^+ \in L^1$, implies that $\sup_{\mathbf{N}} E(X_{\nu_\varepsilon \wedge n}^+) < \infty$; Theorem IV-**1**-2 then implies that the limit $\lim_{n \to \infty} X_n$ exists and is finite on the event $\{\nu_\varepsilon = +\infty\}$. But on this event, $X_n \geqslant Z_n + \varepsilon$ by definition of the stopping time ν_ε, and we therefore have

$$\lim_{n \to \infty} X_n \geqslant \limsup_{n \to \infty} Z_n + \varepsilon \quad \text{a.s.} \quad \text{on } \{\nu_\varepsilon = + \infty\}.$$

Since the limit $\lim_{n \to \infty} X_n$ is finite on $\{\nu_\varepsilon = + \infty\}$, this inequality is only compatible with the result of Lemma VI-**1**-6 above if the event $\{\nu_\varepsilon = +\infty\}$ is null.

We have thus shown that $\nu_\varepsilon < +\infty$ a.s. The inequality $X_{\nu_\varepsilon} \leqslant Z_{\nu_\varepsilon} + \varepsilon$ and Lemma VI-**1**-4 then imply that

$$\sup_{\Lambda_0} E(Z_\nu) = E(X_0) \leqslant E(X_{\nu_\varepsilon}) \leqslant E(Z_{\nu_\varepsilon}) + \varepsilon$$

and this finishes the proof of the proposition. ■

COROLLARY VI-**1**-7. *Let $(Z_n, n \in \mathbf{N})$ be a sequence of integrable r.r.v.'s such that $E(\sup_{\mathbf{N}} Z_n^+) < \infty$. If this sequence converges to $-\infty$ a.s. when $n \uparrow \infty$, then the supremum $\sup_\nu E(Z_\nu)$ taken over the set of all finite stopping times, is attained.*

PROOF. By an argument similar to that given in the third part of the proof of the proposition above, the sequence $(X_{\nu_0 \wedge n}, n \in \mathbf{N})$ is an integrable martingale satisfying $\sup_{\mathbf{N}} E(X_{\nu_0 \wedge n}^+) < \infty$. Consequently, the limit $\lim_{n \to \infty} X_n$ exists and is finite on $\{\nu_0 = +\infty\}$ by Theorem IV-**1**-1; but this contradicts either the hypothesis $\lim_{n \to \infty} Z_n = -\infty$ or the result of Lemma VI-**1**-6, at least when the event $\{\nu_0 = + \infty\}$ is not null. We have thus proved that $\nu_0 < \infty$ a.s. ■

We end this section by remarking that Snell's problem for a finite sequence $(Z_n, 0 \leqslant n \leqslant p)$ of integrable r.r.v.'s is much more simply resolved than in the case of an infinite sequence. Indeed, for every finite sequence $(Z_n, 0 \leqslant n \leqslant p)$ of integrable r.r.v.'s, it is easy to check that the sequence $(X_n, 0 \leqslant n \leqslant p)$ defined using backwards induction by the formulae $X_p = Z_p$ and

$$X_{p-m} = \max(Z_{p-m}, E^{\mathscr{B}_{p-m}}(X_{p-m+1})) \quad \text{for } 0 < m \leqslant p$$

defines the smallest integrable supermartingale dominating the sequence (Z_n). Since $X_p = Z_p$, the formula

$$\nu_0 = \min(n : 0 \leqslant n \leqslant p,\ X_n = Z_n)$$

then defines a stopping time on the entire space Ω such that

$$X_0 = E^{\mathscr{B}_0}(X_{\nu_0}) = E^{\mathscr{B}_0}(Z_{\nu_0})$$

because $(X_{\nu_0 \wedge n}, 0 \leqslant n \leqslant p)$ is a martingale; since on the other hand

$$X_0 \geqslant E^{\mathscr{B}_0}(X_\nu) \geqslant E^{\mathscr{B}_0}(Z_\nu)$$

for every stopping time $\nu : \Omega \to [0, p]$, it is quite clear that the stopping time ν_0 is optimal and that

$$X_0 = \operatorname*{ess\,sup}_{\nu} E^{\mathscr{B}_0}(Z_\nu).$$

By considering only sequences of r.v.'s which begin at the index n, we see in the same way that

$$X_n = \operatorname*{ess\,sup}_{\nu:\nu \geqslant n} E^{\mathscr{B}_n}(Z_\nu) \qquad (0 \leqslant n \leqslant p)$$

This is the solution of Snell's problem in the case of a finite sequence. Now let $p \uparrow \infty$.

Then let $(Z_n, n \in \mathbf{N})$ be an integrable sequence of r.r.v.'s such that $\sup_{\mathbf{N}} Z_n^+ \in L^1$. For every $p \in \mathbf{N}$ denote by $(X_n^p, 0 \leqslant n \leqslant p)$ the smallest supermartingale dominating the sequence $(Z_n, 0 \leqslant n \leqslant p)$; it is easy to see that $X_n^p \leqslant X_n^{p+1}$ if $0 \leqslant n \leqslant p$, which allows us to put $X_n^\infty = \lim\uparrow_p X_n^p$. Taking into account the inequalities $Z_n \leqslant X_n^\infty \leqslant E^{\mathscr{B}_n}(\sup_{\mathbf{N}} Z_p^+)$, it is not hard to show that $(X_n^\infty, n \in \mathbf{N})$ is the smallest integrable supermartingale dominating the sequence $(Z_n, n \in \mathbf{N})$, and that furthermore

$$X_n^\infty = \operatorname*{ess\,sup}_{\Lambda_{n,b}} E^{\mathscr{B}_n}(Z_\nu)$$

where $\Lambda_{n,b}$ denotes the set of all *bounded* stopping times $\nu \geqslant n$.

The supermartingale $(X_n, n \in \mathbf{N})$ of Proposition VI-1-2 thus dominates the supermartingale $(X_n^\infty, n \in \mathbf{N})$. When the sequence $(Z_n, n \in \mathbf{N})$ is positive (or more generally when it is bounded below by an integrable r.r.v.) these two supermartingales coincide; this follows directly from Fatou's inequality

$$E^{\mathscr{B}_n}(Z_\nu) \leqslant \liminf_k E^{\mathscr{B}_n}(Z_{\nu \wedge k}) \leqslant X_n^\infty \qquad (n \in \mathbf{N})$$

valid for every $\nu \in \Lambda_n$. On the other hand, the following example shows that when the sequence $(Z_n, n \in \mathbf{N})$ is not positive, the supermartingales $(X_n, n \in \mathbf{N})$ and $(X_n^\infty, n \in \mathbf{N})$ can be "very" different, and that the optimal stopping times for the finite sequences $(Z_n, 0 \leqslant n \leqslant p)$ need not bear the slightest relation to the optimal stopping times for the infinite sequence $(Z_n, n \in \mathbf{N})$!

EXAMPLE. Let $(Y_n, n \in \mathbf{N}^*)$ be an independent sequence of r.v.'s taking the values ± 1 with probabilities $\frac{1}{2}, \frac{1}{2}$; if $(c_n, n \in \mathbf{N})$ is a sequence of real numbers such that $0 = c_0 < c_1 < c_2 < \ldots < 1$, the formula

$$Z_n = \min(1, Y_1 + \ldots + Y_n) - c_n \qquad (n \in \mathbf{N}) \qquad (Z_0 = 0)$$

defines a supermartingale bounded above by 1; since the sequence $(c_n, n \in \mathbf{N})$ is strictly increasing, the inequalities $Z_n \geqslant E^{\mathscr{B}_n}(Z_{n+1})$ are even a.s. strict. It is therefore clear that $X_n^p = Z_n\ (0 \leqslant n \leqslant p)$, that $X_n^\infty = Z_n (n \in \mathbf{N})$, and that in the time interval $[0, p]$ the stopping time 0 is the unique optimal stopping time. However, the stopping time

$$\nu_0 = \inf(n : Y_1 + \ldots + Y_n = 1)$$

is a.s. finite, the r.v. $Z_{\nu_0} = 1 - c_{\nu_0}$ is strictly positive and such that $Z_n \leqslant Z_{\nu_0}$ a.s. for all $n \in \mathbf{N}$ (because $Z_n \leqslant 0$ if $n < \nu_0$ and $Z_n \leqslant 1 - c_n$ if $n > \nu_0$); thus the stopping time ν_0 is clearly optimal and we have

$$X_0 = \operatorname*{ess\,sup}_{\Lambda_0} E^{\mathscr{B}_0}(Z_\nu) = E^{\mathscr{B}_0}(Z_{\nu_0}) > 0$$

and $X_0^\infty = Z_0 = 0$ a.s. Further, if for all $p \in \mathbf{N}$, ν_p denotes the finite stopping time

$$\nu_p = \inf(n : n \leqslant p, Y_1 + \ldots + Y_n \geqslant 1),$$

it is not hard to show that $X_p = E^{\mathscr{B}_p}(Z_{\nu_p}) = E^{\mathscr{B}_p}(1 - c_{\nu_p})$ and that $\{X_p = Z_p\} = \{Y_1 + \ldots + Y_p \geqslant 1\}$, so that the supermartingales $(X_p^\infty = Z_p, p \in \mathbf{N})$ and $(X_p, p \in \mathbf{N})$ only agree at the times p at which $Y_1 + \ldots + Y_p \geqslant 1$. ■

VI-2. Application to Markov chains

We continue with the notations and definitions concerning Markov chains used in Section III-5. Given a Markov chain $(X_n, n \in \mathbf{N})$ we propose to maximise the expectation $\mathbf{E}_x(f(X_\nu))$ for an arbitrary initial state x, where f is an arbitrary positive function defined on the state space E. In this case the solution of the optimisation problem takes a particularly simple form involving the notion of smallest superharmonic majorant defined below.

PROPOSITION VI-2-8. *Given the canonical Markov chain $(X_n, n \in \mathbf{N})$ with state space E and transition matrix P, every positive function $f : E \to \mathbf{R}_+$ admits a smallest superharmonic majorant, say f^*. This function f^* can be constructed as the limit of the increasing sequence $(f_k, k \in \mathbf{N})$ of functions defined recursively by*

$$f_0 = f, \qquad f_{k+1} = \max(f, Pf_k) \quad (k \in \mathbf{N}),$$

and it satisfies the equality $f^* = \max(f, Pf^*)$.

When $\mathbf{E}_x(\sup_{\mathbf{N}} f(X_n)) < \infty$, the positive supermartingale $(U_n, n \in \mathbf{N})$ that Proposition VI-1-2 associates with the sequence $(f(X_n), n \in \mathbf{N})$ coincides with $(f^*(X_n), n \in \mathbf{N})$. Under this condition the supremum $\sup_\nu \mathbf{E}_x(f(X_\nu))$, which equals $f^*(x)$, is attained if and only if the hitting time of the chain $(X_n, n \in \mathbf{N})$ to the set $\{f = f^*\}$ is a.s. finite; in this case this hitting time is optimal. On the other hand, for all $\varepsilon > 0$ the hitting time of the chain $(X_n, n \in \mathbf{N})$ to the set $\{f^* < f + \varepsilon\}$ is a.s. finite and ε-optimal.

PROOF. The sequence $(f_k, k \in \mathbf{N})$ in the statement of the proposition is increasing. Indeed, we have $f_1 = \max(f, Pf_0) \geqslant f = f_0$ and proceeding inductively we see that if $f_k \geqslant f_{k-1}$, then

$$f_{k+1} = \max(f, Pf_k) \geqslant \max(f, Pf_{k-1}) = f_k \qquad (k \in \mathbf{N}).$$

The limit $f^* = \lim\uparrow_k f_k$ therefore satisfies the equality $f^* = \max(f, Pf^*)$ obtained by passing to the limit in the equations defining the f_k. The two inequalities $f^* \geqslant Pf^*$ and $f^* \geqslant f$ imply that the sequence $(f^*(X_n), n \in \mathbf{N})$ is a positive supermartingale (Lemma III-5-12) dominating the sequence $(f(X_n), n \in \mathbf{N})$.

Every superharmonic function g dominating f also dominates f^*. Indeed, the inequalities $g \geqslant Pg$ and $g \geqslant f_k$ for a fixed $k \in \mathbf{N}$ imply that $g \geqslant Pg \geqslant Pf_k$ and hence that $g \geqslant f_{k+1}$; the result then follows by arguing inductively on k. More generally, every supermartingale $(Y_n, n \in \mathbf{N})$ dominating the sequence

$(f(X_n), n \in \mathbf{N})$ necessarily dominates the supermartingale $(f^*(X_n), n \in \mathbf{N})$. In fact, the inequalities

$$Y_n \geqslant E^{\mathscr{B}_n}(Y_{n+1}), \qquad Y_n \geqslant f_k(X_n) \qquad (n \in \mathbf{N})$$

imply that $Y_n \geqslant E^{\mathscr{B}_n}(f_k(X_{n+1})) = Pf_k(X_n)$ and thus that $Y_n \geqslant f_{k+1}(X_n)$ $(n \in \mathbf{N})$; the result again follows by arguing inductively on k.

The rest of the proposition follows from Snell's theorem. ■

EXAMPLE. For every subset F of the state-space the hitting probability for F, i.e. the function $\phi_F(\cdot) = \mathbf{P}.(\nu_F < \infty)$ already studied in Proposition III-5-14, is the smallest superharmonic majorant of the indicator function 1_F; in fact, it is easily verified inductively that the functions f_k of the proposition equal $f_k = \mathbf{P}.(\nu_F \leqslant k)$ when $f_0 = 1_F$. It therefore follows that

$$\phi_F(x) = \sup_\nu \mathbf{P}_x(X_\nu \in F) \qquad (x \in E),$$

the supremum on the right-hand side being taken over all the a.s. finite stopping times.

Since $\{1_F = \phi_F\} = F + \{\phi_F = 0\}$ (because $\phi_F = 1$ on F), the stopping time ν_0 is the first entry time into the set $F + \{\phi_F = 0\}$; if this first entry time is a.s. finite, the time ν_0 is optimal for initial state x. Similarly the stopping times ν_ε are the first entry times to the sets $F + \{\phi_F \leqslant \varepsilon\}$; we know that they are a.s. finite and ε-optimal. Let us also note that $1_{\limsup_n \{X_n \in F\}} = \lim_n \phi_F(X_n)$ a.s. in this example by Lemma VI-1-6 and that this formula has already been proved in Proposition III-3-12; from this formula follows the property $\nu_\varepsilon < \infty$ a.s. ■

VI-3. Applications to random walks

Let $(Y_n, n \in \mathbf{N}^*)$ be an independent sequence of identically distributed centred r.r.v.'s such that $E(|Y_1| \log^+ |Y_1|) < \infty$ and let $(S_n = \sum_{m=1}^n Y_m, n \in \mathbf{N})$ be the integrable random walk with which it is associated. We intend to study Snell's problem for the sequence $(Z_n = S_n/n, n \in \mathbf{N}^*)$.

The integrability hypothesis made on Y_1 implies that $E(\sup_{\mathbf{N}} |Z_n|) < \infty$ and hence that $Z_\nu \in L^1$ for every stopping time $\nu : \Omega \to [1, \infty[$. Indeed, after we have recalled the canonical case where the Y_n are defined as coordinate mappings on the space $\Omega = \mathbf{R}^{\mathbf{N}^*}$, let us denote by $\mathscr{S}_n$ the σ-field of measurable subsets of $\mathbf{R}^{\mathbf{N}^*}$ invariant under the group of permutations of the first n coordinates; then $E^{\mathscr{S}_n}(Y_m)$ does not depend on m when $m \in [1, n]$ and therefore it is easy to check that

$$E^{\mathcal{S}_n}(Y_1) = E^{\mathcal{S}_n}\left(\frac{Y_1 + \dots + Y_n}{n}\right) = Z_n \qquad (n \in \mathbf{N}^*).$$

The σ-fields $\mathcal{S}_n$ decrease with n ($n \in \mathbf{N}^*$); the lines of the proof of Proposition IV-2-20 applied to the martingale $(Z_n, Z_{n-1}, \dots, Z_1, Z_1, Z_1, \dots)$ adapted to the increasing sequence $(\mathcal{S}_n, \mathcal{S}_{n-1}, \dots, \mathcal{S}_1, \mathcal{S}_1, \mathcal{S}_1, \dots)$ of σ-fields then show that

$$(1 - e^{-1})\, E(\sup_{1 \leqslant m \leqslant n} |Z_m|) \leqslant 1 + E(|Y_1| \log^+ |Y_1|)$$

since $Z_1 = Y_1$. Letting $n \uparrow \infty$ in this inequality, we find that $\sup_{\mathbf{N}^*} |Z_n|$ is integrable. (Davis has shown in [98] that conversely, if $E(|Y_1| \log^+ |Y_1|) = \infty$, there exists an a.s. finite stopping time ν such that $E(Z_\nu^+) = +\infty$.)

Next we note that Snell's problem for the sequence $(Z_n, n \in \mathbf{N}^*)$ is the same as that for the positive sequence $(Z_n^+, n \in \mathbf{N}^*)$. In fact, as the random walk $(S_n, n \in \mathbf{N})$ is recurrent, we know that

$$P(\limsup_n \{S_n \geqslant 0\}) = 1;$$

then for every a.s. finite stopping time ν, the formula

$$\tilde{\nu} = \inf(n : n \geqslant \nu, S_n \geqslant 0)$$

also defines an a.s. finite stopping time and we have $Z_{\tilde{\nu}} \geqslant Z_\nu^+$ because $\tilde{\nu} = \nu$ if $Z_\nu \geqslant 0$ and $Z_{\tilde{\nu}} \geqslant 0$ in any case. This implies that the supermartingales which Proposition VI-1-2 associates with $(Z_n, n \in \mathbf{N}^*)$ and $(Z_n^+, n \in \mathbf{N}^*)$ coincide; let us denote this supermartingale by $(X_n, n \in \mathbf{N}^*)$, which is hence the smallest *positive* supermartingale dominating the sequence $(Z_n, n \in \mathbf{N}^*)$.

Next let $f : \mathbf{N}^* \times \mathbf{R} \to \mathbf{R}_+$ be the function defined by

$$f(n, x) = \sup_\nu E\left(\frac{x + S_\nu}{n + \nu}\right) \qquad (n \in \mathbf{N}^*, x \in \mathbf{R}),$$

where the supremum is taken over all finite stopping times $\nu \geqslant 0$; the function f is positive since the first time $n \geqslant 1$ at which $x + S_n \geqslant 0$ is a finite stopping time. We leave to the reader the task of establishing that Snell's supermartingale is given by the formula

$$X_n = \operatorname*{ess\,sup}_{\nu : n \leqslant \nu < \infty} E^{\mathcal{B}_n}\left(\frac{S_\nu}{\nu}\right) = f(n, S_n) \qquad (n \in \mathbf{N}^*).$$

But the real-valued function $f(n, \cdot)$ is the upper envelope on $\mathbf{R}$ of the lines $x \to xE(1/(n+\nu)) + E(S_\nu/(n+\nu))$ of slope $\leqslant 1/n$; it is therefore clear that the set $\{x : f(n, x) = x/n\}$ is an interval of the form $[c_n, \infty[$ for some constant $c_n \in [0, \infty]$. We have thus found the stopping time ν_0 to be of the form

$$\nu_0 = \inf(n : n \geqslant 1, S_n \geqslant nc_n).$$

A detailed study of these constants c_n has been made by Y.S. Chow and H. Robbins [75], who have shown that ν_0 is finite a.s. and hence optimal.

VI-4. Another application

Let $(Y_n, n \in \mathbf{N}^*)$ be an independent and identically distributed sequence of r.r.v.'s with finite variance, and let $(M_n, n \in \mathbf{N}^*)$ be the increasing sequence of partial maxima of these random variables. If c denotes a real number > 0, we study Snell's problem for the sequence

$$Z_n = M_n - cn,$$

where $M_n = \max_{1 \leqslant m \leqslant n} Y_m$ $(n \in \mathbf{N}^*)$, of integrable r.r.v.'s. The condition $E(Y_1^2) < \infty$ implies that the r.r.v. $\sup_{\mathbf{N}^*} Z_n^+$ is integrable. In fact, we first note the equalities

$$\sup_{\mathbf{N}^*} Z_n = \sup_{\mathbf{N}^*} (M_n - cn) = \sup_{\mathbf{N}^*} (Y_n - cn)$$

(the second follows from $Y_n - cn \leqslant M_n - cn \leqslant \sup_{m \leqslant n}(Y_m - cm)$); these equalities allow us to write

$$E(\sup_{\mathbf{N}^*} Z_n^+) = E(\sup_{\mathbf{N}^*} (Y_n - cn)^+) \leqslant \sum_{\mathbf{N}^*} E((Y_n - cn)^+)$$

$$= \sum_{\mathbf{N}^*} E((Y_1 - cn)^+) \leqslant \frac{1}{2c} E((Y_1^+)^2),$$

the last inequality being a consequence of

$$\sum_{\mathbf{N}^*} (y - cn)^+ \leqslant \int_0^\infty (y - cu)^+ \, du = \frac{(y^+)^2}{2c}$$

for every $y \in \mathbf{R}$. The integrability of $\sup_{\mathbf{N}^*} Z_n^+$ is then established whenever

$$E((Y_1^+)^2) < \infty.$$

(It is also possible to prove that $E(\sup_{\mathbf{N}^*} Z_n^+) = \infty$ when $E((Y_1^+)^2) = \infty$.)

We next observe that $\lim_{n\to\infty} Z_n = -\infty$ a.s. Indeed the r.r.v. $\sup_{\mathbf{N}^*} Z_n^+$ being integrable and thus a.s. finite for every value of the constant $c > 0$, the inequality

$$Z_n = (M_n - \varepsilon n) - (c - \varepsilon) n \leqslant \sup_{\mathbf{N}^*} (M_p - \varepsilon p)^+ - (c - \varepsilon) n,$$

where we take $\varepsilon \in [0, c[$, implies that Z_n tends to $-\infty$ at least as fast as $-(c-\varepsilon)n$. Corollary VI-2-7 therefore implies that *Snell's stopping time* ν_0 *is a.s. finite and hence optimal.* We now go on to determine this stopping time.

We begin by showing that if $f : \mathbf{R} \to \mathbf{R}_+$ denotes the continuous decreasing positive function defined by $f(a) = E[(Y_1 - a)^+]$ for every $a \in \mathbf{R}$, then

$$E^{\mathscr{B}_n}(M_{n+1}) - M_n = f(M_n) \qquad (n \in \mathbf{N}^*)$$

Indeed, the independence and identical distribution of the terms of the sequence $(Y_n, n \in \mathbf{N}^*)$ implies firstly that $E^{\mathscr{B}_n}((Y_{n+1} - a)^+) = f(a)$ for all $a \in \mathbf{R}$; by initially considering the case of step r.v.'s it follows readily that

$$E^{\mathscr{B}_n}((Y_{n+1} - A)^+) = f(A)$$

for every $\mathscr{B}_n$-measurable and integrable r.r.v. A. But we have

$$(Y_{n+1} - M_n)^+ = \max(Y_{n+1}, M_n) - M_n = M_{n+1} - M_n \quad \text{for all } n \in \mathbf{N}^*$$

and the desired formula is therefore the particular case $A = M_n$ of the preceding formula. This formula and the definition of the Z_n imply that

$$E^{\mathscr{B}_n}(Z_{n+1}) - Z_n = f(M_n) - c \qquad (n \in \mathbf{N}^*).$$

But the sequence $(f(M_n), n \in \mathbf{N}^*)$ of positive r.r.v.'s decreases to 0 when $n \uparrow \infty$. Indeed, the fact that the sequence decreases is clear, whilst if we put

$$\sigma = \sup(y : P(Y_1 \leqslant y) \neq 1) \leqslant +\infty,$$

the sequence $(M_n, n \in \mathbf{N})$ increases a.s. to σ when $n \uparrow \infty$ (since $M_n \leqslant y$ a.s. if $y > \sigma$ and since $P(M_n \leqslant y) = [P(Y_1 \leqslant y)^n] \downarrow 0$ if $y < \sigma$); whereas the r.v. $(Y_1 - y)^+$ decreases to 0 when $y \uparrow \sigma$, so that by dominated convergence $f(y) \downarrow 0$ if $y \uparrow \sigma$. The formula

$$\nu^* = \inf(n : f(M_n) \leqslant c)$$

therefore defines an a.s. finite stopping time; let us observe that if $f(0) = E(Y_1^+)$ is bounded above by c, the stopping time is zero everywhere whilst if $f(0) > c$, it can be expressed more simply as $\nu^* = \inf(n : Y_n \geqslant \gamma)$, where γ is the solution of the equation $f(\gamma) = c$.

Relative to the stopping time ν^*, the sequence $(Z_n, n \in \mathbf{N})$ is such that

$$E^{\mathscr{B}_n}(Z_{n+1}) > Z_n \quad \text{on } \{n < \nu^*\},$$
$$E^{\mathscr{B}_n}(Z_{n+1}) \leqslant Z_n \quad \text{on } \{n \geqslant \nu^*\}.$$

It is then neither surprising nor difficult to show that $\nu^* = \nu_0$.

VI-5. Application to sequential statistical analysis

Let us recall the notations and assumptions at the end of Section **III-1**. Suppose also that the expectations $\int \nu \, dP$ and $\int \nu \, dQ$ are taken as measures of the cost of observing the Y_m up to the stopping time ν. The statistician whose strategy consists of choosing a finite stopping time ν and then an event $D \in \mathscr{B}_\nu$ for critical region wants to keep the costs $\int \nu \, dP$ and $\int \nu \, dQ$ and the error probabilities $P(D)$ and $Q(D^c)$ as small as possible; as these wishes conflict he can then try to minimise a linear combination with positive coefficients of these costs and error probabilities. Let

$$aP(D) + a' \int \nu \, dP + \rho[bQ(D^c) + b' \int \nu \, dQ]$$

be such a linear combination $(a, a', b, b', \rho > 0)$; the parameter ρ, whose introduction is not justified here since it is only a multiplier of b and b', will be intended to vary in the sequel while the other constants a, a', b, b' will be kept fixed.

In the preceding expression, ν denotes a finite stopping time and D an event in $\mathscr{B}_\nu$; by Section **III-1** we know that the infimum of this expression when ν is held fixed takes the value

$$a' \int \nu \, dP + \rho b' \int \nu \, dQ + \int \min(a, \rho b f_\nu) \, dP$$
$$= \int (a' \nu + \rho b' \nu f_\nu + \min(a, b f_\nu)) \, dP$$

since $Q = f_\nu \cdot P$ on $\mathscr{B}_\nu$ and ν is $\mathscr{B}_\nu$-measurable. After having put

$$Z_n^{(\rho)} = n(a' + \rho b' f_n) + \min(a, \rho b f_n) \qquad (n \in \mathbf{N}),$$

our problem reduces to minimising the expectation $E(Z_\nu^{(\rho)})$. It amounts to the same to maximise the expectation $E(-Z_\nu^{(\rho)})$ and the problem posed is therefor Snell's problem for the sequence $(-Z_\nu^{(\rho)}, n \in \mathbf{N})$ of negative integrable r.r.v.'s. Also, $Z_n^{(\rho)} > a'n \to +\infty$ as $n \to \infty$.

After changing signs, Proposition VI-**1**-3 and Corollary VI-**1**-7 hence tell us that there exists a largest positive submartingale $(X_n^{(\rho)}, n \in \mathbf{N})$ dominated by the sequence $(Z_n^{(\rho)}, n \in \mathbf{N})$ and such that the stopping time

$$\nu_0^{(\rho)} = \inf(n : Z_n^{(\rho)} = X_n^{(\rho)})$$

is a.s. finite and optimal: $E(Z_{\nu_0^{(\rho)}}^{(\rho)}) = \inf_\nu E(Z_\nu^{(\rho)})$. We now go on to determine the submartingale $(X_n^{(\rho)}, n \in \mathbf{N})$ explicitly.

To this end let us begin by remarking that $X_0^{(\rho)}$ is a constant r.v. since $\mathscr{B}_0 = \{\phi, \Omega\}$ and hence there exists a function $h : \mathbf{R}_+ \to \mathbf{R}_+$ such that $X_0^{(\rho)} = h(\rho)$ a.s. Next let n and p be two integers such that $n \geqslant p \geqslant 0$ and let us write

$$\begin{aligned} Z_n &= n(a' + \rho b' f_n) + \min(a, b\rho f_n) \\ &= p(a' + \rho b' f_n) + (n-p)\left[a' + \rho f_p b' \frac{f_n}{f_p}\right] + \min\left(a, b\rho f_p \frac{f_n}{f_p}\right); \end{aligned}$$

the sum of the second and third terms in the last part can also be obtained by replacing the parameter ρ in the expression for $Z_{n-p}^{(\rho)}$ by ρf_p and the r.v.'s $Y_1, \ldots, Y_{n-p}$ by $Y_{p+1}, \ldots, Y_n$; to see this let us note that

$$\frac{f_n}{f_p} = \prod_{m=p+1}^{n} g(Y_m).$$

Since

$$E^{\mathscr{B}_p}(Z_\nu) = p(a' + \rho b' f_p) + E^{\mathscr{B}_p}\left\{(\nu - p)\, a' + \rho f_p b' \frac{f_\nu}{f_p} + \min\left(a, b\rho f_p \frac{f_\nu}{f_p}\right)\right\}$$

if $\nu \in \Lambda_p$ (note that $E^{\mathscr{B}_p}(f_\nu) = f_p$ if ν is finite), it is then possible to show that

$$X_p^{(\rho)} = \operatorname*{ess\,inf}_{\Lambda_p} E^{\mathscr{B}_p}(Z_\nu^{(\rho)}) = p(a' + \rho b' f_p) + h(\rho f_p).$$

Let us therefore consider the subset I of R_+ defined by

$$I = \{\rho : h(\rho) < \min(a, b\rho)\};$$

the optimal stopping time $\nu_0^{(\rho)}$ is then given by

$$\nu_0^{(\rho)} = \inf(n : Z_n^{(\rho)} = X_n^{(\rho)}) = \inf(n : \rho f_n \notin I)$$

But the function h is increasing and convex, so that I is an open interval R_+; the optimal strategy of the statistician for minimising the expression

$$aP(D) + a' \int v \, dP + bQ(D^c) + b' \int v \, dQ$$

therefore consists of observing the Y_n until the stopping time $v_0^{(1)}$ defined as the first time n for which

$$f_n = \prod_{m=1}^{n} g(Y_m) \leqslant c \text{ or } \geqslant d \qquad (\text{ if } I =]c, d[\,)$$

and then deciding that P (resp. Q) is the probability law governing the observations according as the martingale f exceeds d (resp. is exceeded by c) at the time $v_0^{(1)}$. This test is due to Wald.

VI-6. A stochastic game

The following stochastic game is a slight modification of a game conceived by E. B. Dynkin [129].

On a probability space $(\Omega, \mathscr{A}, P)$ equipped with an increasing sequence $(\mathscr{B}_n, n \in \mathbf{N})$ of sub-σ-fields of $\mathscr{A}$, let us take two adapted sequences of r.r.v.'s, say $(U_n, n \in \mathbf{N})$ and $(V_n, n \in \mathbf{N})$ such that $V_n \leqslant U_n$ for all $n \in \mathbf{N}$; in what follows we will have to suppose that the r.v.'s $\sup_{\mathbf{N}} U_n^-$ and $\sup_{\mathbf{N}} V_n^+$ are integrable. Two players then agree "each to stop the processes" at a time of their choice, the game effectively stopping at the first of the two times chosen. If the game stops at time n, the first player pays the second a total amount (positive or negative!) equal to U_n or V_n according as the stop was decided by the first or by the second player; if the players decide to stop at the same moment n, the sum paid is U_n. Also, the players are allowed the possibility of not stopping the game; if neither of them stops the game, the sum to be paid is zero, which leads us to put $U_\infty = V_\infty = 0$ on Ω.

Each player has at his disposal only the information in $\mathscr{B}_n$ at time n ($n \in \mathbf{N}$), and so the strategies of the two players are their stopping times, say λ and μ, with values in $\overline{\mathbf{N}} = N \cup \{\infty\}$. The sum paid by the first player when the two players adopt the strategies λ and μ respectively equals

$$R(\lambda, \mu) = \begin{cases} U_\lambda & \text{on } \{\lambda \leqslant \mu\}, \\ V_\mu & \text{on } \{\lambda > \mu\}. \end{cases}$$

We will be interested in the expectation $E(R(\lambda, \mu))$ of this r.v., but for this we must suppose that the r.v. $R(\lambda, \mu)$ is integrable; in fact, in what follows we shall restrict λ (resp. μ) to belonging to the class Λ (resp. M) of stopping times for which the r.v. U_λ (resp. V_μ) is integrable. As moreover $U_\lambda^- \leqslant \sup_{\mathbf{N}} U_n^- \in L^1$ by hypothesis, the r.v. U_λ is integrable if and only if U_λ^+ is; similarly V_μ is integrable if and only if V_μ^- is integrable.

The aim of the first player (resp. second player) is to make the expectation $E(R(\lambda, \mu))$ as small (resp. as large) as possible. For a fixed strategy of the first player, the second player is interested in choosing a strategy λ which achieves, or at least approximates, the supremum $\sup_M E[R(\lambda, \mu)]$; if the first player is cautious, he will choose a strategy λ giving (or at least approximating) the infimum

$$\bar{x} = \inf_{\Lambda} \sup_{M} E(R(\lambda, \mu)).$$

This *minimax* ($=\inf\sup$!) strategy assures the first player of an expected loss of not more than $\bar{x}$ whatever the strategy μ adopted by the second player and is optimal from this point of view ($\bar{x}$ is the smallest possible). Reversing the roles of the two players, we also see that a *maximin* strategy, i.e. a strategy μ achieving the supremum

$$\underline{x} = \sup_{M} \inf_{\Lambda} E(R(\lambda, \mu)),$$

gives the second player an expected gain of at least equal to $\underline{x}$ regardless of the strategy adopted by the first player; such a strategy is optimal for the second player from this prudent viewpoint.

It is immediately seen from the definitions of these bounds that $\underline{x} \leqslant \bar{x}$. The aim of this section is to show that $\underline{x} = \bar{x}$ and to find optimal (or ε-optimal) strategies for the two players. To this end, we will study the double essential bounds

$$\begin{aligned} \bar{X}_n &= \operatorname*{ess\,inf}_{\lambda \in \Lambda_n} \operatorname*{ess\,sup}_{\mu \in M_n} E^{\mathscr{B}_n}(R(\lambda, \mu)), \\ \underline{X}_n &= \operatorname*{ess\,sup}_{\mu \in M_n} \operatorname*{ess\,inf}_{\lambda \in \Lambda_n} E^{\mathscr{B}_n}(R(\lambda, \mu)), \end{aligned} \qquad (n \in \mathbf{N}),$$

where Λ_n (resp. M_n) denotes the set of all stopping times λ (resp. μ) with values in $[n, \infty]$ such that $U_\lambda \in L^1$ (resp. $V_\mu \in L^1$); note that the stopping time identically equal to $+\infty$ belongs to both Λ_n and M_n so that these sets are non-empty. The preceding bounds generalise the quantities $\bar{x}$ and $\underline{x}$ introduced previously.

PROPOSITION VI-6-9. *If $(U_n, n \in \mathbf{N})$ and $(V_n, n \in \mathbf{N})$ are two adapted sequences of r.r.v.'s such that*

$$V_n \leqslant U_n \quad (n \in \mathbf{N}), \quad E(\sup_{\mathbf{N}} U_n^-) < \infty, \quad E(\sup_{\mathbf{N}} V_n^+) < \infty,$$

the r.r.v.'s $\overline{X}_n$ and $\underline{X}_n$ introduced above coincide for all $n \in \mathbf{N}$ and constitute the unique sequence, say $(X_n, n \in \mathbf{N})$, of r.r.v.'s simultaneously satisfying the equalities

$$X_n = \begin{cases} U_n & \text{if } E^{\mathscr{B}_n}(X_{n+1}) > U_n, \\ E^{\mathscr{B}_n}(X_n) & \text{if } V_n \leqslant E^{\mathscr{B}_n}(X_{n+1}) \leqslant U_n, \\ V_n & \text{if } E^{\mathscr{B}_n}(X_{n+1}) < V_n \end{cases} \qquad (n \in \mathbf{N})$$

and the inequalities

$$\tilde{U}_n \leqslant X_n \leqslant \tilde{V}_n \qquad (n \in \mathbf{N})$$

where $(\tilde{U}_n, n \in \mathbf{N})$ denotes the greatest negative submartingale dominated by the sequence $(U_n, n \in \mathbf{N})$ and $(\tilde{V}_n, n \in \mathbf{N})$ denotes the smallest positive supermartingale dominating the sequence $(V_n, n \in \mathbf{N})$.

Further, for every $\varepsilon > 0$ the stopping times λ_ε and μ_ε defined by

$$\lambda_\varepsilon = \inf(n : X_n \leqslant V_n + \varepsilon), \qquad \mu_\varepsilon = \inf(n : X_n \geqslant U_n - \varepsilon)$$

satisfy the ε-optimality inequalities

$$E^{\mathscr{B}_0}(R(\lambda_\varepsilon, \mu)) - \varepsilon \leqslant X_0 \leqslant E^{\mathscr{B}_0}(R(\lambda, \mu_\varepsilon)) + \varepsilon.$$

for every λ or μ.

PROOF. (1) If $\lambda \in \Lambda_n$ and $\mu \in M_n$, the r.v. $R(\lambda, \mu)$ is integrable as it is dominated in absolute value by the integrable r.v. $|U_\lambda| + |V_\mu|$.

For every $n \in \mathbf{N}$ and $\lambda \in \Lambda_n$ the family

$$(E^{\mathscr{B}_n}(R(\lambda, \mu)), \mu \in M_n)$$

of integrable r.v.'s is directed upwards and its essential supremum is also integrable. Indeed, on the one hand, if μ_1 and $\mu_2 \in M_n$, the formula

$$\mu = \begin{cases} \mu_1 & \text{if } E^{\mathscr{B}_n}(R(\lambda, \mu_1)) \geqslant E^{\mathscr{B}_n}(R(\lambda, \mu_2)), \\ \mu_2 & \text{if } E^{\mathscr{B}_n}(R(\lambda, \mu_1)) < E^{\mathscr{B}_n}(R(\lambda, \mu_2)) \end{cases}$$

defines a stopping time in M_n (because $\mu = \mu_1$ or μ_2 and $V_\mu = V_{\mu_1}$ or V_{μ_2}, so that $|V_\mu| \leqslant |V_{\mu_1}| + |V_{\mu_2}|$ everywhere) such that

$$E^{\mathscr{B}_n}(R(\lambda, \mu)) = \max(E^{\mathscr{B}_n}(R(\lambda, \mu_1)), E^{\mathscr{B}_n}(R(\lambda, \mu_2))).$$

On the other hand, the inequality $R(\lambda, \mu) \leqslant U_\lambda^+ + \sup_{\mathbf{N}} V_p^+$ valid at least for all $\mu \in M_n$ implies that

$$\operatorname*{ess\,sup}_{M_n} E^{\mathscr{B}_n}(R(\lambda, \mu)) \leqslant E^{\mathscr{B}_n}(U_\lambda^+) + E^{\mathscr{B}_n}(\sup_{\mathbf{N}} V_p^+)$$

and hence this essential supremum is integrable. The properties of the family $(E^{\mathscr{B}_n}[R(\lambda, \mu)], \mu \in M_n)$ that we have just proved imply in particular that for $n \geqslant 1$,

$$E^{\mathscr{B}_{n-1}}(\operatorname*{ess\,sup}_{M_n} E^{\mathscr{B}_n}(R(\lambda, \mu))) = \operatorname*{ess\,sup}_{M_n} E^{\mathscr{B}_{n-1}}(R(\lambda, \mu))$$

by the last part of Proposition VI-**1**-1.

For every $n \in \mathbf{N}$ the family of integrable r.v.'s $\operatorname{ess\,sup}_{M_n} E^{\mathscr{B}_n}(R(\lambda, \mu))$ as λ varies over Λ_n is a decreasing directed set and so the essential infimum, which we denote by $\bar{X}_n$, is again integrable. Indeed, the decreasing property is proved just as in the preceding paragraph; on the other hand, since for every $\lambda \in \Lambda_n$

$$\operatorname*{ess\,sup}_{M_n} E^{\mathscr{B}_n}(R(\lambda, \mu)) \geqslant E^{\mathscr{B}_n}(R(\lambda, +\infty)) = E^{\mathscr{B}_n}(U_\lambda),$$

we see that $\bar{X}_n \geqslant -E^{\mathscr{B}_n}(\sup_{\mathbf{N}} U_p^-)$, which assures the integrability of this r.v. As above, the property just proved implies by Proposition VI-**1**-1 that

$$E^{\mathscr{B}_{n-1}}(\bar{X}_n) = \operatorname*{ess\,inf}_{\Lambda_n} E^{\mathscr{B}_{n-1}}(\operatorname*{ess\,sup}_{M_n} E^{\mathscr{B}_n}(R(\lambda, \mu))).$$

Comparing this formula with the one obtained in the previous paragraph, we obtain the following important formula:

$$E^{\mathscr{B}_{n-1}}(\bar{X}_n) = \operatorname*{ess\,inf}_{\Lambda_n} \operatorname*{ess\,sup}_{M_n} E^{\mathscr{B}_{n-1}}(R(\lambda, \mu)).$$

(2) After these preliminaries we now show that the sequence $(\bar{X}_n, n \in \mathbf{N})$ of r.v.'s satisfies the equalities of the proposition. To this end, we will prove in turn that $V_n \leqslant \bar{X}_n \leqslant U_n$ and

$$\min(U_n, E^{\mathscr{B}_n}(\bar{X}_{n+1})) \leqslant \bar{X}_n \leqslant \max(V_n, E^{\mathscr{B}_n}(\bar{X}_{n+1}))$$

for every $n \in \mathbf{N}$.

Since we may not be able to consider the constant stopping time n which may not belong to Λ_n, for every real $a > 0$ let us put

$$\lambda_{n,a} = \begin{cases} n & \text{if } U_n \leqslant a, \\ \infty & \text{otherwise.} \end{cases}$$

This r.v. $\lambda_{n,a}$ is a stopping time $\geqslant n$ such that $U_{\lambda_{n,a}} \leqslant a$; it therefore belongs to Λ_n. But whatever $\mu \in M_n$, we have $E^{\mathscr{B}_n}(R(\lambda_{n,a}, \mu)) = U_n$ on the event $\{U_n \leqslant a\}$ (which belongs to $\mathscr{B}_n$); hence

$$\bar{X}_n \leqslant \operatorname*{ess\,sup}_{M_n} E^{\mathscr{B}_n}(R(\lambda_{n,a}, \mu)) = U_n \quad \text{on } \{U_n \leqslant a\}.$$

Letting $a \uparrow \infty$, we find that $\bar{X}_n \leqslant U_n$ on $\{U_n < \infty\}$ and hence everywhere. Similarly, for all real numbers $b < 0$ the formula

$$\mu_{n,b} = \begin{cases} n & \text{if } V_n \geqslant b, \\ \infty & \text{otherwise,} \end{cases}$$

defines a stopping time of M_n such that $E^{\mathscr{B}_n}(R(\lambda, \mu_{n,b})) \geqslant V_n$ on $\{V_n \geqslant b\}$ for every $\lambda \in \Lambda_n$; we conclude that

$$\bar{X}_n \geqslant \operatorname*{ess\,inf}_{\Lambda_n} E^{\mathscr{B}_n}(R(\lambda, \mu_{n,b})) \geqslant V_n \quad \text{on } \{V_n \geqslant b\}$$

and hence, letting $b \downarrow -\infty$, that $\bar{X}_n \geqslant V_n$ everywhere.

Let n be an integer $\geqslant 1$, and let λ be a stopping time of Λ_{n-1}. We associate with it a stopping time of Λ_n by putting

$$\lambda^* = \begin{cases} \lambda & \text{if } \lambda \geqslant n, \\ \infty & \text{if } \lambda = n-1. \end{cases}$$

For every $\mu \in M_{n-1}$ the r.v. $E^{\mathscr{B}_{n-1}}(R(\lambda, \mu))$ is equal to $E^{\mathscr{B}_{n-1}}(R(\lambda^*, \mu))$ if $\lambda \geqslant n$ and to U_{n-1} if $\lambda = n-1$ (because $\{\lambda = n-1\} \in \mathscr{B}_{n-1}$); hence

$$\operatorname*{ess\,sup}_{M_{n-1}} E^{\mathscr{B}_{n-1}}(R(\lambda, \mu)) \begin{cases} \geqslant \operatorname*{ess\,sup}_{M_{n-1}} E^{\mathscr{B}_{n-1}}(R(\lambda^*, \mu)) & \text{if } \lambda \geqslant n \\ = U_{n-1} & \text{if } \lambda = n-1. \end{cases}$$

But taking the formula derived in the first part of this proof into account, the right-hand side of the inequality is greater than $\min(E^{\mathscr{B}_{n-1}}(\bar{X}_n), U_{n-1})$. It then follows that

$$\bar{X}_{n-1} = \operatorname*{ess\,inf}_{\Lambda_{n-1}} \operatorname*{ess\,sup}_{M_{n-1}} E^{\mathscr{B}_{n-1}}(R(\lambda, \mu)) \geqslant \min(E^{\mathscr{B}_{n-1}}(\bar{X}_n), U_{n-1}) \quad \text{if } n \geqslant 1.$$

Finally, let n be an integer $\geqslant 1$ and let μ be a stopping time of M_{n-1} which we will associate with the stopping time μ^* of M_n defined by

$$\mu^* = \begin{cases} \mu & \text{if } \mu \geqslant n, \\ +\infty & \text{if } \mu = n-1. \end{cases}$$

For every $\lambda \in \Lambda_n$, the r.v. $E^{\mathscr{B}_{n-1}}(R(\lambda, \mu))$ is then equal to $E^{\mathscr{B}_{n-1}}(R(\lambda, \mu^*))$ if $\mu \geqslant n$ and to V_{n-1} if $\mu = n-1$. It then follows easily that

$$\operatorname*{ess\,sup}_{M_{n-1}} E^{\mathscr{B}_{n-1}}(R(\lambda, \mu)) \leqslant \max(V_{n-1}, \operatorname*{ess\,sup}_{M_n} E^{\mathscr{B}_{n-1}}(R(\lambda, \mu^*)))$$

for every $\lambda \in \Lambda_n$; consequently,

$$\bar{X}_{n-1} \leqslant \operatorname*{ess\,inf}_{\Lambda_n} \operatorname*{ess\,sup}_{M_{n-1}} E^{\mathscr{B}_{n-1}}(R(\lambda, \mu^*))$$

$$\leqslant \max(V_{n-1}, \operatorname*{ess\,inf}_{\Lambda_n} \operatorname*{ess\,sup}_{M_n} E^{\mathscr{B}_{n-1}}(R(\lambda, \mu^*))) = \max(V_{n-1}, E^{\mathscr{B}_{n-1}}(\bar{X}_n))$$

by the formula in the first part of the proof.

(3) It is easy to check that $\tilde{U}_n \leqslant \bar{X}_n \leqslant \tilde{V}_n$ for every $n \in \mathbf{N}$. Indeed, the definition of the $\bar{X}_n$ $(n \in \mathbf{N})$ implies that

$$\operatorname*{ess\,inf}_{\Lambda_n} E^{\mathscr{B}_n}(R(\lambda, \infty)) \leqslant \bar{X}_n \leqslant \operatorname*{ess\,sup}_{M_n} E^{\mathscr{B}_n}(R(\infty, \mu))$$

which can be written more simply as

$$\operatorname*{ess\,inf}_{\Lambda_n} E^{\mathscr{B}_n}(U_\lambda) \leqslant \bar{X}_n \leqslant \operatorname*{ess\,sup}_{M_n} E^{\mathscr{B}_n}(V_\mu).$$

But we have $E^{\mathscr{B}_n}(U_\lambda) \geqslant E^{\mathscr{B}_n}(\tilde{U}_\lambda) \geqslant U_n$ if $\lambda \in \Lambda_n$ and $E^{\mathscr{B}_n}(V_\mu) \leqslant E^{\mathscr{B}_n}(\tilde{V}_\mu) \leqslant \tilde{V}_n$ if $\mu \in M_n$ $(n \in \mathbf{N})$ by Proposition II-2-13.

(4) We have shown above that the sequence $(\bar{X}_n, n \in \mathbf{N})$ satisfies the equalities and inequalities in the statement of the proposition. Now we show that conversely every sequence $(X_n, n \in \mathbf{N})$ satisfying these relations is such that $\bar{X}_n \leqslant X_n \leqslant \underline{X}_n$ for every $n \in \mathbf{N}$; as the definitions imply that $\underline{X}_n \leqslant \bar{X}_n$ $(n \in \mathbf{N})$, this will

show simultaneously that $\overline{X}_n = \underline{X}_n$ for every $n \in \mathbf{N}$ and that these r.v.'s constitute the unique sequence having the properties of the proposition.

First we note that the inequalities $\tilde{U}_n \leqslant X_n \leqslant \tilde{V}_n$ $(n \in \mathbf{N})$ imply that

$$\min(\liminf_n U_n, 0) \leqslant \liminf_n X_n \leqslant \limsup_n X_n \leqslant \max(\limsup_n V_n, 0).$$

Indeed, considering for example the last inequality, the sequence

$$(E^{\mathscr{B}_n}(\sup_{p \geqslant n} V_p^+), n \in \mathbf{N})$$

is a positive supermartingale dominating the sequence $(V_n, n \in \mathbf{N})$ so that

$$X_n \leqslant \tilde{V}_n \leqslant E^{\mathscr{B}_n}(\sup_{p \geqslant n} V_p^+) \quad \text{for every } n \in \mathbf{N};$$

consequently, taking lim sup shows that

$$\limsup_n X_n \leqslant \limsup_n E^{\mathscr{B}_n}(\sup_{p \geqslant n} V_p^+) \leqslant \inf_q \lim_n E^{\mathscr{B}_n}(\sup_{p \geqslant q} V_p^+) = \lim_p \sup V_p^+.$$

For every $\varepsilon > 0$ and every $n \in \mathbf{N}$, let us introduce the stopping times

$$\mu_\varepsilon^n = \inf(p : p \geqslant n, X_p < V_p + \varepsilon) \text{ or } +\infty.$$

The hypotheses then imply that $X_p \leqslant E^{\mathscr{B}_p}(X_{p+1})$ if $n \leqslant p < \mu_\varepsilon^n$; we deduce that

$$X_n \leqslant E^{\mathscr{B}_n}(X_{\lambda \wedge \mu_\varepsilon^n}) \quad \text{for every } \lambda \in \Lambda_n,$$

where by definition $X_{\lambda \wedge \mu_\varepsilon^n} = \lim_{p \to \infty} X_p$ on $\{\lambda \wedge \mu_\varepsilon^n = \infty\}$. The correctness of this inequality is assured by the fact that the sequence $(X_n, n \in \mathbf{N})$ is bounded by two uniformly integrable martingales

$$-E^{\mathscr{B}_n}(\sup_p U_p^-) \leqslant X_n \leqslant E^{\mathscr{B}_n}(\sup_p V_p^+) \qquad (n \in \mathbf{N}),$$

so that the sequence $(X_{p \wedge \mu_\varepsilon^n}, p \geqslant n)$ is a regular submartingale.

We will show that

$$X_{\lambda \wedge \mu_\varepsilon^n} \leqslant R(\lambda, \mu_\varepsilon^n) + \varepsilon \quad \text{if } \lambda \in \Lambda_n, \qquad n \in \mathbf{N}.$$

Indeed, on the event $\{\lambda \leqslant \mu_\varepsilon^n, \lambda < \infty\}$ we have $R(\lambda, \mu_\varepsilon^n) = U_\lambda \geqslant X_\lambda$, whilst on the event $\{\mu_\varepsilon^n < \lambda\}$ we have $R(\lambda, \mu_\varepsilon^n) = V_{\mu_\varepsilon^n} \geqslant X_{\mu_\varepsilon^n} - \varepsilon$ by the definition of μ_ε^n.

Lastly, on $\{\lambda=\mu_\varepsilon^n=\infty\}$ we have $R(\lambda,\mu_\varepsilon^n)=0$ and $X_{\lambda\wedge\mu_\varepsilon^n}=\lim_p X_p$; but on the event $\{\mu_\varepsilon^n=\infty\}=\bigcap_{p\geqslant n}\{X_p\geqslant V_p+\varepsilon\}$, we clearly have

$$\limsup_p X_p \geqslant \limsup_p V_p+\varepsilon$$

and this is only compatible with the inequality $\limsup_p X_p \leqslant (\limsup_p V_p)^+$ proved above if $\limsup_p V_p \leqslant 0$ (since $(\limsup_p V_p)^+$ is integrable and so finite); hence we see that $\limsup_p X_p \leqslant 0$ on $\{\mu_\varepsilon^n=\infty\}$ and as a result,

$$X_{\lambda\wedge\mu_\varepsilon^n}\leqslant R(\lambda,\mu_\varepsilon^n)\quad \text{on}\quad \{\lambda\wedge\mu_\varepsilon^n=\infty\}.$$

We have proved that

$$X_n\leqslant E^{\mathscr{B}_n}(X_{\lambda\wedge\mu_\varepsilon^n})\leqslant E^{\mathscr{B}_n}(R(\lambda,\mu_\varepsilon^n))+\varepsilon$$

for every $\lambda\in\Lambda_n$, every $\varepsilon>0$ and every $n\in\mathbf{N}$. It clearly follows that

$$X_n\leqslant \operatorname*{ess\,inf}_{\Lambda_n} E^{\mathscr{B}_n}(R(\lambda,\mu_\varepsilon^n))+\varepsilon\leqslant \underline{X}_n+\varepsilon\qquad(\varepsilon>0)$$

and hence $X_n\leqslant\underline{X}_n$ for all $n\in\mathbf{N}$. The proof of the inequalities $X_n\geqslant\overline{X}_n$ $(n\in\mathbf{N})$ is given entirely symmetrically using the stopping times

$$\lambda_n^\varepsilon=\inf(p:p\geqslant n, X_p>U_p-\varepsilon)\quad\text{or}\quad=+\infty\qquad(\varepsilon>0,n\in\mathbf{N}).$$

Finally we note that the properties of the stopping times $\lambda_\varepsilon=\lambda_\varepsilon^0$ and $\mu_\varepsilon=\mu_\varepsilon^0$ stated in the last part of the proposition have been established above. ■

CHAPTER VII

DOOB'S DECOMPOSITION OF SUBMARTINGALES AND ITS APPLICATION TO SQUARE-INTEGRABLE MARTINGALES

VII-1. Generalities

The introduction of the notion of increasing process allows us, in this and the following section, to effect decompositions of sub- and super- martingales which are very important in both theory and practice. As before, we take once and for all a probability space $(\Omega, \mathscr{A}, P)$ and an increasing sequence $(\mathscr{B}_n, n \in \mathbf{N})$ of sub-σ-fields of $\mathscr{A}$.

Definition VII-1-1. A sequence $(U_n, n \in \mathbf{N})$ of r.r.v.'s is said to be *predictable* if the r.v. U_0 is $\mathscr{B}_0$-measurable and if for all $n \in \mathbf{N}$ the r.v. U_{n+1} is $\mathscr{B}_n$-measurable. An *increasing process* is defined as a predictable sequence $(A_n, n \in \mathbf{N})$ of finite r.r.v.'s such that

$$0 = A_0 \leqslant A_1 \leqslant A_2 \leqslant \dots \quad \text{a.s.} \quad \text{on } \Omega.$$

It is important to note that for a predictable sequence and in particular for an increasing process, not only the sequence $(U_n, n \in \mathbf{N})$ but also the sequence $(U_{n+1}, n \in \mathbf{N})$ is adapted to the sequence $(\mathscr{B}_n, n \in \mathbf{N})$ of σ-fields. We will soon see the importance of predictable sequences; the interest of increasing processes largely depends on Doob's decomposition theorem stated in the first part of the following proposition. (The second part of this proposition will not be used in the sequel.)

PROPOSITION VIII-1-2. (1) *Every integrable submartingale can be written in a unique way as the sum of an integrable martingale* $(M_n, n \in \mathbf{N})$ *and an increasing process* $(A_n, n \in \mathbf{N})$, *say*

$$X_n = M_n + A_n \qquad (n \in \mathbf{N}).$$

(2) *Further, the condition* $\sup_{\mathbf{N}} E(X_n^+) < \infty$ *(which suffices to ensure the a.s. convergence of the submartingale) is equivalent to the conjunction of the two conditions*

$$\sup_{\mathbf{N}} E(|M_n|) < \infty \quad \text{and} \quad A_\infty \in L^1,$$

whilst the convergence in L^1 *of the submartingale* $(X_n, n \in \mathbf{N})$ *is equivalent to the regularity of the martingale* $(M_n, n \in \mathbf{N})$ *together with the condition* $A_\infty \in L^1$.

For every stopping time ν *regular for the martingale* $(M_n, n \in \mathbf{N})$, *the r.v.* X_ν *is integrable if and only if* $E(A_\nu) < \infty$, *and then*

$$E(X_\nu) = E(M_0) + E(A_\nu).$$

PROOF. (1) The r.r.v.'s M_n and A_n $(n \in \mathbf{N})$ will be defined through their differences by the formulae

$$M_0 = X_0, \quad M_{n+1} - M_n = X_{n+1} - E^{\mathscr{B}_n}(X_{n+1}),$$

$$A_0 = 0, \qquad A_{n+1} - A_n = E^{\mathscr{B}_n}(X_{n+1}) - X_n.$$

These formulae show immediately that the sequence $(M_n, n \in \mathbf{N})$ is an integrable martingale and that the sequence $(A_n, n \in \mathbf{N})$ is an increasing process; since $M_0 + A_0 = X_0$ and the increments of the two sequences

$$(M_n + A_n, n \in \mathbf{N}) \quad \text{and} \quad (X_n, n \in \mathbf{N})$$

coincide, it is clear that $X_n = M_n + A_n$ for every $n \in \mathbf{N}$.

The uniqueness of the Doob decomposition is easy to establish. Indeed, if $X_n = M'_n + A'_n$ $(n \in \mathbf{N})$ is a decomposition of the submartingale $(X_n, n \in \mathbf{N})$ as the sum of a martingale $(M'_n, n \in \mathbf{N})$ and an increasing process $(A'_n, n \in \mathbf{N})$, the equality of the increments

$$A'_{n+1} - A'_n = (X_{n+1} - X_n) - (M'_{n+1} - M'_n) \qquad (n \in \mathbf{N})$$

implies, upon taking the conditional expectation $E^{\mathscr{B}_n}$ of both sides and using the assumptions, that

$$A'_{n+1} - A'_n = E^{\mathscr{B}_n}(X_{n+1}) - X_n = A_{n+1} - A_n.$$

It follows that $A'_n = A_n$ for all $n \in \mathbf{N}$ since $A'_0 = A_0 = 0$, and that

$$M'_n = X_n - A'_n = X_n - A_n = M_n \qquad (n \in \mathbf{N}).$$

(2) The decomposition formula implies that $X_n^+ = (M_n + A_n)^+ \leqslant M_n^+ + A_n$ for every $n \in \mathbf{N}$ and hence that

$$\sup_{\mathbf{N}} E(X_n^+) \leqslant \sup_{\mathbf{N}} E(M_n^+) + E(A_\infty);$$

the left-hand side of this inequality is therefore finite if both terms on the right-hand side are finite. On the other hand, the inequality $X_n \geqslant M_n$ ($n \in \mathbf{N}$) which follows from the positivity of the A_n implies that

$$\sup_{\mathbf{N}} E(M_n^+) \leqslant \sup_{\mathbf{N}} E(X_n^+),$$

whilst the relations $A_n = X_n - M_n \leqslant X_n^+ - M_n$ ($n \in \mathbf{N}$) imply that

$$E(A_\infty) \leqslant \sup_{\mathbf{N}} E(X_n^+) - E(M_0)$$

since $E(M_n) = E(M_0)$ for all $n \in \mathbf{N}$. We have thus proved the equivalence

$$\sup_{\mathbf{N}} E(X_n^+) < \infty \Leftrightarrow \sup_{\mathbf{N}} E(M_n^+) < \infty \text{ and } E(A_\infty) < \infty$$

and we know that the two conditions $\sup_{\mathbf{N}} E(M_n^+) < \infty$ and $\sup_{\mathbf{N}} E(|M_n|) < \infty$ are equivalent for every integrable martingale.

If $A_\infty \in L^1$, the dominated convergence theorem implies that $A_n \to A_\infty$ in L^1 when $n \uparrow \infty$. The regularity of the martingale $(M_n, n \in \mathbf{N})$, i.e. its convergence in L^1, and the condition $A_\infty \in L^1$ therefore imply that the submartingale $(X_n, n \in \mathbf{N})$ converges in L^1. Conversely, if the submartingale $(X_n, n \in \mathbf{N})$ converges in L^1 to a r.v. X_∞, we can write

$$E(A_\infty) = \lim_n \uparrow E(A_n) = \lim_n \uparrow (E(X_n) - E(M_n)) = E(X_\infty) - E(M_0) < +\infty,$$

which shows that $A_\infty \in L^1$ and hence that $A_n \to A_\infty$ in L^1. But then the martingale $(M_n = X_n - A_n, n \in \mathbf{N})$ also converges in L^1. The second part of the proposition is thus proved.

If ν is a stopping time regular for the martingale $(M_n, n \in \mathbf{N})$, the r.v. M_ν (equal to $\lim_{n\to\infty} M_n$ a.s. on $\{\nu = \infty\}$) exists a.s., is integrable and by Proposition IV-3-12 satisfies $E(M_\nu) = E(M_0) = E(X_0)$. Consequently, the r.v. X_ν (equal to $\lim_{n\to\infty} X_n$ a.s. on $\{\nu = \infty\}$) exists and equals $M_\nu + A_\nu$ a.s.; hence it is integrable if and only if A_ν is, and when it is integrable, it satisfies $E(X_\nu) = E(M_\nu) + E(A_\nu)$. The proposition is thus completely proved. ■

REMARK. Let $(X_n = M_n + A_n, n \in \mathbf{N})$ be the Doob decomposition of an integrable submartingale $(X_n, n \in \mathbf{N})$. Then the condition $\sup_{\mathbf{N}} E(|M_n|) < \infty$, which by the preceding proposition is weaker than $\sup_{\mathbf{N}} E(X_n^+) < \infty$, already suffices to imply the a.s. convergence of the submartingale $(X_n = M_n + A_n, n \in \mathbf{N})$ in $]-\infty, +\infty]$ since it implies that the martingale $(M_n, n \in \mathbf{N})$ converges a.s. to a finite limit and since $A_n \uparrow A_\infty \leqslant +\infty$ when $n \uparrow +\infty$. But this observation is not very useful! ■

VII-2. Asymptotic behaviour of a square-integrable martingale

For every square-integrable martingale $(X_n, n \in \mathbf{N})$ the sequence $(X_n^2, n \in \mathbf{N})$ is a positive integrable submartingale since Schwartz's inequality for conditional expectations shows that

$$E^{\mathscr{B}_n}(X_{n+1}^2) \geqslant [E^{\mathscr{B}_n}(X_{n+1})]^2 = X_n^2 \qquad (n \in \mathbf{N}).$$

This section is devoted to a study of the martingale $(X_n, n \in \mathbf{N})$ through the increasing process $(A_n, n \in \mathbf{N})$ obtained from Doob's decomposition

$$X_n^2 = M_n + A_n \qquad (n \in \mathbf{N})$$

of the submartingale $(X_n^2, n \in \mathbf{N})$. Note that by definition this increasing process is given by the formula

$$A_{n+1} - A_n = E^{\mathscr{B}_n}(X_{n+1}^2) - X_n^2 \qquad (n \in \mathbf{N});$$

by the "conditional variance" identity,

$$E^{\mathscr{B}}((Y - E^{\mathscr{B}}(Y))^2) = E^{\mathscr{B}}(Y^2) - [E^{\mathscr{B}}(Y)]^2,$$

this formula can also be written in the very useful form

$$A_{n+1} - A_n = E^{\mathscr{B}_n}((X_{n+1} - X_n)^2) \qquad (n \in \mathbf{N}).$$

In the case where the martingale is formed by the partial sums $X_n = \sum_{m=1}^n Y_m$ $(n \in \mathbf{N})$ of a sequence $(Y_n, n \in \mathbf{N}^*)$ of independent r.v.'s with zero mean and finite variance, the increasing process reduces to the increasing sequence $(a_n, n \in \mathbf{N})$ of real numbers defined by

$$a_n = \sum_{m=1}^{n} E(Y_m^2) = E(X_n^2) \qquad (n \in \mathbf{N}).$$

The proposition below simplifies considerably in this case; nevertheless it establishes the following two results:

(a) the sequence $(X_n = \sum_{m=1}^n Y_m, n \in \mathbf{N})$ converges a.s. and in L^2 if $\sum_{\mathbf{N}} E(Y_n^2) < \infty$,

(b) the stopping time ν is regular for the martingale $(X_n, n \in \mathbf{N})$ whenever $E(\sqrt{a_\nu}) < \infty$.

Let us return to the general case. Replacing the martingale $(X_n, n \in \mathbf{N})$ by the martingale $(X_n - X_0, n \in \mathbf{N})$, which does not change the increasing process $(A_n, n \in \mathbf{N})$, without loss of generality we can suppose in what follows that $X_0 = 0$.

PROPOSITION VII-**2**-3. *If $(X_n, n \in \mathbf{N})$ is a square-integrable martingale such that $X_0 = 0$ and $(A_n, n \in \mathbf{N})$ denotes the increasing process associated with the submartingale $(X_n^2, n \in \mathbf{N})$ by the Doob decomposition, then:*

(a) *if $E(A_\infty) < \infty$, the martingale $(X_n, n \in \mathbf{N})$ converges in L^2 and is therefore regular; further, $E(\sup_{\mathbf{N}} X_n^2) \leqslant 4E(A_\infty)$;*

(b) *if $E(\sqrt{A_\infty}) < \infty$, the martingale $(X_n, n \in \mathbf{N})$ is regular and such that $E(\sup_{\mathbf{N}} |X_n|) \leqslant 3E(\sqrt{A_\infty})$; more generally, a stopping time v is regular for $(X_n, n \in \mathbf{N})$ whenever $E(\sqrt{A_v}) < \infty$ and then $E(\sup_{n \leqslant v} |X_n|) \leqslant 3E(\sqrt{A_v})$;*

(c) *in every case the martingale $(X_n, n \in \mathbf{N})$ converges a.s. to a finite limit on the event $\{A_\infty < \infty\}$.*

PROOF. Since $M_0 = X_0 = 0$, the martingale $(M_n, n \in \mathbf{N})$ in Doob's decomposition of $(X_n^2, n \in \mathbf{N})$ is centred and the identity $E(X_n^2) = E(A_n)$ $(n \in \mathbf{N})$ is therefore valid; it implies that

$$\sup_{\mathbf{N}} E(X_n^2) = E(A_\infty).$$

The first part of the proposition is therefore only a reformulation in terms of the increasing process of Propositions IV-**2**-7 and IV-**2**-8 in the case $p = 2$.

Let us remark next that for every stopping time v, the stopped increasing process $(A_{v \wedge n}, n \in \mathbf{N})$ is the same as the one which by Doob's decomposition is associated with the square of the stopped martingale $(X_{v \wedge n}, n \in \mathbf{N})$. Indeed, for every $n \in \mathbf{N}$ we have

$$\begin{aligned} E^{\mathscr{B}_n}((X_{v \wedge (n+1)} - X_{v \wedge n})^2) &= E^{\mathscr{B}_n}(1_{\{v > n\}}(X_{n+1} - X_n)^2) \\ &= 1_{\{v > n\}}(A_{n+1} - A_n) \\ &= A_{v \wedge (n+1)} - A_{v \wedge n}. \end{aligned}$$

We apply this remark to the stopping time v_a ($a > 0$ real) defined by

$$v_a = \begin{cases} \min(n : A_{n+1} > a^2) & \\ \infty & \text{if } A_\infty \leqslant a^2; \end{cases}$$

note that the r.v. v_a is indeed a stopping time since the sequence $(A_{n+1}, n \in \mathbf{N})$ is adapted. As $A_{v_a} \leqslant a^2$, the first part of the proposition shows that the stopping time ν_a is regular for the martingale $(X_n, n \in \mathbf{N})$; consequently the limit $\lim_n X_n$ exists and is a.s. finite on the event $\{\nu_a = \infty\} = \{A_\infty \leqslant a^2\}$. Letting $a \uparrow \infty$ through integer values, we obtain the third part of the proposition.

To prove the second part of the proposition, let us first write down the inequality

$$P(\sup_{\mathbf{N}} |X_n| > a) \leqslant P(\nu_a < \infty) + P(\nu_a = \infty, \sup_{\mathbf{N}} |X_n| > a)$$

$$\leqslant P(\nu_a < \infty) + P(\sup_{\mathbf{N}} |X_{\nu_a \wedge n}| > a),$$

and then, applying Lemma IV-2-9 to the positive submartingale $(X^2_{\nu_a \wedge n}, n \in \mathbf{N})$:

$$P(\sup_{\mathbf{N}} X^2_{\nu_a \wedge n} > a^2) \leqslant a^{-2} \lim_n\uparrow E(X^2_{\nu_a \wedge n}) = a^{-2} E(A_{\nu_a}).$$

Since the r.v. A_{ν_a} is bounded above by both A_∞ and a^2 and as $\{\nu_a < \infty\} = \{A_\infty > a^2\}$, we have proved that

$$P(\sup_{\mathbf{N}} |X_n| > a) \leqslant P(A_\infty > a^2) + a^{-2} E(\min(A_\infty, a^2))$$

Integrating out the variable a on both sides with respect to Lebesgue measure on $\mathbf{R}_+$, we obtain by Fubini's theorem

$$E(\sup_{\mathbf{N}} |X_n|) = \int_{\mathbf{R}^+} P(\sup_{\mathbf{N}} |X_n| > a)\,da$$

$$\leqslant \int_{\mathbf{R}^+} P(A_\infty > a^2)\,da + \int_{\mathbf{R}^+} a^{-2} E(\min(A_\infty, a^2))\,da$$

$$= 3E(\sqrt{A_\infty}).$$

By Proposition IV-2-3(c) the condition $E(\sqrt{A_\infty}) < \infty$ implies that the martingale $(X_n, n \in \mathbf{N})$ is regular; the proposition is therefore proved. ∎

EXAMPLE. In the case of a centred square-integrable random walk $(X_n = \sum_{m=1}^{n} Y_m, n \in \mathbf{N})$, where the Y_m are independent identically distributed r.r.v.'s with zero mean and finite variance, the preceding proposition implies that $E(X_\nu) = 0$ whenever $E(\sqrt{\nu}) < \infty$. It follows, for example, that the stopping time $\nu_a = \min(n : X_n \geqslant a)$ $(a > 0$ real), which is a.s. finite because the random walk is recurrent, must be such that $E(\sqrt{\nu_a}) = +\infty$ since the a.s. inequality $X_{\nu_a} \geqslant a$ makes the equality $E(X_{\nu_a}) = 0$ impossible. ∎

REMARK. A minor change in the proof of property (b) of the preceding proposition allows us to establish more generally that

$$E(\sup_{\mathbf{N}} |X_n|^p) \leqslant c_p E(A_\infty^{p/2})$$

for every real $p \in]0, 2[$, the constant c_p in this inequality only depending on p. ■

On the event $\{A_\infty = \infty\}$ we cannot expect the martingale $(X_n, n \in \mathbf{N})$ to remain bounded; nonetheless, the following proposition shows that when $n \uparrow \infty$ the martingale $(X_n, n \in \mathbf{N})$ remains an order of magnitude smaller than $A_n^{1/2+\varepsilon}$ or even $A_n^{1/2}(\log^+ A_n)^{1/2+\varepsilon}$ if $\varepsilon > 0$.

PROPOSITION VII-2-4. *If $(X_n, n \in \mathbf{N})$ is a square-integrable martingale such that $X_0 = 0$ and if $(A_n, n \in \mathbf{N})$ denotes the increasing process associated with the submartingale $(X_n^2, n \in \mathbf{N})$ by the Doob decomposition, then*

$$X_n = \mathrm{o}(f(A_n)) \quad \text{a.s.} \quad \text{on } \{A_\infty = \infty\}$$

for every increasing function $f : \mathbf{R}_+ \to \mathbf{R}_+$ increasing sufficiently rapidly at infinity that $\int_0^\infty (1 + f(t))^{-2} \mathrm{d}t < \infty$. In particular, the functions $f_\alpha(t) = t^\alpha$ and $f'_\alpha(t) = t^{1/2}(\log^+ t)^\alpha$ are suitable if $\alpha > \frac{1}{2}$ (in both cases).

PROOF. Let $f : \mathbf{R}_+ \to \mathbf{R}_+$ be an increasing function such that $\int_0^\infty (1 + f(t))^{-2} \mathrm{d}t$ is finite; then $\lim_{t \uparrow \infty} f(t) = +\infty$. We shall show that the formula

$$Z_n = \sum_{m<n} \frac{X_{m+1} - X_m}{1 + f(A_{m+1})} \qquad (n \in \mathbf{N})$$

defines a martingale bounded in L^2 (this formula provides the first example of a transformation of martingales which we shall be studying in the next chapter). The sequence $(Z_n, n \in \mathbf{N})$ is a martingale because

$$E^{\mathscr{B}_n}(Z_{n+1} - Z_n) = E^{\mathscr{B}_n}\left(\frac{X_{n+1} - X_n}{1 + f(A_{n+1})}\right) = 0,$$

since $1 + f(A_{n+1})$ is $\mathscr{B}_n$-measurable and $E^{\mathscr{B}_n}(X_{n+1} - X_n) = 0$. The increments of the increasing process $(B_n, n \in \mathbf{N})$ in Doob's decomposition of $(Z_n^2, n \in \mathbf{N})$ are obtained analogously as

$$B_{n+1} - B_n = E^{\mathscr{B}_n}[(Z_{n+1} - Z_n)^2] = E^{\mathscr{B}_n}\left[\frac{(X_{n+1} - X_n)^2}{(1 + f(A_{n+1}))^2}\right] = \frac{A_{n+1} - A_n}{[1 + f(A_{n+1})]^2}$$

for every $n \in \mathbf{N}$. But

$$\sum_{\mathbf{N}} \frac{A_{n+1} - A_n}{[1 + f(A_{n+1})]^2} \leqslant \sum_{\mathbf{N}} \int_{A_n}^{A_{n+1}} \frac{\mathrm{d}t}{[1 + f(t)]^2} \leqslant \int_0^{\infty} \frac{\mathrm{d}t}{[1 + f(t)]^2};$$

denoting by c the value of the integral on the right-hand side, we have therefore proved that $B_\infty \leqslant c$ a.s. By Proposition VII-2-3 (a) the martingale $(Z_n, n \in \mathbf{N})$ converges a.s. to a finite limit when $n \uparrow \infty$; Kronecker's lemma given below then implies that

$$\frac{X_n - X_0}{1 + f(A_n)} \to 0 \quad \text{a.s.} \quad \text{on } \{f(A_n) \uparrow \infty\} = \{A_\infty = \infty\},$$

which is equivalent to the result stated in the proposition.

LEMMA VII-2-5 (Kronecker). *If* $(u_n, n \geqslant 1)$ *is an increasing sequence of positive real numbers tending to* $+\infty$ *and the sequence* $(y_n, n \geqslant 1)$ *of real numbers is such that* $\lim_n \sum_{m=1}^n (y_m/u_m)$ *exists and is finite, then* $\lim_n (u_n^{-1} \sum_{m=1}^n y_m) = 0$.

PROOF. The real numbers $v_n = u_n - u_{n-1}$ ($= u_1$ if $n = 1$) are positive and such that $\sum_{m=1}^n v_m = u_n \uparrow \infty$. On the other hand, the $z_n = \sum_{m=1}^n (y_m/u_m)$ converge to a finite limit when $n \uparrow \infty$ and satisfy the equality

$$\sum_{m=1}^n y_m = \sum_{m=1}^n u_m(z_m - z_{m-1}) = \sum_{m=1}^n v_m(z_n - z_{m-1}).$$

Consequently, if $p < n$,

$$\left| \frac{1}{u_n} \sum_{m=1}^n y_m \right| \leqslant \frac{1}{u_n} \left| \sum_{m=1}^p v_m(z_n - z_{m-1}) \right| + \frac{1}{u_n} \left(\sum_{m=p+1}^n v_m \right) \sup_{p \leqslant m \leqslant n} |z_n - z_m|;$$

letting $n \uparrow \infty$ and then $p \uparrow \infty$, we obtain

$$\lim_{n \to \infty} \frac{1}{u_n} \sum_{m=1}^n y_m = 0. \blacksquare$$

Here is a corollary of the two preceding propositions; this corollary generalises the elementary Borel–Cantelli lemma.

COROLLARY VII-2-6. *For every adapted sequence* $(B_n, n \in \mathbf{N}^*)$ *of events, the a.s. equality*

$$\{\sum_{\mathbf{N}^*} 1_{B_n} < \infty\} = \{\sum_{\mathbf{N}^*} P^{\mathscr{B}_{n-1}}(B_n) < \infty\}$$

is valid, and further

$$\frac{\sum_{1 \leqslant m \leqslant n} 1_{B_m}}{\sum_{1 \leqslant m \leqslant n} P^{\mathscr{B}_{m-1}}(B_m)} \to 1 \quad \text{a.s. on} \left\{ \sum_{\mathbf{N}^*} P^{\mathscr{B}_{n-1}}(B_n) = \infty \right\}.$$

PROOF. Consider the martingale $(X_n, n \in \mathbf{N})$ defined by the formula

$$X_n - X_{n-1} = 1_{B_n} - P^{\mathscr{B}_{n-1}}(B_n) \qquad (n \in \mathbf{N}^*)$$

with $X_0 = 0$. For this martingale the increments of the increasing process $(A_n, n \in \mathbf{N})$ associated with the submartingale $(X_n^2, n \in \mathbf{N})$ are given by

$$A_n - A_{n-1} = E^{\mathscr{B}_{n-1}}((X_n - X_{n-1})^2) = P^{\mathscr{B}_{n-1}}(B_n) - [P^{\mathscr{B}_{n-1}}(B_n)]^2 \qquad (n \in \mathbf{N}^*).$$

Note that by this inequality $A_n - A_{n-1} \leqslant P^{\mathscr{B}_{n-1}}(B_n)$ for every $n \in \mathbf{N}^*$.

On the event $\{\sum_{\mathbf{N}^*} P^{\mathscr{B}_{n-1}}(B_n) < \infty\}$ the r.v. $A_\infty = \sum_{\mathbf{N}^*}(A_n - A_{n-1})$ is a.s. finite, and consequently the martingale $(X_n, n \in \mathbf{N})$ converges there to a finite limit (Proposition VII-**2**-3(c)); it follows that $\sum_{\mathbf{N}^*} 1_{B_n} < \infty$ on this event.

On the event $\{A_\infty < \infty\} \cap \{\sum_{\mathbf{N}^*} P^{\mathscr{B}_{n-1}}(B_n) = \infty\}$ the martingale $(X_n, n \in \mathbf{N})$ again has an a.s. *finite* limit and so when $n \to \infty$

$$\frac{\sum_{1 \leqslant m \leqslant n} 1_{B_m}}{\sum_{1 \leqslant m \leqslant n} P^{\mathscr{B}_{m-1}}(B_m)} - 1 = \frac{X_n}{\sum_{1 \leqslant m \leqslant n} P^{\mathscr{B}_{m-1}}(B_m)} \to 0.$$

Finally, on the event $\{A_\infty = \infty\}$ the inequalities $A_n \leqslant \sum_{1 \leqslant m \leqslant n} P^{\mathscr{B}_{m-1}}(B_m)$ imply that a.s.

$$\left| \frac{\sum_{1 \leqslant m \leqslant n} 1_{B_m}}{\sum_{1 \leqslant m \leqslant n} P^{\mathscr{B}_{m-1}}(B_m)} - 1 \right| \leqslant \frac{|X_n|}{A_n} \to 0 \qquad (n \uparrow \infty)$$

by the preceding proposition (with $f(t) = t$). ■

REMARK. In this corollary the sequence of indicator r.v.'s 1_{B_n} $(n \in \mathbf{N})$ can be replaced by an arbitrary adapted sequence $(Z_n, n \in \mathbf{N})$ taking their values in the interval [0, 1]; neither the statement nor the proof of the corollary requires further changes. ■

The class of functions f in Proposition VII-2-4 cannot be enlarged without putting further assumptions on the square-integrable martingale $(X_n, n \in \mathbf{N})$. Petrov [231] has shown on this point that if $f : \mathbf{R}_+ \to \mathbf{R}_+$ is a function such that $f(t)/\sqrt{t}$ is increasing and $\int_0^\infty (1+f(t))^{-2} dt = +\infty$, there exists at least one square-integrable martingale $(X_n, n \in \mathbf{N})$ such that

$$\limsup_{n\to\infty} \frac{X_n}{f(A_n)} > 0 \text{ a.s.}$$

On the other hand, for a martingale $(X_n, n \in \mathbf{N})$ whose differences are dominated by the same finite constant, the behaviour of the martingale $(X_n, n \in \mathbf{N})$ on $\{A_\infty = \infty\}$ can be described in a very precise way by a "law of the iterated logarithm".

PROPOSITION VII-2-7. *Let $(X_n, n \in \mathbf{N})$ be a square-integrable martingale such that* $\sup_{\mathbf{N}} |X_{n+1} - X_n| \leqslant c$ *a.s. for a finite constant c. If $(A_n, n \in \mathbf{N})$ denotes the increasing process associated with the submartingale $(X_n^2, n \in \mathbf{N})$, we have*

$$\left.\begin{aligned} &\limsup_n ([2A_n \log\log A_n]^{-1/2} X_n) = 1, \\ &\liminf_n ([2A_n \log\log A_n]^{-1/2} X_n) = -1, \end{aligned}\right\} \text{ a.s. on } \{A_\infty = \infty\}.$$

In particular this proposition applies to the sequence $(X_n = \sum_{m=1}^n Y_m, n \in \mathbf{N})$ of partial sums of a sequence of independent zero-mean r.v.'s dominated in absolute value by the same constant c and such that $\sum_{\mathbf{N}^*} E(Y_n^2) = +\infty$.

PROOF. We shall show that every square-integrable martingale $(X_n, n \in \mathbf{N})$ such that $X_{n+1} - X_n \leqslant c$ a.s. for all $n \in \mathbf{N}$ satisfies the inequality

$$\limsup_{n\to\infty} ([2A_n \log\log A_n]^{-1/2} X_n) \leqslant 1 \text{ a.s.} \qquad \text{on } \{A_\infty = \infty\}.$$

The assumption that all the increments $X_{n+1} - X_n$ $(n \in \mathbf{N})$ are bounded by a constant c is only needed here to obtain the lemma below. [The preceding inequality will therefore be satisfied for every square-integrable martingale satisfying the conclusions of this lemma, for example for every gaussian martingale $(X_n, n \in \mathbf{N})$ for which the sequences $(\exp[\lambda X_n - \frac{1}{2}\lambda^2 E(X_n^2)], n \in \mathbf{N})$ are necessarily martingales.]

LEMMA VII-2-8. *For every real number $c > 0$, let $\phi_c : \mathbf{R}_+ \to \mathbf{R}_+$ be the function defined by $\phi_c(\lambda) = c^{-2}[\exp(\lambda c) - 1 - \lambda c]$; it satisfies $\phi_c(\lambda) = \frac{1}{2}\lambda^2 + o(\lambda^2)$ at the neighbourhood of the origin. For every square-integrable martingale $(X_n, n \in \mathbf{N})$*

such that $\sup(X_{n+1} - X_n) \leqslant c$ *a.s., the sequence*

$$\exp(\lambda X_n - \phi_c(\lambda) A_n) \qquad (n \in \mathbf{N})$$

of r.v.'s is then a positive supermartingale for every $\lambda \in \mathbf{R}_+$.

We will be using the following elementary inequality, valid for every real number $y \in]-\infty, c]$ and every real $c > 0$:

$$\exp(\lambda y) \leqslant 1 + \lambda y + \phi_c(\lambda) y^2.$$

This inequality is easily proved from the series expansion of the exponential function when $y \in [0, c]$, viz.

$$\exp(\lambda y) = 1 + \lambda y + \sum_{n \geqslant 2} \frac{\lambda^n y^n}{n!} \leqslant 1 + \lambda y + y^2 \sum_{n \geqslant 2} \frac{\lambda^n c^{n-2}}{n!}$$

$$= 1 + \lambda y + \lambda^2 \phi_c(y).$$

To prove the above inequality when $y \leqslant 0$ we integrate the inequality $\exp(\lambda y) \leqslant 1$ twice over the interval $[y, 0]$; we find that

$$\exp(\lambda y) - (1 + \lambda y) \leqslant \tfrac{1}{2}\lambda^2 y^2 \qquad (y \leqslant 0),$$

but by definition, $\frac{1}{2}\lambda^2 \leqslant \phi_c(\lambda)$.

The inequality that we have just proved implies that under the assumptions of the lemma

$$E^{\mathscr{B}_n}(\exp[\lambda(X_{n+1} - X_n]) \leqslant E^{\mathscr{B}_n}(1 + \lambda(X_{n+1} - X_n) + \phi_c(\lambda)(X_{n+1} - X_n)^2)$$

$$= 1 + \phi_c(\lambda)(A_{n+1} - A_n) \leqslant \exp[\phi_c(\lambda)(A_{n+1} - A_n)]$$

for every $n \in \mathbf{N}$. But this can also be written

$$E^{\mathscr{B}_n}(\exp[\lambda X_{n+1} - \phi_c(\lambda) A_{n+1}]) \leqslant \exp[\lambda X_n - \phi_c(\lambda) A_n] \qquad (n \in \mathbf{N})$$

and the lemma is therefore proved. ■

From now on we will suppose, without loss of generality, that $X_0 = 0$. Proposition II-**2**-7 applied to the positive supermartingale in the above lemma shows that

$$P(\bigcup_{\mathbf{N}} \{\exp[\lambda X_n - \phi_c(\lambda) A_n] > \exp(\lambda a)\}) \leqslant \exp(-\lambda a)$$

if λ and $a \in \mathbf{R}_+$, as $\exp[\lambda X_0 - \phi_c(\lambda) A_0] = 1$. This inequality can also be written

$$P(\bigcup_{\mathbf{N}} \{X_n > a + \lambda^{-1} \phi_c(\lambda) A_n\}) \leqslant \exp(-\lambda a) \qquad (a > 0, \lambda > 0).$$

Let θ be a real number >1 and δ a real number >0. Put

$$a_k = \tfrac{1}{2}(1+\delta) h(\theta^k), \qquad \lambda_k = \theta^{-k} h(\theta^k),$$

denoting by h the function $h(t) = (2t \log\log t)^{1/2}$ defined on $[e, \infty[$, and letting k vary over the integers for which $\theta^k \geqslant e$; then we have

$$\exp(-\lambda_k a_k) = [k \log \theta]^{-(1+\delta)}.$$

For the same values of k let us denote by v_k the stopping times

$$v_k = \begin{cases} \min(n : A_{n+1} > \theta^k) & \\ +\infty & \text{if } A_\infty \leqslant \theta^k; \end{cases}$$

the following bounds are valid on $\{v_k < n \leqslant v_{k+1}\}$:

$$a_k + \frac{\phi_c(\lambda_k)}{\lambda_k} A_n \leqslant a_k + \frac{\phi_c(\lambda_k)}{\lambda_k} \theta^{k+1} = C_k(\theta, \delta) h(\theta^k) \leqslant C_k(\theta, \delta) h(A_n)$$

subject to putting

$$C_k(\theta, \delta) = \tfrac{1}{2}(1+\delta) + \theta \frac{\phi_c(\lambda_k)}{\lambda_k^2}$$

The inequality at the beginning hence implies that

$$P(X_n > C_k(\theta, \delta) h(A_n) \text{ for some } n \in]v_k, v_{k+1}]) \leqslant [k \log \theta]^{-(1+\delta)}.$$

Since the series $\sum_k k^{-(1+\delta)}$ is convergent, the classical Borel–Cantelli lemma implies that on $\{A_\infty = \infty\} = \{v_k < \infty \text{ for all } k \text{ and } v_k \uparrow \infty\}$,

$$\limsup_n \frac{X_n}{h(A_n)} \leqslant \lim_k C_k(\theta, \delta) = \tfrac{1}{2}(1 + \delta + \theta).$$

Letting $\delta \downarrow 0$ and $\theta \downarrow 1$, we find that

$$\limsup_n \frac{X_n}{h(A_n)} \leqslant 1.$$

For the proof of the inequality

$$\limsup\,([2A_n \log\log A_n]^{-1/2}\, X_n) \geqslant 1 \text{ p.s.}$$

the reader is referred to the original article of W. Stout [277]. ■

VII-3. Martingales with integrable pth power

For every martingale $(X_n, n \in \mathbf{N})$ with integrable p^{th} power $(p \geqslant 1)$, the sequence $(|X_n|^p, n \in \mathbf{N})$ is an integrable submartingale by virtue of the convexity inequality $E^{\mathscr{B}}(|X|^p) \geqslant |E^{\mathscr{B}}(X)|^p$. The increasing process $(A_n^{(p)}, n \in \mathbf{N})$ in the Doob decomposition, say

$$|X_n|^p = M_n^{(p)} + A_n^{(p)} \qquad (n \in \mathbf{N}),$$

is defined by the formula

$$A_{n+1}^{(p)} - A_n^{(p)} = E^{\mathscr{B}_n}(|X_{n+1}|^p) - |X_n|^p \qquad (n \in \mathbf{N}).$$

Note that apart from the case $p = 2$ this expression does not equal

$$E^{\mathscr{B}_n}(|X_{n+1} - X_n|^p).$$

If $p > 1$ and $E(\mathrm{A}_\infty^{(p)}) < \infty$, it is not hard to see that the martingale $(X_n, n \in \mathbf{N})$ is regular and convergent in L^p (the case $p = 2$ has moreover already been considered in Proposition VII-**2**-3); more generally, if $p > 1$ and $E(\mathrm{A}_\nu^{(p)}) < \infty$, the stopping time ν is regular and $X_\nu \in L^p$. But this section is devoted to another type of result; we will show that under suitable conditions a martingale which does not converge oscillates unboundedly between $-\infty$ and $+\infty$.

PROPOSITION VII-**3**-9. *For every integrable martingale $(X_n, n \in \mathbf{N})$ satisfying the condition $E(\sup_{\mathbf{N}} |X_{n+1} - X_n|) < \infty$, we have*

$$\limsup_n X_n = +\infty \qquad \liminf_n X_n = -\infty$$

a.s. outside the set on which $\lim_n X_n$ exists and is finite.

More precisely and more generally, let p be a real number $\geqslant 1$ and let $(X_n, n \in \mathbf{N})$ be a martingale with integrable p*th power. Then the r.r.v.* $\lim_n X_n$ *exists and is finite a.s. on $\{A_\infty^{(p)} < \infty\}$. Further, if $E(\sup_n |X_{n+1} - X_n|^p) < \infty$, we have*

$$\limsup_n X_n = +\infty \quad \textit{and} \quad \liminf_n X_n = -\infty \quad \text{on } \{A_\infty^{(p)} = \infty\}.$$

In particular for $p=2$ the second part of the above result shows that

$$\{X_n \to\} = \{A_\infty < \infty\} \quad \text{a.s.}$$

if $\sup_{\mathbf{N}} |X_n - X_{n-1}| \in L^2$, denoting by $(A_n = A_n^{(2)}, n \in \mathbf{N})$ the increasing process in Doob's decomposition of the submartingale $(X_n^2, n \in \mathbf{N})$. This result is more precise than Proposition VII-**2**-3(c) which only establishes the inclusion $\{A_\infty < \infty\} \subset \{X_n \to\}$ for every square-integrable martingale, but it also requires a further hypothesis.

PROOF. Since the first part of the proposition follows from the second, it clearly suffices to establish the second part. Let p be a real number $\geqslant 1$ and let $(X_n, n \in \mathbf{N})$ be a martingale with integrable p^{th} power such that $X_0 = 0$. Then $((X_n^+)^p, n \in \mathbf{N})$ is a positive integrable submartingale because

$$E^{\mathscr{B}_n}[(X_{n+1}^+)^p] \geqslant [E^{\mathscr{B}_n}(X_{n+1}^+)]^p \geqslant (X_n^+)^p.$$

Let us denote by $(A'_n, n \in \mathbf{N})$ the increasing process in the Doob decomposition of this submartingale. We shall show that

(a′) $\lim_n X_n$ exists and is finite a.s. on $\{A'_\infty < \infty\}$,

(b′) $\limsup_n X_n = +\infty$ a.s. on $\{A'_\infty = \infty\}$.

This will be enough to prove the proposition, for upon applying this result to the martingale $(-X_n, n \in \mathbf{N})$ we find that

(a″) $\lim_n X_n$ exists and is finite a.s. on $\{A''_\infty < \infty\}$,

(b″) $\liminf_n X_n = -\infty$ a.s. on $\{A''_\infty = \infty\}$,

if $(A''_n, n \in \mathbf{N})$ denotes the increasing process in the Doob decomposition of the submartingale $((X_n^-)^p, n \in \mathbf{N})$. Comparing the two results (a′), (b′) and (a″), (b″), we obtain $\{A'_\infty < \infty\} = \{A''_\infty < \infty\}$ (for when the limit $\lim_n X_n$ exists in $\mathbf{R}$, $\limsup_n X_n$ and $\liminf_n X_n$ cannot be infinite). Finally, upon remarking that the increasing process $(A_n^{(p)}, n \in \mathbf{N})$ in the Doob decomposition of $(|X_n|^p, n \in \mathbf{N})$ is simply the sum of the two increasing processes $(A'_n, n \in \mathbf{N})$ and $A''_n, n \in \mathbf{N})$, we see that A_∞^p is finite a.s. at the same time as A'_∞ and A''_∞. The proposition will thus be proved.

(a′) For every real $a > 0$, let ν_a be the stopping time defined by the formula

$$\nu_a = \begin{cases} \min(n : A'_{n+1} > a) & \\ +\infty & \text{if } A'_\infty \leqslant a. \end{cases}$$

Then $A'_{\nu_a} \leqslant a$ and consequently

$$\lim_n \uparrow E((X^+_{\nu_a \wedge n})^p) = \lim_n \uparrow E(A'_{\nu_a \wedge n}) = E(A'_{\nu_a}) \leqslant a.$$

Thus $\lim\uparrow_n E(X^+_{\nu_a \wedge n}) < \infty$ by Hölder's inequality; Theorem IV-**1**-2 then implies that the martingale $(X_{\nu_a \wedge n}, n \in \mathbf{N})$ converges a.s. to a finite limit. It then follows that the limit $\lim_n X_n$ exists and is finite a.s. on the event $\{\nu_a = \infty\} = \{A'_\infty \leqslant a\}$ and hence also, letting $a \uparrow \infty$, on the event $\{A'_\infty < \infty\}$.

(b′) For every real number $a > 0$, let ν'_a be the stopping time defined by the formula

$$\nu'_a = \begin{cases} \min(n : X_n > a) & \\ +\infty & \text{if } \sup_{\mathbf{N}} X_n \leqslant a. \end{cases}$$

The condition $\sup_{\mathbf{N}}(X_{n+1} - X_n)^+ \in L^p$ then suffices to infer that the expression

$$E(A'_{\nu'_a}) = \lim_m \uparrow E((X^+_{\nu'_a \wedge m})^p)$$

is finite. Indeed, $X_{\nu'_a \wedge m} \leqslant a$ on $\{\nu'_a > m\}$, whilst on the complementary event $\{\nu'_a \leqslant m\}$ we have

$$X_{\nu'_a \wedge m} \leqslant a + (X_{\nu'_a} - X_{\nu'_a - 1})^+ \leqslant a + \sup_{\mathbf{N}}(X_{n+1} - X_n)^+;$$

it follows that

$$\|X^+_{\nu'_a \wedge m}\|_p \leqslant a + \|\sup_{\mathbf{N}}(X_{n+1} - X_n)^+\|_p < \infty \qquad (m \in \mathbf{N}).$$

The positive r.r.v. $X'_{\nu'_a}$ therefore belongs to L^1, and is hence a.s. finite; we have thus shown that

$$A'_\infty < \infty \text{ a.s.} \quad \text{on } \{\nu' = \infty\} = \{\sup_{\mathbf{N}} X_n \leqslant a\}.$$

Letting $a \uparrow \infty$, we see that $A'_\infty < \infty$ a.s. on $\{\sup_{\mathbf{N}} X_n < \infty\}$ or, equivalently, that $\sup_{\mathbf{N}} X_n = \infty$ on $\{A'_\infty = \infty\}$; to complete the proof it remains to observe that

$$\{\sup_{\mathbf{N}} X_n = \infty\} = \{\limsup_n X_n = \infty\}.$$

since $X_n < \infty$ a.s. for all $n \in \mathbf{N}$. ■

REMARK. If we had contented ourselves with applying the line of reasoning which gave (a′) and (b′) above to the submartingale $(|X_n|^p, n \in \mathbf{N})$ and its associated increasing process $(A_n^{(p)}, n \in \mathbf{N})$, we would only have obtained

$$\limsup_n |X_n| = +\infty \text{ a.s.} \quad \text{on } \{A_\infty^{(p)} = \infty\},$$

which is weaker than the result given in the proposition.

Supplement

The increasing process $(A_n^{(p)}, n \in \mathbf{N})$ in the Doob decomposition of the submartingale $(|X_n|^p, n \in \mathbf{N})$ associated with a martingale $(X_n, n \in \mathbf{N})$ with integrable pth power is defined by the formula

$$A_{n+1}^{(p)} - A_n^{(p)} = E^{\mathscr{B}_n}(|X_{n+1}|^p - |X_n|^p) \qquad (n \in \mathbf{N})$$

and is distinct, if $p \neq 2$, from the increasing process $(B_n^{(p)}, n \in \mathbf{N})$ defined by

$$B_{n+1}^{(p)} - B_n^{(p)} = E^{\mathscr{B}_n}(|X_{n+1} - X_n|^p) \qquad (n \in \mathbf{N}).$$

Nevertheless, the following is a proposition relating these two increasing processes when p is an *even integer*!

PROPOSITION VII-3-10. *Let p be an even integer $\geqslant 2$. Then $E(A_\infty^{(p)}) < \infty$, which implies that the martingale $(X_n, n \in \mathbf{N})$ is regular and convergent in L^p, whenever $B_\infty^{(q)} \in L_{p/q}$ for every integer $q = 2, 3, \ldots, p$.*

By considering the martingale $(X_{\nu \wedge n}, n \in \mathbf{N})$ stopped at a stopping time ν, we deduce from this proposition that $E(A_\nu^{(p)}) < \infty$ and that ν is regular for $(X_n, n \in \mathbf{N})$ whenever $B_\nu^{(q)} \in L_{p/q}$ for $q = 2, 3, \ldots, p$.

PROOF. Since p is an integer, the binomial formula applied to $[X_m + (X_{m+1} - X_m)]^p$ shows that

$$\begin{aligned} E^{\mathscr{B}_m}(X_{m+1}^p) &= \sum_{q=0}^{p} \binom{p}{q} X_m^{p-q} E^{\mathscr{B}_m}((X_{m+1} - X_m)^q) \\ &= X_m^p + \sum_{q=2}^{p} \binom{p}{q} X_m^{p-q} E^{\mathscr{B}_m}((X_{m+1} - X_m)^q), \end{aligned}$$

and when p is even it follows that

$$A_{m+1}^{(p)} - A_m^{(p)} = E^{\mathscr{B}_m}(|X_{m+1}|^p) - |X_m|^p \leqslant \sum_{q=2}^{p} \binom{p}{q} |X_m|^{p-q} [B_{m+1}^{(q)} - B_m^{(q)}].$$

Consequently,

$$E(A_n^{(p)}) \leqslant \sum_{q=2}^{p} \binom{p}{q} E\left(\sum_{m=0}^{n-1} |X_m|^{p-q} [B_{m+1}^{(q)} - B_m^{q}] \right).$$

But $|X_m|^{p-q} \leqslant E^{\mathscr{B}_m}(|X_n|^{p-q})$ if $m < n$ and as $B_{m+1}^{(q)} - B_m^{(q)}$ is $\mathscr{B}_m$-measurable, we see that

$$E(|X_m|^{p-q} [B_{m+1}^{(q)} - B_m^{(q)}]) \leqslant E(|X_n|^{p-q} [B_{m+1}^{(q)} - B_m^{(q)}])$$

if $0 \leqslant m < n$. Hence finally

$$E(A_n^{(p)}) \leqslant \sum_{q=2}^{p} \binom{p}{q} E(|X_n|^{p-q} B_n^{(q)}) \qquad (n \in \mathbf{N}).$$

Hölder's inequality shows that

$$E(|X_n|^{p-q} B_n^{(q)}) \leqslant \| |X_n|^{p-q} \|_{p/(p-q)} \|B_n^{(q)}\|_{p/q} = \|X_n\|_p^{p-q} \|B_n^{(q)}\|_{p/q}.$$

Using the equality $E(|X_n|^p) = E(A_n^{(p)})$, we finally see that

$$E(A_n^{(p)}) \leqslant \sum_{q=2}^{p} \binom{p}{q} [E(A_n^{(p)})]^{p-q/p} \|B_n^{(q)}\|_{p/q}$$

and hence that

$$1 \leqslant \sum_{q=2}^{p} \binom{p}{q} \frac{\|B_\infty^{(q)}\|_{p/q}}{[E(A_n^{(p)})]^{q/p}}$$

bounding the norm of $B_n^{(q)}$ by that of $B_\infty^{(q)}$. If we then had $\lim\uparrow_n E(A_n^{(p)}) = +\infty$, when $n \uparrow \infty$ the last inequality would show that $1 \leqslant 0$! ■

For a random walk $X_n = \sum_{m=1}^{n} Y_m$ in L^p (p an even integer), the preceding proposition implies that $E(|X_\nu|^p) = E(A_\nu^{(p)})$ for every stopping time ν such that $E(\nu^{p/2}) < \infty$ since the increasing processes $(B_n^{(q)}, n \in \mathbf{N})$ are all proportional to n.

Corollary VII-3-11. *For every stopping time ν, the condition $E(A_\nu^{(4)}) < \infty$ is valid whenever $E(\nu B_\nu^{(4)}) < \infty$.*

Proof. The condition $E(\nu B_\nu^{(4)}) < \infty$ implies that simultaneously $E(B_\nu^{(4)}) < \infty$, $E([B_\nu^{(3)}]^{4/3}) < \infty$ and $E([B_\nu^{(2)}]^2) < \infty$ by Hölder's inequality and the fact that $B_0^{(\cdot)} = 0$. ■

VII-4. Gundy's condition

The aim of this section is to provide converses to Proposition VII-2-3(a)–(c). Concerning result (*c*), according to which every square-integrable martingale $(X_n, n \in \mathbf{N})$ converges to a finite limit a.s. on the event $\{A_\infty < \infty\}$, we have already found in Proposition VII-3-9 that when $\sup_{\mathbf{N}}|X_{n+1}-X_n| \in L^2$ we have

$$\limsup_{n\to\infty} X_n = +\infty, \qquad \liminf_{n\to\infty} X_n = -\infty \quad \text{a.s. on } \{A_\infty = \infty\}.$$

[A stronger result of the same kind is even given by the law of the iterated logarithm when $\sup_{\mathbf{N}}|X_{n+1}-X_n| \in L^\infty$, (Proposition VII-2-7).] Another hypothesis, due to Gundy, lets us obtain an analogous result; we begin with a lemma which is elementary although its proof is rather long, and which is intended to make one understand this hypothesis.

Definition VII-4-12. A square-integrable martingale $(X_n, n \in \mathbf{N})$ is said to *satisfy Gundy's condition* if it satisfies one of the following equivalent conditions:

(1) There exist two real numbers $a > 0$ and $\alpha > 0$ such that

$$P^{\mathscr{B}_n}(|X_{n+1}-X_n| \geqslant a\sqrt{(A_{n+1}-A_n)}) \geqslant \alpha \ a.s. \quad \text{for every } n \in \mathbf{N}.$$

(2) There exists a real number $b > 0$ such that

$$E^{\mathscr{B}_n}(|X_{n+1}-X_n|) \geqslant b\sqrt{(A_{n+1}-A_n)} \ a.s. \quad \text{for every } n \in \mathbf{N}.$$

(3) There exist two real numbers $c > 0$ and $\gamma > 0$ such that

$$E^{\mathscr{B}_n}((X_{n+1}-X_n)^2 1_{\{(X_{n+1}-X_n)^2 \geqslant c(A_{n+1}-A_n)\}}) \leqslant (1-\gamma)(A_{n+1}-A_n) \text{ a.s.} \quad \text{for every } n \in \mathbf{N}.$$

The equivalence of these three conditions results directly from applying the following lemma to the positive r.r.v.'s $|X_{n+1}-X_n|$ $(n \in \mathbf{N})$ and the respective σ-fields $\mathscr{B}_n$.

LEMMA VII-4-13. *Let Y be a positive r.r.v., and let $\mathscr{B}$ be a sub-σ-field of $\mathscr{A}$. Suppose that $E^{\mathscr{B}}(Y^2) < \infty$, and let us put $S^2 = E^{\mathscr{B}}(Y^2)$. Then for all constants $a,c > 0$ and $\alpha,b,\gamma \in]0,1[$, the following implications are valid:*

(1) $E^{\mathscr{B}}(Y) \geqslant bS \Rightarrow P^{\mathscr{B}}(Y \geqslant \tfrac{1}{2}bS) \geqslant \tfrac{1}{4}b^2$.
(2) $P^{\mathscr{B}}(Y \geqslant aS) \geqslant \alpha \Rightarrow E^{\mathscr{B}}(Y^2 1_{\{Y^2 \geqslant (2/\alpha)S^2\}}) \leqslant (1-\tfrac{1}{2}a^2\alpha)S^2$.
(3) $E^{\mathscr{B}}(Y^2 1_{\{Y^2 \geqslant cS^2\}}) \leqslant (1-\gamma)S^2 \Rightarrow E^{\mathscr{B}}(Y) \geqslant (\gamma/\sqrt{c})S$.

PROOF. (1) By Schwartz's inequality,

$$E^{\mathscr{B}}(Y 1_{\{Y \geqslant aS\}}) \leqslant S\sqrt{P^{\mathscr{B}}(Y \geqslant aS)}$$

for every $a \in \mathbf{R}_+$. On the other hand, it is clear that $E^{\mathscr{B}}(Y 1_{\{Y < aS\}}) \leqslant aS$ so that assumption (1) implies that

$$bS \leqslant E^{\mathscr{B}}(Y) \leqslant aS + S\sqrt{P^{\mathscr{B}}(Y \geqslant aS)}$$

Taking $a = \frac{1}{2}b$, we find that $P^{\mathscr{B}}(Y \geqslant aS) \geqslant \frac{1}{4}b^2$ on $\{S > 0\}$. On the other hand, on the event $\{S = 0\}$, which is $\mathscr{B}$-measurable, we clearly have $P^{\mathscr{B}}(Y \geqslant aS) = 1 \geqslant \frac{1}{4}b^2$.

(2) We remark that by Bienaymé's inequality $P^{\mathscr{B}}(Y^2 \geqslant (2/\alpha)S^2) \leqslant \frac{1}{2}\alpha$, so that the assumption implies that

$$P^{\mathscr{B}}(a^2 S^2 \leqslant Y^2 < (2/\alpha) S^2) \geqslant \alpha - \tfrac{1}{2}\alpha = \tfrac{1}{2}\alpha.$$

Then we write

$$E^{\mathscr{B}}(Y^2 1_{\{a^2 S^2 \leqslant Y^2 < (2/\alpha) S^2\}}) \geqslant a^2 S^2 P^{\mathscr{B}}(a^2 S^2 \leqslant Y^2 < (2/\alpha) S^2) \geqslant \tfrac{1}{2} a^2 \alpha S^2,$$

which implies relation (2).

(3) The following bounds are clear:

$$E^{\mathscr{B}}(Y^2 1_{\{Y^2 < cS^2\}}) \leqslant \sqrt{c}\, S E^{\mathscr{B}}(Y 1_{\{Y^2 < cS^2\}}) \leqslant \sqrt{c}\, S E^{\mathscr{B}}(Y).$$

Assumption (3) thus implies that

$$S^2 = E^{\mathscr{B}}(Y^2) \leqslant (1-\gamma) S^2 + \sqrt{c}\, S E^{\mathscr{B}}(Y),$$

which can be rewritten $\gamma S \leqslant (\sqrt{c}) E^{\mathscr{B}}(Y)$. ■

PROPOSITION VII-4-14. *Let $(X_n, n \in \mathbf{N})$ be a square-integrable martingale satisfying Gundy's condition. Then*

$$\limsup_{n\to\infty} |X_n| = +\infty \text{ a.s.} \quad \text{on } \{A_\infty = \infty\}.$$

PROOF. (1) We begin by showing that

$$\limsup_{n\to\infty} (A_{n+1} - A_n) < \infty \text{ a.s.} \quad \text{on } \{\limsup_{n\to\infty} |X_n| < \infty\}.$$

To this end, note that Gundy's condition in the form VII-4-12(1) implies that

$$P^{\mathscr{B}_n}(|X_{n+1} - X_n| > au) \geqslant 1_{\{\sqrt{(A_{n+1} - A_n)} > u\}} P^{\mathscr{B}_n}(|X_{n+1} - X_n| > a\sqrt{(A_{n+1} - A_n)})$$
$$\geqslant \alpha\, 1_{\{\sqrt{(A_{n+1} - A_n)} > u\}}$$

for every fixed real number $u > 0$. Thus on the event $\limsup_n\{\sqrt{(A_{n+1} - A_n)} > u\}$ the series $\sum_{\mathbf{N}} P^{\mathscr{B}_n}(|X_{n+1} - X_n| > au)$ is a.s. finite; by the generalised Borel–Cantelli lemma (Corollary VII-2-6) this implies that

$$\limsup_{n\to\infty}\{\sqrt{(A_{n+1} - A_n)} > u\} \subset \limsup_{n\to\infty}\{|X_{n+1} - X_n| > au\}$$

for every real number $u > 0$. Letting $u \uparrow \infty$, we find that

$$\{\limsup_{n\to\infty}(A_{n+1} - A_n) = +\infty\} \subset \{\limsup_{n\to\infty}|X_{n+1} - X_n| = +\infty\}.$$

But since the inequality $|X_{n+1} - X_n| > v$ implies that either $|X_{n+1}|$ or $|X_n|$ exceeds $\frac{1}{2}v$, it is clear that

$$\{\limsup_{n\to\infty}|X_{n+1} - X_n| = +\infty\} \subset \{\limsup_{n\to\infty}|X_n| = +\infty\};$$

we have thus obtained the stated result.

(2) For every real number $u > 0$ let ν_u be the stopping time defined by

$$\nu_u = \inf(n : n > 0,\ |X_n| + \sqrt{c(A_{n+1} - A_n)} > u),$$

where c denotes the positive constant in the form VII-4-12(3) of Gundy's condition. We immediately see that this definition implies that

$$\lim_{u\uparrow\infty}\uparrow\{\nu_u = +\infty\} = \{\sup_{\mathbf{N}}|X_n| < \infty,\ \sup_{\mathbf{N}}|A_{n+1} - A_n| < \infty\}$$

and the right-hand side is a.s. equal to $\{\sup_{\mathbf{N}}|X_n| < \infty\}$ by the first part of the proof.

On the other hand, on the event $\{0 \leqslant n < \nu_u\}$ the following inequalities are valid:

$$\begin{aligned}|X_{n+1}| &\leqslant |X_n| + |X_{n+1} - X_n| \leqslant u - \sqrt{c(A_{n+1} - A_n)} + |X_{n+1} - X_n| \\ &\leqslant u + |X_{n+1} - X_n|\,1_{\{|X_{n+1} - X_n|^2 \geqslant c(A_{n+1} - A_n)\}}.\end{aligned}$$

Hence for all $p \in \mathbf{N}$

$$|X_{\nu_u \wedge p}| \leqslant u + \{\sum_{n < \nu_u \wedge p} |X_{n+1} - X_n|^2\, 1_{\{|X_{n+1} - X_n|^2 \geqslant c(A_{n+1} - A_n)\}}\}^{1/2}$$

and then

$$\{E(X^2_{\nu_u\wedge p})\}^{1/2} \leqslant u + \Big\{E\big(\sum_{n<p} 1_{\{\nu_u>n\}}|X_{n+1}-X_n|^2 1_{\{|X_{n+1}-X_n|^2\geqslant c(A_{n+1}-A_n)\}}\big)\Big\}^{1/2}$$

$$= u + E\Big\{\sum_{n<p} 1_{\{\nu_u>n\}} E^{\mathscr{B}_n}\big(|X_{n+1}-X_n|^2 1_{\{|X_{n+1}-X_n|^2>c(A_{n+1}-A_n)\}}\big)\Big\}^{1/2}$$

By Gundy's condition this shows that

$$\|X_{\nu_u\wedge p}\|_2 \leqslant u + \sqrt{(1-\gamma)\,E(A_{\nu_u\wedge p})}.$$

But $E(X^2_{\nu_u\wedge p}) = E(A_{\nu_u\wedge p}) < \infty$; the preceding inequality can thus also be written

$$E(A^2_{\nu_u\wedge p}) \leqslant \{1-\sqrt{(1-\gamma)}\}^{-2}u^2.$$

As the right-hand side does not depend on p, this implies that $E(A_{\nu_u}) < \infty$ and hence that $A_\infty < \infty$ on $\{\nu_u = \infty\}$. We have thus proved that $A_\infty < \infty$ on $\lim\uparrow_u\{\nu_u=\infty\} = \{\sup_{\mathbf{N}}|X_n| < \infty\}$. ■

REMARK. By a different method, Davis [96] has obtained the more precise result that whenever Gundy's condition is satisfied,

$$\limsup_n X_n = +\infty, \qquad \liminf_n X_n = -\infty \text{ a.s.} \quad \text{on}\,\{A_\infty = \infty\}. \blacksquare$$

We now give a converse to part (b) of Proposition VII-**2**-3 which is also based on Gundy's condition.

PROPOSITION VII-**4**-15. *Let $(X_n, n\in\mathbf{N})$ be a square-integrable martingale with $X_0 = 0$ and satisfying Gundy's condition. Then there exists a constant k depending only on the constants involved in this condition such that*

$$E(\sqrt{A_\infty}) \leqslant kE(\sup_{\mathbf{N}}|X_n|).$$

PROOF. We begin by showing that

$$E(\sup_{\mathbf{N}}\sqrt{(A_{n+1}-A_n)}) \leqslant (2/a\alpha)\,E(\sup_{\mathbf{N}}|X_n|),$$

where α and a are the constants in the first form of Gundy's condition. (Definition VII-**4**-12(1)). To this end, for all $u\in\mathbf{R}_+$ we denote by ν_u the stopping time defined by

$$\nu_u = \begin{cases} \min(n : A_{n+1}-A_n > u^2) & \\ +\infty & \text{if } \sup_{\mathbf{N}}\sqrt{(A_{n+1}-A_n)} \leqslant u. \end{cases}$$

For any $p \in \mathbf{N}$, Gundy's condition allows us to write

$$P^{\mathscr{B}_p}(|X_{p+1} - X_p| > au) \geqslant P^{\mathscr{B}_p}(|X_{p+1} - X_p| > a\sqrt{(A_{p+1} - A_p)}) \geqslant \alpha$$

on the event $\{\nu_u = p\}$; this implies that

$$\begin{aligned} P(\sup_{\mathbf{N}} |X_{n+1} - X_n| > au) &\geqslant P(\nu_u < \infty, |X_{\nu_u+1} - X_{\nu_u}| > au) \\ &= E(\sum_{\mathbf{N}} 1_{\{\nu_u = p\}} P^{\mathscr{B}_p}(|X_{p+1} - X_p| > au)) \\ &\geqslant \alpha\, P(\nu_u < \infty) = \alpha P(\sup_{\mathbf{N}} \sqrt{(A_{n+1} - A_n)} > u) \end{aligned}$$

for every $u \in \mathbf{R}_+$. The inequality stated in the beginning is then obtained by integrating the two extreme terms in the preceding inequality with respect to Lebesgue measure on $\mathbf{R}_+$.

Next let us define a stopping time ν'_u for every $u \in \mathbf{R}_+$ by putting

$$\nu'_u = \min(n : |X_n| + \sqrt{c(A_{n+1} - A_n)} > u) \quad \text{or} + \infty,$$

taking for the constant c that figuring in the third form of Gundy's condition. For every integer $p \geqslant 1$, on the event $\{\nu'_u = p\}$ we can then write

$$\begin{aligned} |X_p| &\leqslant |X_{p-1}| + \sqrt{c(A_p - A_{p-1})} + |X_p - X_{p-1}| - \sqrt{c(A_p - A_{p-1})} \\ &\leqslant \min(u, \sup_{\mathbf{N}}(|X_n| + \sqrt{c(A_{n+1} - A_n)})) \\ &\quad + |X_p - X_{p-1}|\, 1_{\{|X_p - X_{p-1}|^2 \geqslant c(A_p - A_{p-1})\}}. \end{aligned}$$

As $X_0 = 0$, this implies that

$$\begin{aligned} |X_{\nu'_u \wedge p}| &\leqslant \min(u, \sup_{\mathbf{N}}(|X_n| + \sqrt{c(A_{n+1} - A_n)})) \\ &\quad + \left[\sum_{p \leqslant \nu'_u} (X_p - X_{p-1})^2 1_{\{(X_p - X_{p-1})^2 > c(A_p - A_{p-1})\}}\right]^{1/2} \end{aligned}$$

and upon taking the quadratic norms of both sides, we obtain

$$\begin{aligned} \sqrt{E(A_{\nu'_u \wedge p})} = \|X_{\nu'_u \wedge p}\|_2 &\leqslant \|\min(u, \sup_{\mathbf{N}}(|X_n| + \sqrt{c(A_{n+1} - A_n)}))\|_2 \\ &\quad + \left[E\left(\sum_{p \leqslant \nu'_u} E^{\mathscr{B}_{p-1}}(X_p - X_{p-1})^2\, 1_{\{(X_p - X_{p-1})^2 > c(A_p - A_{p-1})\}}\right)\right]^{1/2} \end{aligned}$$

But by Gundy's condition the last term is dominated by

$$[(1-\gamma)\,E(A_{\nu'_u\wedge p})]^{1/2};$$

the inequality which we have obtained can thus be rewritten in the following form, also letting $p \uparrow \infty$:

$$(1-\sqrt{(1-\gamma)})\sqrt{E(A_{\nu'_u})} = \lim_p\uparrow (1-\sqrt{(1-\gamma)})\sqrt{E(A_{\nu'_u\wedge p})}$$
$$\leqslant \|\min(u, \sup_{\mathbf{N}}(|X_n|, \sqrt{c(A_{n+1}-A_n)}))\|_2.$$

Since $1-\sqrt{(1-\gamma)} > \frac{1}{2}\gamma$, we have therefore shown that

$$E(A_{\nu'_u}) \leqslant 4\gamma^{-2}\,E(\min(u^2, \sup_{\mathbf{N}}\{|X_n| + \sqrt{c(A_{n+1}-A_n)}\}^2)).$$

To complete the proof, we then write

$$P(A_\infty > u^2) \leqslant P(\nu'_u < \infty) + P(\nu'_u = \infty,\ A_{\nu'_u} > u^2)$$
$$\leqslant P(\sup_{\mathbf{N}}(|X_n| + \sqrt{c(A_{n+1}-A_n)}) > u) + u^{-2}\,E(A_{\nu'_u}).$$

Using the inequality obtained above, an integration in u with respect to Lebesgue measure on $\mathbf{R}_+$ turns this last inequality into

$$E(\sqrt{A_\infty}) = \int_0^\infty P(A_\infty > u^2)\,\mathrm{d}u$$
$$\leqslant E(\sup_{\mathbf{N}}(|X_n| + \sqrt{c(A_{n+1}-A_n)}))$$
$$+ 8\gamma^{-2}\,E(\sup_{\mathbf{N}}(|X_n| + \sqrt{c(A_{n+1}-A_n)})).$$

Taking the first part of the proof into account, this shows that

$$E(\sqrt{A_\infty}) \leqslant (1+8\gamma^{-2})(1+2(a\alpha)^{-1}\sqrt{c})\,E(\sup_{\mathbf{N}}|X_n|).\ \blacksquare$$

VII-5. Exercises

VII-1. Let $(B_n, n \in \mathbf{N})$ be an adapted sequence of events in the space $[\Omega, \mathscr{A}, P, (\mathscr{B}_n, n \in \mathbf{N})]$ such that $P(\limsup_{n\to\infty} B_n) = 0$. Show that the formula

$$A_{n+1} - A_n = 1_{B_n} P^{\mathscr{B}_n}((\bigcup_{k>n} B_k)^c) \qquad (n \in \mathbf{N};\ A_0 = 0)$$

defines an increasing process $(A_n, n \in \mathbf{N})$ dominated by the constant 1 whose potential $(X_n = E^{\mathscr{B}_n}(A_\infty - A_n), n \in \mathbf{N})$ is

$$X_n = P^{\mathscr{B}_n}(\bigcup_{k \geq n} B_k) \qquad (n \in \mathbf{N}).$$

As a consequence, this increasing process has the following two properties:

(a) $A_{n+1} - A_n = 0$ on B_n^c,
(b) $X_n = 1$ on B_n $\quad (n \in \mathbf{N})$.

Conversely, if $(B_n, n \in \mathbf{N})$ is an adapted sequence of events and if there exists an increasing process having properties (a) and (b), show that

$$P(\limsup_{n \to \infty} B_n) = 0.$$

[Deduce that $X_n \geq P^{\mathscr{B}_n}(\bigcup_{n \leq k < n+j} B_k)$ from (a) and (b) by arguing inductively on j.]

VII-2 (a) Let $(X_n, n \in \mathbf{N})$ be a positive integrable submartingale defined on

$$[\Omega, \mathscr{A}, P, (\mathscr{B}_n, n \in \mathbf{N})]$$

and let $(c_n, n \in \mathbf{N})$ be a decreasing sequence in $\mathbf{R}_+$. Put

$$Z_n = c_0 X_0 + \sum_{m=1}^{n} c_m(X_m - X_{m-1})$$

for every $n \in \mathbf{N}$ and observe that this definition implies that $Z_n \geq c_n X_n$ for all $n \in \mathbf{N}$. Show that

$$P(\sup_{m \leq n} c_m X_m > a) \leq a^{-1} E(Z_n) \qquad (n \in \mathbf{N})$$

for every real number $a > 0$.

(b) If $(Y_n, n \in \mathbf{N})$ is a square-integrable martingale and $(c_n, n \in \mathbf{N})$ is a decreasing sequence in $\mathbf{R}_+$, show that

$$a^2 P(\sup_{\mathbf{N}} c_n |Y_n| > a) \leq c_0^2 E(Y_0^2) + \sum_{\mathbf{N}} c_m^2 E((Y_m - Y_{m-1})^2)$$

for every real $a > 0$ and conclude by a passage to the limit that

$$P(\limsup_{n \to \infty} c_n |Y_n| = 0) = 1$$

whenever $\sum_{\mathbf{N}^*} c_m^2 E((Y_m - Y_{m-1})^2) < \infty$ and $\lim\downarrow_p c_p = 0$.

VII-3. Let $(X_n, n \in \mathbf{N})$ be a martingale with integrable pth power $(p > 1)$ and let $|X_n|^p = M_n^{(p)} + A_n^{(p)}$ be the Doob decomposition of the submartingale $(|X_n|^p, n \in \mathbf{N})$. Show that every stopping time ν for which $E(A_\nu^{(p)}) < \infty$ is regular for both the martingales $(X_n, n \in \mathbf{N})$ and $(M_n^{(p)}, n \in \mathbf{N})$ and conclude that

$$E(|X_\nu|^p) = E(|X_0|^p) + E(|A_\nu|^p).$$

[Recall the case $\nu = \infty$. Then show that $\sup_{\mathbf{N}} |X_n| \in L^p$ if $E(A_\infty^{(p)}) < \infty$ and then under this same hypothesis that $\sup_{\mathbf{N}} |M_n^{(p)}| \leqslant \sup_{\mathbf{N}} |X_n|^p + A_\infty^{(p)} \in L^1$.]

VII-4. Let $(Y_n, n \in \mathbf{N})$ be an adapted sequence of real-valued r.v.'s defined on the space $[\Omega, \mathscr{A}, P, (\mathscr{B}_n, n \in \mathbf{N})]$ such that $|Y_n| \leqslant 1$ a.s. for every $n \in \mathbf{N}$. Show that for every $\lambda \in \mathbf{R}$, the sequence

$$Z_n^{(\lambda)} = \prod_{m=1}^{n} \frac{\exp(\lambda Y_m)}{E^{\mathscr{B}_{m-1}}(\exp(\lambda Y_m))} \qquad (n \in \mathbf{N})$$

of r.r.v.'s converges a.s. to a finite limit $Z_\infty^{(\lambda)}$ say. Then show that

$$\{Z_\infty^{(\lambda)} > 0\} = \left\{ \sum_{\mathbf{N}^*} V_n < \infty \right\},$$

where $V_n = E^{\mathscr{B}_{n-1}}([Y_n - E^{\mathscr{B}_{n-1}}(Y_n)]^2)$ for every $n \in \mathbf{N}$. [Use the inequality

$$1 \leqslant E^{\mathscr{B}}(\exp(\lambda U)) \leqslant \exp[\phi_c(\lambda)\, E^{\mathscr{B}}(U^2)]$$

valid for every r.r.v. such that $|U| \leqslant c$ and for every real number λ, if

$$\phi_c(\lambda) = c^{-2}(\exp(\lambda c) - 1 - \lambda c).]$$

How do the preceding results specialise in the case of an independent sequence $(Y_n, n \in \mathbf{N})$?

VII-5. Let $(X_n, n \in \mathbf{N})$ be a square-integrable martingale such that $X_0 = 0$, and let $(A_n, n \in \mathbf{N})$ be the increasing process in the Doob decomposition of the submartingale $(X_n^2, n \in \mathbf{N})$. If $h: \mathbf{R} \times \mathbf{R}_+ \to \bar{\mathbf{R}}_+$ is a Borel-measurable function satisfying the inequality

$$h(x, a) \geqslant \int_{\mathbf{R}} h(x + y, a + b)\, \mathrm{d}\mu(y)$$

for all $x \in \mathbf{R}$, all a,b $\in \mathbf{R}_+$ and all probability measures μ on $\mathbf{R}$ such that $\int y\, \mathrm{d}\mu(y) = 0$ and $\int y^2\, \mathrm{d}\mu(y) = b$, show that for every fixed $(x, a) \in \mathbf{R} \times \mathbf{R}_+$ the sequence

$$(h(x + X_n, a + A_n), n \in \mathbf{N})$$

is a positive supermartingale.

If $g : \mathbf{R}_+ \to \mathbf{R}_+$ is a positive decreasing function, the function

$$h(x,a) = x^2 g(a) + \int_a^\infty g(u)\, du$$

satisfies the above assumptions; in particular the function $h_0(x,a) = (a^{-1}x)^2 + a^{-1}$ satisfies this hypothesis. Using the supermartingale associated with h_0, show that

$$P(\sup_{\mathbf{N}}(|X_n| > \alpha A_n + \beta)) \leqslant (\alpha\beta)^{-1} \quad \text{if } \alpha,\beta > 0 \text{ in } \mathbf{R},$$

and deduce that $\lim_{n\to\infty}(X_n/A_n) = 0$ a.s. on $\{A_\infty = \infty\}$. By taking $g = 1/f^2$, where f satisfies the assumptions of Proposition VII-**2**-4 in the above, one can improve this last result and recover Proposition VII-2-4.

VII-6. (a) Let $M = (M_n, n \in \mathbf{N})$ be a square-integrable martingale and let $(A_n, n \in \mathbf{N})$ be the increasing process in the Doob decomposition of the submartingale $(M_n^2, n \in \mathbf{N})$. Denote by $\Lambda^2(M)$ the class of all adapted sequences $Y = (Y_n, n \in \mathbf{N})$ such that

$$E\left(\sum_{\mathbf{N}} Y_n^2(A_{n+1} - A_n)\right) < \infty;$$

subject to identifying two sequences Y and Y' whose differences $\Delta_n = Y_n' - Y_n\ (n \in \mathbf{N})$ are such that $\sum_{\mathbf{N}} \Delta_n^2(A_{n+1} - A_n) = 0$ a.s., show that $\Lambda^2(M)$ is a real Hilbert space for the inner product

$$\langle Y, Y' \rangle = E\left(\sum_{\mathbf{N}} Y_n\, Y_n'(A_{n+1} - A_n)\right).$$

Show also that the formula

$$M(Y) = \sum_{\mathbf{N}} Y_n(M_{n+1} - M_n) \qquad (Y \in \Lambda^2(M))$$

(where the right-hand side converges in L^2) defines an isometry of $\Lambda^2(M)$ into $L^2(\Omega, \mathscr{A}, P)$ whose image will be denoted by H_M.

(b) Two square-integrable martingales $(M_n, n \in \mathbf{N})$ and $(M_n', n \in \mathbf{N})$ are said to be orthogonal if $H_M \perp H_{M'}$ in $L^2(\Omega, \mathscr{A}, P)$; show that this orthogonality condition obtains if and only if the sequence $(M_n M_n', n \in \mathbf{N})$ of integrable r.r.v.'s is a martingale. If M and M' are two arbitrary square-integrable martingales, show that there exists a unique element $Y \in \Lambda^2(M)$ such that the difference

$$(M_n' - E^{\mathscr{B}_n}(M(Y)), \quad n \in \mathbf{N})$$

is a martingale orthogonal to M; moreover, Y is given by

$$Y_n(A_{n+1} - A_n) = E^{\mathscr{B}_n}(M_{n+1} M_{n+1}') - M_n M_n' \qquad (n \in \mathbf{N}).$$

CHAPTER VIII

DOOB'S DECOMPOSITION OF POSITIVE SUPERMARTINGALES

VIII-1. Generalities

PROPOSITION VIII-1-1. *Every finite positive supermartingale* $(X_n, n \in \mathbf{N})$ *can be written in one and only one way as the difference between a finite positive martingale* $(M_n, n \in \mathbf{N})$ *and an increasing process* $(A_n, n \in \mathbf{N})$,

$$X_n = M_n - A_n \qquad (n \in \mathbf{N}).$$

We have $M_0 = X_0$ *and* $E^{\mathscr{B}_0}(A_\infty) \leqslant X_0$. *Further,* $M_n \geqslant E^{\mathscr{B}_n}(A_\infty)$ *for every* $n \in \mathbf{N}$.

PROOF. Let us define the sequences $(M_n, n \in \mathbf{N})$ and $(A_n, n \in \mathbf{N})$ of r.r.v.'s through their increments by the formulae

$$\begin{aligned} M_0 &= X_0, \qquad M_{n+1} - M_n = X_{n+1} - E^{\mathscr{B}_n}(X_{n+1}), \\ A_0 &= 0, \qquad A_{n+1} - A_n = X_n - E^{\mathscr{B}_n}(X_{n+1}) \end{aligned} \qquad (n \in \mathbf{N}).$$

Note that the hypothesis $0 \leqslant E^{\mathscr{B}_n}(X_{n+1}) \leqslant X_n$ implies that the r.v.'s $E^{\mathscr{B}_n}(X_{n+1})$ are finite. It is then immediately seen from the preceding formulae that the sequence $(A_n, n \in \mathbf{N})$ is an increasing process. The decomposition formula $X_n = M_n - A_n$ $(n \in \mathbf{N})$ follows directly from the equality $X_0 = M_0 - A_0$ and the equality of the increments of the two sequences $(X_n, n \in \mathbf{N})$ and $(M_n - A_n, n \in \mathbf{N})$. This formula shows that $M_n \geqslant A_n \geqslant 0$ for every $n \in \mathbf{N}$; finally it is readily seen from the definition of the increments that the sequence $(M_n, n \in \mathbf{N})$ is a martingale.

The uniqueness of the Doob decomposition of the supermartingale $(X_n, n \in \mathbf{N})$ is easy to establish. Indeed, if $X_n = M'_n - A'_n$ $(n \in \mathbf{N})$ is a second decomposition of $(X_n, n \in \mathbf{N})$ of the same type, the equality

$$A'_{n+1} - A'_n = (M'_{n+1} - M'_n) - (X_{n+1} - X_n)$$

implies, if we take the conditional expectation of both sides, that

$$A'_{n+1} - A'_n = X_n - E^{\mathscr{B}_n}(X_{n+1}) = A_{n+1} - A_n;$$

since $A'_0 = 0 = A_0$ this implies that $A'_n = A_n$ and hence also, by adding X_n, that $M'_n = M_n$ $(n \in \mathbf{N})$.

The inequality $A_p \leqslant M_p$ $(p \in \mathbf{N})$ which follows from the positivity of X_p $(p \in \mathbf{N})$ implies that

$$E^{\mathscr{B}_n}(A_p) \leqslant E^{\mathscr{B}_n}(M_p) = M_n \quad \text{if } n \leqslant p \quad (n, p \in \mathbf{N}),$$

and hence, letting $p \uparrow \infty$, that $E^{\mathscr{B}_n}(A_\infty) \leqslant M_n$ for every $n \in \mathbf{N}$. ∎

The following proposition introduces the notion of potential of an increasing process and gives some consequences of Proposition VIII-**1**-1.

PROPOSITION VIII-**1**-2. *For every increasing process* $(A_n, n \in \mathbf{N})$ *such that* $E^{\mathscr{B}_0}(A_\infty) < \infty$ *a.s., the formula*

$$X_n = E^{\mathscr{B}_n}(A_\infty) - A_n \qquad (n \in \mathbf{N})$$

defines a finite positive supermartingale $(X_n, n \in \mathbf{N})$ *which is called the potential of the increasing process* $(A_n, n \in \mathbf{N})$. *This potential* $(X_n, n \in \mathbf{N})$ *determines the increasing process* $(A_n, n \in \mathbf{N})$ *uniquely.*

For a finite positive supermartingale $(X_n, n \in \mathbf{N})$ *to be the potential of an increasing process* $(A_n, n \in \mathbf{N})$ *such that* $E^{\mathscr{B}_0}(A_\infty) < \infty$ *a.s., it is necessary and sufficient that* $\lim\downarrow_{n\uparrow\infty} E^{\mathscr{B}_0}(X_n) = 0$ *a.s.*

PROOF. For every increasing sequence $(B_n, n \in \mathbf{N})$ of positive r.r.v.'s such that $E^{\mathscr{B}_0}(B_\infty) < \infty$ a.s. (where $B_\infty = \lim\uparrow_n B_n$), in particular for every increasing process $(A_n, n \in \mathbf{N})$ such that $E^{\mathscr{B}_0}(A_\infty) < \infty$, the formula

$$X_n = E^{\mathscr{B}_n}(B_\infty - B_n) \qquad (n \in \mathbf{N}),$$

which can also be written as $X_n = E^{\mathscr{B}_n}(A_\infty) - A_n$ in the case of an increasing process, defines a supermartingale since

$$E^{\mathscr{B}_n}(X_{n+1}) = E^{\mathscr{B}_n}(B_\infty - B_{n+1}) \leqslant E^{\mathscr{B}_n}(B_\infty - B_n) = X_n \qquad (n \in \mathbf{N});$$

note that $E^{\mathscr{B}_n}(B_\infty) < \infty$ a.s. for all $n \in \mathbf{N}$ since the conditional expectation $E^{\mathscr{B}_0}(E^{\mathscr{B}_n}(B_\infty)) = E^{\mathscr{B}_0}(B_\infty)$ is a.s. finite by assumption. This supermartingale is positive and finite; further

$$E^{\mathscr{B}_0}(X_n) = E^{\mathscr{B}_0}(B_\infty) - E^{\mathscr{B}_0}(B_n) \downarrow 0 \quad \text{when } n \uparrow \infty.$$

Conversely, let $(X_n, n \in \mathbf{N})$ be a finite positive supermartingale such that $\lim\downarrow_n E^{\mathscr{B}_0}(X_n) = 0$ a.s. Let us write its Doob decomposition $X_n = M_n - A_n$ $(n \in \mathbf{N})$ as in Proposition VIII-**1**-1; then we have

$$E^{\mathscr{B}_0}(X_n) = M_0 - E^{\mathscr{B}_0}(A_n) \quad \text{for every } n \in \mathbf{N},$$

so that the assumption $\lim\downarrow_n E^{\mathscr{B}_0}(X_n)=0$ a.s. implies, by passing to the limit, that $M_0=E^{\mathscr{B}_0}(A_\infty)$. Since by Proposition VII-**1**-1 the martingale $(M_n-E^{\mathscr{B}_n}(A_\infty),\ n\in\mathbf{N})$ is positive, the initial r.v. $M_0-E^{\mathscr{B}_0}(A_\infty)$ of this martingale can only be a.s. null if the martingale is identically zero; hence

$$X_n=M_n-A_n=E^{\mathscr{B}_n}(A_\infty)-A_n.$$

Finally, by the uniqueness of the Doob decomposition, there cannot be a further increasing process distinct from $(A_n,\ n\in\mathbf{N})$ for which $(X_n,\ n\in\mathbf{N})$ is the potential. ■

The first part of the preceding proof also establishes the following result:

COROLLARY VIII-**1**-3. *For every increasing sequence $(B_n,\ n\in\mathbf{N})$ of positive r.r.v.'s such that $E^{\mathscr{B}_0}(B_\infty)<\infty$ a.s., where $B_\infty=\lim\uparrow_n B_n$, there exists a unique increasing process $(A_n,\ n\in\mathbf{N})$ such that $E^{\mathscr{B}_0}(A_\infty)<\infty$ a.s. and for which*

$$E^{\mathscr{B}_n}(B_\infty-B_n)=E^{\mathscr{B}_n}(A_\infty-A_n)\qquad(n\in\mathbf{N}).$$

By the explicit formulae in the proof of Proposition VIII-**1**-1, the increasing process $(A_n,\ n\in\mathbf{N})$ of this corollary, which is the increasing process in the Doob decomposition of the supermartingale

$$(E^{\mathscr{B}_n}(B_\infty-B_n),\ n\in\mathbf{N}),$$

is given by the simple formula

$$A_{n+1}-A_n=E^{\mathscr{B}_n}(B_{n+1}-B_n)\qquad(n\in\mathbf{N}).$$

The introduction of potentials of increasing processes suggests writing the Doob decomposition of a finite positive supermartingale in a slightly different way. Putting $M_n'=M_n-E^{\mathscr{B}_n}(A_\infty)$ for every $n\in\mathbf{N}$, we see that *every finite positive supermartingale $(X_n,\ n\in\mathbf{N})$ can be written as the sum of a finite positive martingale and the potential of an increasing process, say*

$$X_n=M_n'+E^{\mathscr{B}_n}(A_\infty-A_n)\qquad(n\in\mathbf{N}).$$

This decomposition is called the *Riesz decomposition* of the supermartingale, and like Doob's decomposition it is unique.

EXAMPLE. Consider a Markov chain with state space E and transition matrix P and also a finite positive superharmonic function f on E (see Section III-**5**). The increasing process $(A_n,\ n\in\mathbf{N})$ in the Doob and Riesz decompositions of the finite positive supermartingale $(f(X_n),\ n\in\mathbf{N})$ is given by the formulae

$$A_0=0,\qquad A_{n+1}-A_n=E^{\mathscr{B}_n}(f(X_n)-f(X_{n+1}))\quad(n\in\mathbf{N});$$

defining the function $u : E \to \mathbf{R}_+$ by $u = f - Pf$, we thus see that

$$A_n = \sum_{m<n} u(X_m), \qquad A_\infty = \sum_{\mathbf{N}} u(X_n) \text{ a.s.},$$

whatever the initial state of the Markov chain. Consequently,

$$E^{\mathscr{B}_n}(A_\infty - A_n) = E^{\mathscr{B}_n}\left(\sum_{m \geqslant n} u(X_m)\right) = Uu(X_n),$$

where $U = \sum_{\mathbf{N}} P_n$ denotes the potential matrix of the chain, and

$$Uu(x) = \mathbf{E}_x^{\mathscr{B}_0}(A_\infty) \leqslant f(x) \qquad (x \in E).$$

The positive function $f' = f - Uu$ is harmonic since it corresponds to the martingale

$$f'(X_n) = f(X_n) - Uu(X_n) = M_n - E^{\mathscr{B}_n}(A_\infty) = M'_n,$$

if $f(X_n) = M_n - A_n$ represents the Doob decomposition of the supermartingale $(f(X_n), n \in \mathbf{N})$. ■

We conclude this section by an elementary result comparing the L^p-norms of the final r.r.v. A_∞ of an increasing process $(A_n, n \in \mathbf{N})$ and the supremum $\sup_{\mathbf{N}} X_n$ of its potential.

PROPOSITION VIII-1-4. *The potential* $(X_n = E^{\mathscr{B}_n}(A_\infty - A_n), n \in \mathbf{N})$ *of every increasing process* $(A_n, n \in \mathbf{N})$ *satisfies the two inequalities*

$$\|\sup X_n\|_p \leqslant \frac{p}{p-1} \|A_\infty\|_p, \qquad \|A_\infty\|_p \leqslant p \|\sup_{\mathbf{N}} X_n\|_p,$$

for any real number $p > 1$. *More generally, for every positive r.r.v.* Z *such that* $X_n \leqslant E^{\mathscr{B}_n}(Z)$ *for every* $n \in \mathbf{N}$, *we have*

$$\|A_\infty\|_p \leqslant p \|Z\|_p \quad \text{if } p \geqslant 1.$$

In particular, if $X_n \leqslant c$ $(n \in \mathbf{N})$ *for a constant* c, *the pth power of the r.v.* A_∞ *is integrable for every* $p < \infty$; *furthermore, we then have*

$$E^{\mathscr{B}_n}(\exp \lambda(A_\infty - A_n)) \leqslant (1 - \lambda c)^{-1} \qquad (n \in \mathbf{N})$$

for every real $\lambda \in]0, c^{-1}[$.

PROOF. Since $0 \leqslant X_n \leqslant E^{\mathscr{B}_n}(A_\infty)$ for every $n \in \mathbf{N}$, the first inequality of the proposition is an immediate consequence of the formula of Proposition IV-**2**-8.

Let Z be a positive r.r.v. whose associated martingale $(E^{\mathscr{B}_n}(Z), n \in \mathbf{N})$ dominates the potential $(X_n = E^{\mathscr{B}_n}(A_\infty - A_n), n \in \mathbf{N})$; in particular, $Z = \sup_{\mathbf{N}} X_n$ is one such r.v. To show that $\|A_\infty\|_p \leqslant p\|Z\|_p$, we can suppose that the r.v. $A_\infty \in L^p$; indeed, if this inequality is proved when $A_\infty \in L^p$, it will be true in the general case because applying it to the truncated increasing process $(\min(A_n, a), n \in \mathbf{N})$ whose potential is also dominated by $(E^{\mathscr{B}_n}(Z), n \in \mathbf{N})$ since $\min(A_\infty, a) - \min(A_n, a) \leqslant A_\infty - A_n$, we find that $\|\min(A_\infty, a)\|_p \leqslant p\|Z\|_p$ and hence, letting $a \uparrow \infty$, that $\|A_\infty\|_p \leqslant p\|Z\|_p$.

Suppose that $A_\infty \in L^p$, and for every $u \in \mathbf{R}_+$ let v_u denote the stopping time defined by

$$v_u = \begin{cases} \inf(n : A_{n+1} > u), & \\ +\infty & \text{if } A_\infty \leqslant u. \end{cases}$$

Then $A_{v_u} \leqslant u$ on Ω and consequently, since the event $\{A_\infty > u\} = \{v_u < \infty\}$ belongs to $\mathscr{B}_{v_u}$, we can write

$$\begin{aligned} E^{\mathscr{B}_0}((A_\infty - u)\,1_{\{A_\infty > u\}}) &\leqslant E^{\mathscr{B}_0}((A_\infty - A_{v_u})\,1_{\{v_u < \infty\}}) \\ &= E^{\mathscr{B}_0}(E^{\mathscr{B}_{v_u}}(A_\infty - A_{v_u})\,1_{\{v_u < \infty\}}) \\ &\leqslant E^{\mathscr{B}_0}(E^{\mathscr{B}_{v_u}}(Z)\,1_{\{v_u < \infty\}}) = E^{\mathscr{B}_0}(Z\,1_{\{A_\infty > u\}}). \end{aligned}$$

for every $u \in \mathbf{R}_+$. Integrate the first and last terms of these inequalities with respect to the positive measure $(p-1)u^{p-2}du$ over $\mathbf{R}_+$; by Fubini's theorem and an elementary calculation, we find that

$$\begin{aligned} \frac{1}{p} E(A_\infty^p) &= \int_0^\infty (p-1)\,u^{p-2}\,E((A_\infty - u)\,1_{\{A_\infty > u\}})\,du \\ &\leqslant \int_0^\infty (p-1)\,u^{p-2}\,E(Z\,1_{\{A_\infty > u\}})\,du = E(Z\,A_\infty^{p-1}). \end{aligned}$$

By Hölder's inequality, the last term is also dominated by

$$E(Z\,A_\infty^{p-1}) \leqslant \|Z\|_p \|A_\infty^{p-1}\|_{p/p-1} = \|Z\|_p \|A_\infty\|_p^{p-1};$$

hence we have proved that

$$\|A_\infty\|_p^p = E(A_\infty^p) \leqslant p\|Z\|_p \|A_\infty\|_p^{p-1}.$$

Dividing both sides by the *finite* number $\|A_\infty\|_p^{p-1}$ gives us the desired inequality.

Finally, let us suppose that the potential $X_n = E^{\mathscr{B}_n}(A_\infty - A_n)$ is dominated by a constant c for each $n \in \mathbf{N}$. Then the results just proved imply that $\|A_\infty\|_p \leqslant pc$ for every real number $p \in [1, \infty[$; this does not imply that the r.v. A_∞ is bounded, but it is not hard to deduce from the preceding argument that $E^{\mathscr{B}_0}(\exp(\lambda A_\infty)) \leqslant (1 - \lambda c)^{-1}$ when $\lambda \in]0, 1/c[$. It suffices to integrate the inequality

$$E^{\mathscr{B}_0}((A_\infty - u)1_{\{A_\infty > u\}}) \leqslant E^{\mathscr{B}_0}(c\,1_{\{A_\infty > u\}}) \qquad (u \in \mathbf{R}_+)$$

obtained above with respect to the positive measure $\lambda^2 e^{\lambda u}\, du$ on $\mathbf{R}_+$; we find that

$$E^{\mathscr{B}_0}(\exp(\lambda A_\infty) - \lambda A_\infty - 1) \leqslant c\lambda\, E^{\mathscr{B}_0}(\exp(\lambda A_\infty) - 1)$$

or, equivalently,

$$(1 - \lambda c)\, E^{\mathscr{B}_0}(\exp(\lambda A_\infty)) \leqslant 1 + \lambda[E^{\mathscr{B}_0}(A_\infty) - c],$$

at least when $E^{\mathscr{B}_0}(\exp(\lambda A_\infty)) < \infty$. Since $E^{\mathscr{B}_0}(A_\infty) = X_0 \leqslant c$, we have thus established that $(1 - \lambda c)E^{\mathscr{B}_0}(\exp(\lambda A_\infty)) \leqslant 1$ if the conditional expectation in the left-hand side is finite; but the latter restriction can easily be lifted by reasoning with the processes $(\min(A_n, a), n \in \mathbf{N})$ $(a \in \mathbf{R}_+)$. Finally, the last inequality in the statement,

$$(1 - \lambda c)\, E^{\mathscr{B}_n}(\exp \lambda(A_\infty - A_n)) \leqslant 1 \quad (n \in \mathbf{N}, \lambda \in]0, c^{-1}[\,),$$

is obtained by a shift of time in the result already obtained. ■

A potential $(X_n, n \in \mathbf{N})$ can often be dominated by martingales other than the martingale $(E^{\mathscr{B}_n}(\sup_{\mathbf{N}} X_n), n \in \mathbf{N})$ considered in the preceding proof. For example, if $(B_n, n \in \mathbf{N})$ is an increasing sequence of positive r.r.v.'s (not necessarily adapted, nor a fortiori predictable) such that $B_\infty \in L^p$, Corollary VIII-**1**-3 shows that $(X_n = E^{\mathscr{B}_n}(B_\infty - B_n), n \in \mathbf{N})$ is the potential of an increasing process $(A_n, n \in \mathbf{N})$; this potential is clearly dominated by the martingale $(E^{\mathscr{B}_n}(B_\infty), n \in \mathbf{N})$ and the preceding proposition shows in this case that $\|A_\infty\|_p \leqslant p\|B_\infty\|_p$.

The following corollary is an immediate application of the preceding proposition.

Corollary VIII-**1**-5. *For every sequence $(Z_n, n \in \mathbf{N})$ of positive r.r.v.'s and for every real number $p \in [1, \infty[$, the following inequality holds:*

$$\|\sum_{\mathbf{N}} E^{\mathscr{B}_n}(Z_n)\|_p \leqslant p\|\sum_{\mathbf{N}} Z_n\|_p.$$

PROOF. The formula $A_n = \sum_{m<n} E^{\mathscr{B}_n}(Z_m)$, where n varies over $\mathbf{N}$, defines an increasing process whose potential satisfies the inequality

$$E^{\mathscr{B}_n}(A_\infty - A_n) = E^{\mathscr{B}_n}\Big(\sum_{m\geqslant n} E^{\mathscr{B}_m}(Z_m)\Big)$$

$$= E^{\mathscr{B}_n}\Big(\sum_{m\geqslant n} Z_m\Big)$$

$$\leqslant E^{\mathscr{B}_n}(Z) \quad \text{if } Z = \sum_{\mathbf{N}} Z_m.$$

The preceding proposition then shows that $\|A_\infty\|_p \leqslant p\|Z\|_p$, which is precisely the result stated in the corollary. ■

Another example of the application of the preceding proposition is furnished by Doob's decomposition of square-integrable martingales, as the following Corollary shows:

COROLLARY VIII-**1**-6. *Let p be a real number >1 and let Y be a r.v. in $L^{2p}(\Omega, \mathscr{A}, P)$; put $Y_n = E^{\mathscr{B}_n}(Y)$ for every $n \in \mathbf{N}$. The increasing process $(A_n, n \in \mathbf{N})$ in Doob's decomposition of the submartingale $(Y_n^2, n \in \mathbf{N})$ then satisfies the inequality*

$$\|A_\infty\|_p \leqslant p\| Y^2\|_p \qquad (= p\| Y\|_{2p}^2)$$

and in particular, $A_\infty \in L^p$.

The result of this corollary should be compared with that of Proposition VII-**2**-3b and the remark which follows it; by these results the norms $\|\sqrt{A_\infty}\|_q$ and $\| Y\|_q$ satisfy an inequality of the form $\| Y\|_q \leqslant c_q\|\sqrt{A_\infty}\|_q$ if $q \leqslant 2$ and an inequality in the other direction, $\| Y\|_q \geqslant c_q'\|\sqrt{A_\infty}\|_q$ if $q \geqslant 2$.

PROOF. By Section VII-**2** the increasing process $(A_n, n \in \mathbf{N})$ is given by the formula

$$A_{n+1} - A_n = E^{\mathscr{B}_n}(Y_{n+1}^2) - Y_n^2 \qquad (n \in \mathbf{N}).$$

Consequently, for every $m \in \mathbf{N}$ we have

$$E^{\mathscr{B}_m}(A_\infty - A_m) = \sum_{n\geqslant m} E^{\mathscr{B}_m}(A_{n+1} - A_n) = \sum_{n\geqslant m} [E^{\mathscr{B}_m}(Y_{n+1}^2) - E^{\mathscr{B}_m}(Y_n^2)]$$

$$= E^{\mathscr{B}_m}(Y_\infty^2) - Y_m^2$$

because $E^{\mathscr{B}_m}(Y_n^2) \to E^{\mathscr{B}_m}(Y_\infty^2)$ in L^1 (since $Y_n \to Y_\infty = E^{\mathscr{B}_\infty}(Y)$ in L^2) when $n\uparrow\infty$. Following this inequality, the potential

$$(X_m = E^{\mathscr{B}_m}(A_\infty - A_m), m \in \mathbf{N})$$

is dominated by the martingale $(E^{\mathscr{B}_m}(Y_\infty^2), m \in \mathbf{N})$ and thus a fortiori by the martingale $(E^{\mathscr{B}_m}(Y^2), m \in \mathbf{N})$; to complete the proof it only remains to apply the preceding proposition. ■

VIII-2. Supplement; A remarkable duality

There exists a remarkable duality relation between the space of martingales $(X_n, n \in \mathbf{N})$ for which $E(\sqrt{A_\infty}) < \infty$ which we have already encountered in Proposition VII-2-3, and the space of martingales $(Y_n, n \in \mathbf{N})$ such that $\sup_{\mathbf{N}} E^{\mathscr{B}_n}(B_\infty - B_n) \in L^\infty$; here we denote by $(A_n, n \in \mathbf{N})$ (resp. $(B_n\ n \in \mathbf{N})$) the increasing process in Doob's decomposition of the submartingale $(X_n^2, n \in \mathbf{N})$ (resp. $(Y_n^2, n \in \mathbf{N})$). The study of this duality is the object of the following proposition.

Before giving the statement of the proposition let us observe that the condition $E(\sqrt{A_\infty}) < \infty$ can perhaps be satisfied by martingales which are not square-integrable. More precisely, if $(X_n, n \in \mathbf{N})$ is an adapted sequence of finite r.r.v.'s, the conditional expectations $E^{\mathscr{B}_n}((X_{n+1} - X_n)^2)$ of positive r.v.'s always make sense and we can thus introduce the condition

$$E\left(\left\{\sum_{\mathbf{N}} E^{\mathscr{B}_n}((X_{n+1} - X_n)^2)\right\}^{1/2}\right) < \infty;$$

since every r.r.v. Z satisfies the chain of inequalities

$$E(|Z|) = E(E^{\mathscr{B}}(|Z|)) \leqslant E\{\sqrt{(E^{\mathscr{B}}(Z^2))}\} \leqslant \sqrt{\{E(E^{\mathscr{B}}(Z^2))\}} = \sqrt{\{E(Z^2)\}} \leqslant +\infty$$

the preceding condition implies that the r.v.'s $X_{n+1} - X_n (n \in \mathbf{N})$ are integrable, but not necessarily square integrable. If furthermore $X_0 = 0$ and $E^{\mathscr{B}_n}(X_{n+1} - X_n) = 0$ for all $n \in \mathbf{N}$, we say that $(X_n, n \in \mathbf{N})$ is an integrable martingale, null at 0 and satisfying the condition $E(\sqrt{A_\infty}) < \infty$.

PROPOSITION VIII-2-7. *The integrable martingales $X = (X_n, n \in \mathbf{N})$ which are null at* 0 *and which satisfy the condition $E(\sqrt{A_\infty}) < \infty$ form a Banach space, say G, for the norm defined by this expectation (subject to identifying two a.s. equal martingales).*

Every martingale $Y=(Y_n, n\in\mathbf{N})$ bounded in L^2, null at 0 and such that $\sigma^2=\|sup_{\mathbf{N}} E^{\mathscr{B}_n}(B_\infty - B_n)\|_\infty<\infty$ defines a continuous linear functional on the space G by the formula

$$(X, Y)=E\left(\sum_{\mathbf{N}}(X_{n+1}-X_n)(Y_{n+1}-Y_n)\right).$$

Whenever X is bounded in L^2 we have $(X, Y)=E(X_\infty Y_\infty)$. Further, the norm of this linear functional, say $\|Y\|_G'=\sup((X, Y): \|X\|_G\leqslant 1)$ satisfies the double inequality $\sigma\leqslant\|Y\|_G'\leqslant\sigma\sqrt{2}$.

Conversely, every continuous linear functional on the space G is of the preceding form for a uniquely determined martingale Y.

PROOF. (1) The proof of the first part of the proposition uses nothing more than classical ideas from integration theory. If X' and X'' are two integrable martingales in G and if $a', a''\in\mathbf{R}$, the formula $X_n=a'X_n'+a''X_n''$ defines an integrable martingale which is null at the origin. The sublinearity of the functional $\sqrt{\{E^{\mathscr{B}}(Z^2)\}}$ then implies that

$$\sqrt{(A_{n+1}-A_n)}\leqslant|a'|\sqrt{(A_{n+1}'-A_n')}+|a''|\sqrt{(A_{n+1}''-A_n'')}\qquad(n\in\mathbf{N})$$

and the sublinearity of the norm in $l^2(\mathbf{N})$ applied to expressions of the form $\sqrt{A_\infty}=[\sum_{\mathbf{N}}\{\sqrt{(A_{n+1}-A_n)}\}^2]^{1/2}$ implies next that

$$\sqrt{A_\infty}\leqslant|a'|\sqrt{A_\infty'}+|a''|\sqrt{A_\infty''}.$$

It now follows that the space G is a vector space normed by $\|X\|_G=E(\sqrt{A_\infty})$, since it is clear that null martingales are the only ones for which $E(\sqrt{A_\infty})=0$.

To show that a normed vector space is complete it suffices (as is well known and easy to show) to prove that a series $\sum_{\mathbf{N}} X_n$ converges in norm in the space whenever the series $\sum_{\mathbf{N}}\|X_n\|$ of norms converges in R. Hence the following lemma shows that the space G is a Banach space.

LEMMA VIII-2-8. *For every sequence $(X^{(k)}, k\in\mathbf{N})$ of martingales in the space G such that $\sum_{\mathbf{N}}\|X^{(k)}\|_G<\infty$, the series $X_n=\sum_{k\in\mathbf{N}}X_n^{(k)}$ $(n\in\mathbf{N})$ converge in L^1 and a.s.; they define a martingale $X=(X_n, n\in\mathbf{N})$ belonging to the space G such that $\|X\|_G\leqslant\sum_{\mathbf{N}}\|X^{(k)}\|_G$. Furthermore, the series $\sum_{\mathbf{N}}X^{(k)}$ converges to X in G.*

PROOF. The inequalities

$$E(|X_{n+1}^{(k)}-X_n^{(k)}|)\leqslant E(\sqrt{(A_{n+1}^{(k)}-A_n^{(k)})})\leqslant E(\sqrt{A_\infty^{(k)}})$$

imply that $\sum_{k\in\mathbf{N}} E(|X_{n+1}^{(k)} - X_n^{(k)}|) < \infty$ for all $n \in \mathbf{N}$; the series $\sum_{k\in\mathbf{N}} [X_{n-1}^{(k)} - X_n^{(k)}]$ thus converges in L^1 and a.s. ($n \in \mathbf{N}$) and, as $X_0^{(k)} = 0$ for all $k \in \mathbf{N}$, the series $X_n = \sum_{k\in\mathbf{N}} X_n^{(k)}$ is also convergent. It is clear that $(X_n, n \in \mathbf{N})$ is then an integrable martingale which is null at 0. The monotonicity and σ-sub-additivity of the functional $\sqrt{\{E^{\mathscr{B}}(Z^2)\}}$ then implies that

$$\sqrt{E^{\mathscr{B}_n}((X_{n+1} - X_n)^2)} \leqslant \sqrt{E^{\mathscr{B}_n}(\sum_k (X_{n+1}^{(k)} - X_n^{(k)}))^2} \leqslant \sum_k \sqrt{E^{\mathscr{B}_n}((X_{n+1}^{(k)} - X_n^{(k)})^2)}$$

and the sub-σ-additivity of the norm in $l^2(\mathbf{N})$ finally gives

$$\sqrt{A_\infty} \leqslant \sum_{k\in\mathbf{N}} \sqrt{A_\infty^{(k)}}.$$

Our hypotheses thus imply that $E(\sqrt{A_\infty}) < \infty$, i.e. that $X \in G$, and furthermore, that $\|X\|_G \leqslant \sum_k \|X^{(k)}\|_G$. Finally, the convergence of $\sum_k X^{(k)}$ in G to X is an immediate consequence of the preceding inequality because

$$\|X - \sum_{k\leqslant l} X^{(k)}\|_G = \|\sum_{k>l} X^{(k)}\|_G \leqslant \sum_{k>l} \|X^{(k)}\|_G \downarrow 0 \text{ when } l \uparrow \infty.$$

The lemma is thus proved. ■

(2) By Proposition VII-**2**-3, every martingale $(X_n, n \in \mathbf{N})$ in the space G converges in L^1 (and hence a.s.) to a r.v. X_∞ and satisfies

$$X_n = E^{\mathscr{B}_n}(X_\infty) \quad (n \in \mathbf{N}), \qquad E(|X_\infty|) \leqslant E(\sup_{\mathbf{N}} |X_n|) \leqslant 3E(\sqrt{A_\infty}).$$

We next observe that for such a martingale the r.v. $\sum_{\mathbf{N}} (X_{n+1} - X_n)^2/\sqrt{A_{n+1}}$ is integrable; indeed, the inequality

$$E^{\mathscr{B}_n}((X_{n+1} - X_n)^2/\sqrt{A_{n+1}}) = (A_{n+1} - A_n)/\sqrt{A_{n+1}} \leqslant 2(\sqrt{A_{n+1}} - \sqrt{A_n}),$$

valid for all $n \in \mathbf{N}$, implies that

$$E\left(\sum_{\mathbf{N}} (X_{n+1} - X_n)^2/\sqrt{A_{n+1}}\right) \leqslant 2E(\sqrt{A_\infty}) < \infty.$$

Now let $(Y_n, n \in \mathbf{N})$ be a martingale which is null at 0 and bounded in L^2 for which there exists a constant $\sigma < \infty$ such that $E^{\mathscr{B}_n}(B_\infty - B_n) \leqslant \sigma^2$ a.s. for all

$n \in \mathbf{N}$. The r.v. $\sum_{\mathbf{N}} (Y_{n+1} - Y_n)^2 \sqrt{A_{n+1}}$ is integrable for such a martingale since

$$E\left(\sum_{\mathbf{N}} (Y_{n+1} - Y_n)^2 \sqrt{A_{n+1}}\right) = E\left(\sum_{\mathbf{N}} (B_{n+1} - B_n) \sqrt{A_{n+1}}\right)$$

$$= E\left(\sum_{\mathbf{N}} (B_\infty - B_n)(\sqrt{A_{n+1}} - \sqrt{A_n})\right)$$

$$= E\left(\sum_{\mathbf{N}} E^{\mathscr{B}_n}(B_\infty - B_n)(\sqrt{A_{n+1}} - \sqrt{A_n})\right)$$

$$\leqslant \sigma^2 E(\sqrt{A_\infty}) < \infty,$$

since the r.v.'s A_{n+1} and A_n are $\mathscr{B}_n$-measurable. The integrability of the two r.v.'s $\sum_{\mathbf{N}} (X_{n+1} - X_n)^2/\sqrt{A_{n+1}}$ and $\sum_{\mathbf{N}} (Y_{n+1} - Y_n)^2 \sqrt{A_{n+1}}$ just proved implies that the r.v.'s $(X_{n+1} - X_n)(Y_{n+1} - Y_n)$ $(n \in \mathbf{N})$ are integrable and that the series of their L^1 norms is convergent; the series

$$\sum_{\mathbf{N}} (X_{n+1} - X_n)(Y_{n+1} - Y_n)$$

therefore converges in L^1 and a.s. to a r.v. such that

$$E\left(\sum_{\mathbf{N}} (X_{n+1} - X_n)(Y_{n+1} - Y_n)\right) \leqslant E\left(\sum_{\mathbf{N}} |X_{n+1} - X_n|\,|Y_{n+1} - Y_n|\right)$$

$$\leqslant [2\, E(\sqrt{A_\infty})\, \sigma^2\, E(\sqrt{A_\infty})]^{1/2}$$

$$= (\sqrt{2})\, \sigma\, E(\sqrt{A_\infty}).$$

Thus we have constructed from the martingale $(Y_n, n \in \mathbf{N})$ a continuous linear functional, say

$$X \to E\left(\sum_{\mathbf{N}} (X_{n+1} - X_n)(Y_{n+1} - Y_n)\right),$$

whose norm is not greater than $(\sqrt{2})\sigma$. Even though the r.v.'s $(X_{n+1} - X_n), (Y_{n+1} - Y_n)$ $(n \in \mathbf{N})$ are integrable, it is not generally true that $X_n Y_n$ (see Exercise VIII-6); hence it is not possible in general to write the preceding linear functional as $\lim_{n \to \infty} E(X_n Y_n)$ nor as $E(X_\infty Y_\infty)$.

(3) Let Φ be a continuous linear functional on the Banach space G. As every r.v. $X \in L^2(\mathscr{B}_\infty)$ defines a martingale $(X_n = E^{\mathscr{B}_n}(X), n \in \mathbf{N})$ in G such that

$$\|X\|_G = E(\sqrt{A_\infty}) \leqslant \sqrt{\{E(A_\infty)\}} = \|X\|_{L^2(\mathscr{B}_\infty)},$$

the mapping $X \to \Phi((X_n, n \in \mathbf{N}))$ defines a continuous linear functional on the Hilbert space $L^2(\mathscr{B}_\infty)$; consequently, there exists a r.v. Y in this space such that

$$\Phi((X_n, n \in \mathbf{N})) = E(XY) \quad \text{if } X \in L^2(\mathscr{B}_\infty).$$

We shall denote by $(Y_n = E^{\mathscr{B}_n}(Y), n \in \mathbf{N})$ the L^2-bounded martingale associated with Y; we know that

$$E(XY) = E\left(\sum_{\mathbf{N}} (X_{n+1} - X_n)(Y_{n+1} - Y_n)\right) \quad \text{if } X \in L^2(\mathscr{B}_\infty).$$

Also, we denote by $(B_n, n \in \mathbf{N})$ the increasing process associated with the submartingale $(Y_n^2, n \in \mathbf{N})$.

In order to show that $E^{\mathscr{B}_n}(B_\infty - B_n) \leqslant (\|\Phi\|_G')^2$ a.s. for every $n \in \mathbf{N}$, associate with every integer $p \in \mathbf{N}$ and every fixed r.v. $U \in L_+^\infty(\mathscr{B}_p)$ the martingale given by

$$Y_n^{p,U} = \begin{cases} U(Y_n - Y_p) & \text{if } n \geqslant p, \\ 0 & \text{otherwise.} \end{cases}$$

This martingale is L^2-bounded and its increasing process is given by an analogous formula:

$$B_n^{p,U} = \begin{cases} U^2(B_n - B_p) & \text{if } n \geqslant p, \\ 0 & \text{otherwise} \end{cases}$$

The inequality $\Phi(Y^{p,U}) \leqslant \|\Phi\|_G' \| Y^{p,U}\|_G$ can therefore be written explicitly as

$$E(U(Y_\infty - Y_p)\, Y_\infty) \leqslant \|\Phi\|_G'\, E(U\sqrt{(B_\infty - B_p)})$$

or again, since $E^{\mathscr{B}_p}((Y_\infty - Y_p)\, Y_\infty) = E^{\mathscr{B}_p}((Y_\infty - Y_p)^2) = E^{\mathscr{B}_p}(B_\infty - B_p)$,

$$E(UE^{\mathscr{B}_p}(B_\infty - B_p)) \leqslant \|\Phi\|_G'\, E(UE^{\mathscr{B}_p}(\sqrt{(B_\infty - B_p)})).$$

But the latter inequality can only hold for every positive $\mathscr{B}_p$-measurable r.r.v. U if

$$E^{\mathscr{B}_p}(B_\infty - B_p) \leqslant \|\Phi\|_G'\, E^{\mathscr{B}_p}(\sqrt{(B_\infty - B_p)}) \quad \text{a.s.}$$

for every $p \in \mathbf{N}$. Taking Schwartz's inequality

$$[E^{\mathscr{B}_p}(\sqrt{(B_\infty - B_p)})]^2 \leqslant E^{\mathscr{B}_p}(B_\infty - B_p)$$

into account, we find that $E^{\mathscr{B}_p}(\sqrt{(B_\infty - B_p)}) \leqslant \|\Phi\|_G'$, and hence finally that

$$E^{\mathscr{B}_p}(B_\infty - B_p) \leqslant (\|\Phi\|_G')^2.$$

It only remains to prove that the original identity

$$\Phi((X_n, n \in \mathbf{N})) = E\left(\sum_{\mathbf{N}} (X_{n+1} - X_n)(Y_{n+1} - Y_n)\right)$$

is valid for every martingale in G and not just for L^2-bounded martingales in G. But on the one hand, both sides of this identity are continuous on G: the left-hand side by assumption, and the right-hand side also since $\sup_{\mathbf{N}} E^{\mathscr{B}_n}(B_\infty - B_n) \in L^\infty$ has just been proved; on the other hand, the L^2-bounded martingales form a dense subset of G, for if $(X_n, n \in \mathbf{N})$ is a martingale in G, the stopped martingales $(X_{\nu_a \wedge n}, n \in \mathbf{N})$ with stopping times $\nu_a = \inf(n: A_{n+1} > a)$ are L^2-bounded as $E(A_{\nu_a}) \leqslant a^2$, whilst

$$\|X - X_{\nu_a}\|_G = E(\sqrt{(A_\infty - A_{\nu_a})}) \downarrow 0 \quad \text{a.s. } a \uparrow \infty$$

by dominated convergence. The proposition is thus proved. ∎

We conclude this section with a representation theorem for martingales belonging to the space G. A r.r.v. Y such that $E(\sqrt{\{E^{\mathscr{B}_p}(Y^2)\}}) < \infty$ for some fixed $p \in \mathbf{N}$ is necessarily integrable since $E^{\mathscr{B}_p}(|Y|) \leqslant \sqrt{\{E^{\mathscr{B}_p}(Y^2)\}}$. If for the same integer p the r.v. Y satisfies $E^{\mathscr{B}_p}(Y) = 0$, the sequence $(Y_n = E^{\mathscr{B}_n}(Y), n \in \mathbf{N})$ is an integrable martingale which is null for the indices $n \leqslant p$ and thus such that

$$E^{\mathscr{B}_p}(B_\infty) = \sum_{n \geqslant p} E^{\mathscr{B}_p}[(Y_{n+1} - Y_n)^2] = E^{\mathscr{B}_p}(Y^2);$$

this martingale then belongs to the space G since the inequality $E^{\mathscr{B}_p}(\sqrt{B_\infty}) \leqslant \sqrt{\{E^{\mathscr{B}_p}(B_\infty)\}}$ implies that

$$E(\sqrt{B_\infty}) \leqslant E(\sqrt{\{E^{\mathscr{B}_p}(Y^2)\}}).$$

This simple result possesses the following remarkable converse.

PROPOSITION VIII-2-9. *For every martingale* $(X_n = E^{\mathscr{B}_n}(X_\infty),\ n \in \mathbf{N})$ *in the Banach space G there exists at least one sequence* $(Y^{(p)}, p \in \mathbf{N})$ *of integrable r.r.v.'s such that*

$$E^{\mathscr{B}_p}(Y^{(p)}) = 0 \quad (p \in \mathbf{N}), \qquad E\left(\sum_{\mathbf{N}} \sqrt{\{E^{\mathscr{B}_p}(Y^{(p)})^2\}}\right) \leqslant 4E(\sqrt{A_\infty})$$

giving the representation $X_\infty = \sum_{\mathbf{N}} Y^{(p)}$, *the series on the right-hand side converging in the space G (thus a fortiori in* L^1*) and a.s.*

PROOF. Let $(\nu_i, -\infty \leqslant i < +\infty)$ be the increasing sequence of stopping times defined by

$$\nu_i = \begin{cases} \min(n : A_{n+1} > 4^i) & \\ +\infty & \text{if } A_\infty \leqslant 4^i \end{cases}$$

(with $4^i = 0$ if $i = -\infty$). At almost every point of Ω the stopping times ν_i are equal to $+\infty$ if i is sufficiently large (as $A_\infty < \infty$ a.s. by assumption) and are equal to $\nu_{-\infty}$ for sufficiently small i; further, $X_{\nu_{-\infty}} = 0$ when $A_{\nu_{-\infty}} = 0$ and $X_0 = 0$. We can therefore write

$$X_\infty = \sum_{i \in \mathbf{Z}} (X_{\nu_{i+1}} - X_{\nu_i})\, 1_{\{\nu_i < \infty\}} = \sum_{i \in \mathbf{Z}} \sum_{p \in \mathbf{N}} (X_{\nu_{i+1}} - X_p)\, 1_{\{\nu_i = p\}},$$

the series in the last two expressions reducing at almost every point to finite sums.

The integrable r.v.'s $Y^{(p,i)} = (X_{\nu_{i+1}} - X_p)\, 1_{\{\nu_i = p\}}$ $(i \in \mathbf{Z},\ p \in \mathbf{N})$ are such that $E^{\mathscr{B}_p}(Y^{(p,i)}) = 0$ and

$$E^{\mathscr{B}_p}((Y^{(p,i)})^2) = E^{\mathscr{B}_p}((A_{\nu_{i+1}} - A_p)\, 1_{\{\nu_i = p\}}) \leqslant 4^{i+1}\, 1_{\{\nu_i = p\}}$$

since $A_{\nu_{i+1}} \leqslant 4^{\,i+1}$ everywhere. On the other hand, on each of the events $\{\nu_j < \infty = \nu_{j+1}\} = \{4^j < A_\infty \leqslant 4^{j+1}\}$ $(j \in \mathbf{Z})$ we can write

$$\sum_{i \in \mathbf{Z}} \sum_{p \in \mathbf{N}} 2^{i+1}\, 1_{\{\nu_i = p\}} = \sum_{i \in \mathbf{Z}} 2^{i+1}\, 1_{\{\nu_i < \infty\}} = \sum_{i \leqslant j} 2^{i+1} = 4 \cdot 2^j \leqslant 4\sqrt{A_\infty}.$$

As the events $\{\nu_j < \infty = \nu_{j+1}\}$ $(j \in \mathbf{Z})$ together with the event $\{\nu^j = \infty$ for all $j \in \mathbf{Z}\} = \{A_\infty = 0\}$ form a partition of Ω and as the inequality just obtained is trivially true on the latter event, we see that

$$\sum_{i \in \mathbf{Z}} \sum_{p \in \mathbf{N}} \sqrt{\{E^{\mathscr{B}_p}(Y^{(p,i)})^2\}} \leqslant \sum_{i \in \mathbf{Z}} \sum_{p \in \mathbf{N}} 2^{i+1}\, 1_{\{\nu_i = p\}} \leqslant 4\sqrt{A_\infty} \quad \text{a.s.}$$

We then put

$$Y^{(p)} = \sum_{i \in \mathbf{Z}} Y^{(p,i)} = \sum_{i \in \mathbf{Z}} (X_{\nu_{i+1}} - X_p) 1_{\{\nu_i = p\}} \text{ on } \Omega$$

(the last series contains no more than one non-zero term at any given point of Ω). As the inequality just proved implies that

$$\sum_{i \in \mathbf{Z}} \sum_{p \in \mathbf{N}} E(|Y^{(p,i)}|) \leqslant 4E(\sqrt{A_\infty}) < \infty$$

since $E(|Y|) \leqslant E(\sqrt{\{E^{\mathscr{B}}(Y^2)\}})$, the series $\sum_{i \in \mathbf{Z}} Y^{(p,i)}$ also converges in L^1 to $Y^{(p)}$ ($p \in \mathbf{N}$), which implies that $E^{\mathscr{B}_p}(Y^{(p)}) = \sum_{i \in \mathbf{Z}} E^{\mathscr{B}_p}(Y^{(p,i)}) = 0$. Since the functional $\sqrt{\{E^{\mathscr{B}}(Y^2)\}}$ is sub-σ-additive in Y we have

$$\sum_{p \in \mathbf{N}} \sqrt{\{E^{\mathscr{B}_p}((Y^{(p)})^2)\}} \leqslant 4\sqrt{A_\infty} \quad \text{a.s.}$$

It follows that each of the r.v.'s $Y^{(p)}$ belongs to the space G and that $\sum_{p \in \mathbf{N}} \|Y^{(p)}\|_G \leqslant 4E(\sqrt{A_\infty})$ by the results preceding the statement of the proposition; the series $\sum_{\mathbf{N}} Y^{(p)}$ therefore converges in G. This convergence is then also in the sense of L^1 and since $\sum_{\mathbf{N}} Y^{(p)} = X_\infty$ a.s. on Ω, the series $\sum_{\mathbf{N}} Y^{(p)}$ can only converge in G to X_∞. ■

VIII-3. Quadratic variation

Definition VIII-3-10. The *quadratic variation* of a sequence $(X_n, n \in \mathbf{N})$ of r.r.v.'s is the positive r.r.v. V defined by the formula

$$V = X_0^2 + \sum_{\mathbf{N}} (X_{m+1} - X_m)^2.$$

In this section we shall also be making constant use of the sequence $(V_n, n \in \mathbf{N})$ defined by

$$V_n = X_0^2 + \sum_{m<n} (X_{m+1} - X_m)^2 \quad (n \in \mathbf{N});$$

this sequence increases towards V when $n \uparrow \infty$.

Recall that a sequence $(x_n, n \in \mathbf{N})$ of real numbers is convergent whenever it satisfies the condition $\sum_{\mathbf{N}} |x_{n+1} - x_n| < \infty$, but on the other hand the quadratic condition $\sum_{\mathbf{N}} (x_{n+1} - x_n)^2 < \infty$ is not sufficient to imply convergence. Nevertheless, if the sequence $(x_n, n \in \mathbf{N})$ is convergent, the condition $\sum_{\mathbf{N}} (x_{n+1} - x_n)^2 < \infty$, which is weaker than the condition $\sum_{\mathbf{N}} |x_{n+1} - x_n| < \infty$ since $(x_{n+1} - x_n)^2$ is much smaller than $|x_{n+1} - x_n|$ given that $x_{n+1} - x_n \to 0$, assures a certain regularity of this convergence; for example, Schwartz's inequality implies that for such a sequence

$$|x_p - x_q|^2 \leqslant |p - q| v \quad \text{if } v = \sum_n (x_{n+1} - x_n)^2 \qquad (p, q \in \mathbf{N}).$$

It is a remarkable fact that martingales (resp. sub- or supermartingales) have finite quadratic variation whenever they are convergent. We begin by studying the case of positive supermartingales; the following proposition constitutes the basic result of this section.

PROPOSITION VIII-3-11. *The quadratic variation V of every finite positive supermartingale $(X_n, n \in \mathbf{N})$ is a.s. finite. More precisely, the quadratic variation satisfies the following inequalities:*

(a) *when the supermartingale $(X_n, n \in \mathbf{N})$ is bounded by a constant c $(c \in \mathbf{R}_+)$,*

$$E^{\mathscr{B}_0}(V) \leqslant 2cX_0;$$

(b) *in general,*

$$P^{\mathscr{B}_0}(V > a^2) \leqslant 3a^{-1} X_0 \quad \textit{for every } a \in \mathbf{R}_+.$$

PROOF. (a) We begin by remarking that the r.v.'s

$$V_n = X_0^2 + \sum_{m<n} (X_{m+1} - X_m)^2 \qquad (n \in \mathbf{N})$$

which tend monotonically upwards to V when $n \uparrow \infty$, can also be written

$$V_n = X_n^2 + 2 \sum_{m<n} X_m(X_{m+1} - X_m) \qquad (n \in \mathbf{N}).$$

This way of writing V_n implies that

$$E^{\mathscr{B}_0}(V_n) = E^{\mathscr{B}_0}\left(X_n^2 + 2 \sum_{m<n} E^{\mathscr{B}_m}(X_m(X_m - X_{m+1}))\right)$$

$$= E^{\mathscr{B}_0}\left(X_n^2 + 2 \sum_{m<n} X_m(A_{m+1} - A_m)\right)$$

where $(A_n, n \in \mathbf{N})$ denotes the increasing process in Doob's decomposition of the supermartingale $(X_n, n \in \mathbf{N})$. If $0 \leqslant X_m \leqslant c$ $(m \in \mathbf{N})$, then

$$E^{\mathscr{B}_0}(V_n) \leqslant cE^{\mathscr{B}_0}(X_n + 2A_n) \leqslant 2cE^{\mathscr{B}_0}(X_n + A_n) = 2cX_0$$

for $(X_n + A_n, n \in \mathbf{N})$ is a martingale whose initial term is X_0. It remains to let $n \uparrow \infty$, and the inequality (a) of the proposition is proved.

(b) If $(X_n, n \in \mathbf{N})$ is a positive supermartingale, then $(\min(X_n, c), n \in \mathbf{N})$ is a positive supermartingale bounded above by c. By the first part of the proof, the quadratic variation $U^{(c)}$ of this sequence satisfies the inequality

$$E^{\mathscr{B}_0}(U^{(c)}) \leqslant 2c \min(X_0, c) \leqslant 2cX_0.$$

On the other hand, on the event $\{\sup_{\mathbf{N}} X_n \leqslant c\}$ the two sequences $(X_n, n \in \mathbf{N})$ and $(\min(X_n, c), n \in \mathbf{N})$ coincide, and thus the same is true of the r.v.'s V and $U^{(c)}$; this allows us to write

$$\begin{aligned} P^{\mathscr{B}_0}(V > a^2) &\leqslant P^{\mathscr{B}_0}(\sup_{\mathbf{N}} X_n > c) + P^{\mathscr{B}_0}(\sup_{\mathbf{N}} X_n \leqslant c, U^{(c)} > a^2) \\ &\leqslant P^{\mathscr{B}_0}(\sup_{\mathbf{N}} X_n > c) + a^{-2} E^{\mathscr{B}_0}(U^{(c)}) \leqslant c^{-1} X_0 + 2a^{-2} c X_0 \end{aligned}$$

by the inequality of Proposition II-2-7. Taking $c = a$, we obtain the second inequality of the proposition. ■

Let us note in passing the following consequence of the first part of the preceding proposition.

COROLLARY VIII-3-12. *Let $(X_n, n \in \mathbf{N})$ be a positive supermartingale bounded above by a constant c. The martingale $(M_n, n \in \mathbf{N})$ in the Doob decomposition $X_n = M_n - A_n$ $(n \in \mathbf{N})$ of the supermartingale $(X_n, n \in \mathbf{N})$ is then bounded in L^2. More precisely,*

$$\lim_n \uparrow E^{\mathscr{B}_0}(M_n^2) \leqslant 2\, cX_0 \leqslant 2c^2.$$

PROOF. Since $A_{n+1} - A_n = - E^{\mathscr{B}_n}(X_{n+1} - X_n)$, we have

$$M_{n+1} - M_n = (X_{n+1} - X_n) - E^{\mathscr{B}_n}(X_{n+1} - X_n).$$

It follows that

$$\begin{aligned} E^{\mathscr{B}_n}((M_{n+1} - M_n)^2) &= E^{\mathscr{B}_n}((X_{n+1} - X_n)^2) - [E^{\mathscr{B}_n}(X_{n+1} - X_n)]^2 \\ &\leqslant E^{\mathscr{B}_n}((X_{n+1} - X_n)^2). \end{aligned}$$

Since on the other hand

$$E^{\mathscr{B}_n}((M_{n+1} - M_n)^2) = E^{\mathscr{B}_n}(M_{n+1}^2) - M_n^2,$$

we see that

$$\lim_p \uparrow E^{\mathscr{B}_0}(M_p^2) = E^{\mathscr{B}_0}\left(M_0^2 + \sum_{\mathbf{N}} [E^{\mathscr{B}_n}(M_{n+1}^2) - M_n^2]\right)$$

$$\leqslant E^{\mathscr{B}_0}\left(X_0^2 + E^{\mathscr{B}_n} \sum_{\mathbf{N}} (X_{n+1} - X_n)^2\right) = E^{\mathscr{B}_0}(V).$$

But the preceding proposition has precisely shown that $E^{\mathscr{B}_0}(V) \leqslant 2cX_0$. ∎

We now collect several results concerning the quadratic variation of a martingale into a single proposition.

PROPOSITION VIII-3-13. *The quadratic variation V of an integrable martingale $(X_n, n \in \mathbf{N})$ has the following properties:*

(a) *It is integrable if the martingale $(X_n, n \in \mathbf{N})$ is bounded in L^2; more precisely, $E(V) = \sup_{\mathbf{N}} E(X_n^2)$.*

(b) *It is a.s. finite and satisfies the more precise inequality*

$$P(V > a^2) \leqslant 6a^{-1} \sup_{\mathbf{N}} E(|X_n|)$$

if the martingale $(X_n, n \in \mathbf{N})$ is bounded in L^1.

(c) *It satisfies the inequality $E(\sqrt{V}) \leqslant (\sqrt{2}) E(\sqrt{A_\infty})$.*

(d) *It is a.s. finite on the event $\{A_\infty < \infty\}$ if $(A_n, n \in \mathbf{N})$ denotes the increasing process associated with the submartingale $(X_n^2, n \in \mathbf{N})$.*

PROOF. Properties (a) and (d) are easy to prove directly; on the other hand, the proof of property (b) uses Proposition VIII-3-11 above.

(a) and (d). The identity $E((X_{n+1} - X_n)^2) = E(X_{n+1}^2) - E(X_n^2)$ $(n \in \mathbf{N})$ satisfied by every square-integrable martingale immediately implies that

$$E(V) = E\left(X_0^2 + \sum_{\mathbf{N}} (X_{n+1} - X_n)^2\right)$$

$$= E(X_0^2) + \sum_{\mathbf{N}} [E(X_{n+1}^2) - E(X_n^2)] = \lim_n \uparrow E(X_n^2).$$

Let us note that the last term is also equal to $E(X_0^2 + A_\infty)$ if $(A_n, n \in \mathbf{N})$ denotes the increasing process in the Doob decomposition of the submartingale $(X_n^2, n \in \mathbf{N})$.

We apply this result to the martingale $(X_{\nu_a \wedge n}, n \in \mathbf{N})$ stopped at the stopping time

$$\nu_a = \begin{cases} \inf(n : A_{n+1} > a^2), & \\ +\infty & \text{if } A_\infty \leqslant a^2. \end{cases}$$

As the quadratic variation of this martingale equals V on the event $\{\nu_a = \infty\}$ and is everywhere positive, we see that

$$\int_{\{\nu_a=\infty\}} V \, \mathrm{d}P \leqslant E(X_0^2 + A_{\nu_a}) \leqslant E(X_0^2) + a < \infty,$$

which shows that $V < \infty$ a.s. on the event $\{\nu_a = \infty\} = \{A_\infty < a^2\}$. It only remains to let $a \uparrow \infty$ to complete the proof of the fourth part of the proposition.

(b) We use the Krickeberg decomposition (proved in Theorem IV-**1**-2) of the martingale $(X_n, n \in \mathbf{N})$ which we are now supposing to be bounded in L^1; let $X_n = X_n^{(+)} - X_n^{(-)}$ $(n \in \mathbf{N})$ be this decomposition of the martingale $(X_n, n \in \mathbf{N})$ into the difference of the two positive martingales such that we further have

$$E(X_0^{(+)}) + E(X_0^{(-)}) = \sup_{\mathbf{N}} E(|X_n|).$$

We then note that by the triangle inequality satisfied by the norm in the space $l^2(\mathbf{N})$, the quadratic variations V, $V^{(+)}$ and $V^{(-)}$ of the three martingales $(X_n, n \in \mathbf{N})$, $(X_n^{(+)}, n \in \mathbf{N})$ and $(X_n^{(-)}, n \in \mathbf{N})$ are related by the inequality

$$\sqrt{V} \leqslant \sqrt{V^{(+)}} + \sqrt{V^{(-)}}.$$

Hence by the second part of Proposition VIII-**3**-11 we can write

$$\begin{aligned} P(V > a^2) &\leqslant P(V^{(+)} > \tfrac{1}{4}a^2) + P(V^{(-)} > \tfrac{1}{4}a^2) \\ &\leqslant 6a^{-1} E(X_0^{(+)}) + 6a^{-1} E(X_0^{(-)}) = 6a^{-1} \sup_{\mathbf{N}} E(|X_n|). \end{aligned}$$

(c) The inequality involving $E(\sqrt{V})$ can be proved in the same way as that involving $E(\sup_{\mathbf{N}} |X_n|)$ in Proposition VII-**2**-3 (with the constant 3 instead of $\sqrt{2}$), but the following direct proof is simpler.

Firstly, Schwartz's inequality gives

$$[E(\sqrt{V})]^2 \leqslant E(V/\sqrt{A_\infty})\, E(\sqrt{A_\infty});$$

next, using $E^{\mathscr{B}_n}(V_{n+1} - V_n) = A_{n+1} - A_n$, we can see that

$$E(V/\sqrt{A_\infty}) \leqslant E\left(\sum_{\mathbf{N}} (V_{n+1} - V_n)/\sqrt{A_{n+1}}\right)$$
$$= E\left(\sum_{\mathbf{N}} (A_{n+1} - A_n)/\sqrt{A_{n+1}}\right)$$
$$\leqslant 2\,E\left(\sum_{\mathbf{N}} (\sqrt{A_{n+1}} - \sqrt{A_n})\right) = 2E(\sqrt{A_\infty}),$$

and it then follows that $E(\sqrt{V})^2 \leqslant 2E(\sqrt{A_\infty})^2$. ■

The following theorem identifies the dual of the Banach space of martingales $(X_n, n \in \mathbf{N})$ such that $E(\sqrt{V}) < \infty$ (equipped with the norm defined by this expectation); we will then deduce from this theorem that the two integrability conditions $E(\sqrt{V}) < \infty$ and $E(\sup_{\mathbf{N}} |X_n|) < \infty$ are equivalent for martingales. Hence by the results of Section IV-**2**, an L^1-bounded martingale does not necessarily satisfy $E(\sqrt{V}) < \infty$ although V is a.s. finite by the preceding proposition.

THEOREM VIII-3-14. *The integrable martingales $X = (X_n, n \in \mathbf{N})$ which are null at 0 and such that $E(\sqrt{V}) < \infty$ form a Banach space, say H, for the norm defined by this expectation (i.e. $\|X\|_H = E(\sqrt{V})$).*

On the other hand, the martingales $Y = (Y_n, n \in \mathbf{N})$ which are null at zero, L^2-bounded and such that $\sup_{\mathbf{N}} E^{\mathscr{B}_{n+1}}((Y_\infty - Y_n)^2) \in L^\infty$ form a Banach space, denoted by BMO, *for the norm*

$$\|Y\|_{\mathrm{BMO}} = \|\sup_{\mathbf{N}} \sqrt{\{E^{\mathscr{B}_{n+1}}((Y_\infty - Y_n)^2)\}}\|_\infty.$$

This Banach space BMO *is dual to the Banach space H under the bilinear functional $(X, Y) = \lim_{n\to\infty} E(X_n Y_n)$; however, the norms*

$$\|Y\|'_H = \sup((X, Y) : \|X\|_H \leqslant 1), \qquad \|Y\|_{\mathrm{BMO}}$$

are not equal but merely equivalent, in the sense that

$$\|Y\|'_H \leqslant (\sqrt{2})\|Y\|_{\mathrm{BMO}}, \qquad \|Y\|_{\mathrm{BMO}} \leqslant 2\|Y\|'_H \qquad (Y \in \mathrm{BMO}).$$

The martingales $Y = (Y_n, n \in \mathbf{N})$ in the space BMO *coincide with those which admit a representation of the form*

$$Y_n = \sum_{m < n} (E^{\mathscr{B}_{m+1}} U_m - E^{\mathscr{B}_m} U_m)$$

for a sequence $(U_n, n \in \mathbf{N})$ *of real-valued r.v.'s such that* $\sum_{\mathbf{N}} U_n^2 \in L^\infty$. *Furthermore, every martingale* $Y \in$ BMO *satisfies*

$$\|Y\|'_H = \min \left\| \sqrt{\{\sum_{\mathbf{N}} U_n^2\}} \right\|_\infty,$$

the lower bound on the right being taken over all the possible such representations of Y, *and always being attained.*

In the above statement it is convenient to identify two martingales which are a.s. equal so that the norms introduced above are actually norms.

PROOF. (1) Let us consider the Banach space $L^1_{l^2(\mathbf{N})}(\Omega, \mathscr{A}, P)$; it is formed from (equivalence classes of) integrable functions $f : \Omega \to l^2(\mathbf{N})$, i.e. from (equivalence classes of) sequences of measurable real-valued functions such that $E(\sqrt{\{\sum_{\mathbf{N}} f_n^2\}}) < \infty$, and this expectation is precisely the norm of $f = (f_n, n \in \mathbf{N})$ in $L^1_{l^2(\mathbf{N})}$. In this Banach space, the vector subspace Λ consisting of the $f = (f_n, n \in \mathbf{N})$ such that for every $n \in \mathbf{N}$, f_n is $\mathscr{B}_{n+1}$-measurable and satisfies $E^{\mathscr{B}_n}(f_n) = 0$ is clearly closed. But the mapping Δ which associates with every martingale $X = (X_n, n \in \mathbf{N})$ the sequence $\Delta X = (X_{n+1} - X_n, n \in \mathbf{N})$ establishes a linear bijection from the space H of the proposition *onto* the subspace Λ of $L^1_{l^2(\mathbf{N})}$ just described such that $E(\sqrt{V}) = \|\Delta X\|_\Lambda$; the space H is therefore a Banach space for the norm $\|X\|_H = E(\sqrt{V})$, isomorphic to the closed vector subspace Λ of $L^1_{l^2(\mathbf{N})}$ under the mapping Δ.

The space BMO of martingales introduced in the theorem is a normed vector space by virtue of the sublinearity of the functional $\sqrt{\{E^{\mathscr{B}}(Z^2)\}}$ and the fact that the null martingales are the only martingales in BMO for which $E^{\mathscr{B}_{n+1}}((Y_\infty - Y_n)^2) = 0$ $(n \in \mathbf{N})$. We will show that this space is complete by showing that it is, up to equivalence of norms, the strong dual of the Banach space H.

(2) Let $(Y_n, n \in \mathbf{N})$ be a martingale in the space BMO. For any martingale $(X_n, n \in \mathbf{N})$ in H the two positive r.v.'s

$$\sum_{\mathbf{N}} (X_{n+1} - X_n)^2 / \sqrt{V_{n+1}}, \qquad \sum_{\mathbf{N}} (Y_{n+1} - Y_n)^2 \sqrt{V_{n+1}}$$

are then integrable. Indeed, the inequality

$$(X_{n+1} - X_n)^2 / \sqrt{V_{n+1}} = (V_{n+1} - V_n)/\sqrt{V_{n+1}} \leqslant 2(\sqrt{V_{n+1}} - \sqrt{V_n})$$

implies that

$$E\Big(\sum_{\mathbf{N}}(X_{n+1}-X_n)^2/\sqrt{V_{n+1}}\Big)\leqslant 2E(\sqrt{V})=2\|X\|_H;$$

on the other hand, we have

$$\begin{aligned}
&E\Big(\sum_{n\in\mathbf{N}}(Y_{n+1}-Y_n)^2\sqrt{V_{n+1}}\Big)\\
&=E\Big(\sum_{m\in\mathbf{N}}\Big\{\sum_{n\geqslant m}(Y_{n+1}-Y_n)^2\Big\}(\sqrt{V_{m+1}}-\sqrt{V_m})\Big) \qquad \text{(summing by parts)}\\
&=E\Big(\sum_{m\in\mathbf{N}}E^{\mathscr{B}_{m+1}}((Y_\infty-Y_m)^2)(\sqrt{V_{m+1}}-\sqrt{V_m})\Big)\\
&\leqslant\|\sup_{\mathbf{N}}E^{\mathscr{B}_{m+1}}((Y_\infty-Y_m)^2)\|_\infty E(\sqrt{V})\\
&=\|X\|_H\|Y\|^2_{\mathrm{BMO}},
\end{aligned}$$

taking into account the fact that

$$E^{\mathscr{B}_{m+1}}\Big(\sum_{n\geqslant m}(Y_{n+1}-Y_n)^2\Big)=E^{\mathscr{B}_{m+1}}((Y_\infty-Y_m)^2).$$

This double integrability just shown implies that the series of r.v.'s $(X_{n+1}-X_n)(Y_{n+1}-Y_n)$ converges in L^1 and a.s. and that, furthermore,

$$\begin{aligned}
\Big|E\Big(\sum_{\mathbf{N}}(X_{n+1}-X_n)(Y_{n+1}-Y_n)\Big)\Big|^2&\leqslant\Big[E\Big(\sum_{\mathbf{N}}|X_{n+1}-X_n|\,|Y_{n+1}-Y_n|\Big)\Big]^2\\
&\leqslant E\Big(\sum_{\mathbf{N}}(X_{n+1}-X_n)^2/\sqrt{V_{n+1}}\Big)\\
&\quad\times E\Big(\sum_{\mathbf{N}}(Y_{n+1}-Y_n)^2\sqrt{V_{n+1}}\Big)\\
&\leqslant 2\|X\|^2_H\|Y\|^2_{\mathrm{BMO}}.
\end{aligned}$$

We have thus constructed a continuous linear functional on H, say $(\cdot,\ Y)$ from the martingale $Y\in\mathrm{BMO}$, where

$$(X,Y)=E\Big(\sum_{\mathbf{N}}(X_{n+1}-X_n)(Y_{n+1}-Y_n)\Big)\qquad(X\in H)$$

whose norm $\|Y\|'_H$ is no larger than $(\sqrt{2})\|Y\|_{\mathrm{BMO}}$. Let us also note that the terms Y_n $(n \in \mathbf{N})$ of a martingale $Y \in \mathrm{BMO}$ belong to L^∞ as $Y_0 = 0$ and

$$(Y_{n+1} - Y_n)^2 \leqslant E^{\mathscr{B}_{n+1}}\left(\sum_{m \geqslant n} (Y_{m+1} - Y_m)^2\right) = E^{\mathscr{B}_{n+1}}((Y_\infty - Y_n)^2) \leqslant \|Y\|^2_{\mathrm{BMO}} \quad \text{a.s.}$$

for every $n \in \mathbf{N}$ (on the other hand, the sequence $\{\|Y_n\|_\infty\}$ is not generally bounded and so Y_∞ does not generally belong to L^∞!); this implies that for every integrable martingale and in particular for every martingale $X \in H$,

$$E\left(\sum_{n<p} (X_{n+1} - X_n)(Y_{n+1} - Y_n)\right) = E(X_p Y_p),$$

and thus also that

$$(X, Y) = \lim_{p \to \infty} E(X_p Y_p).$$

(3) We will show that every continuous linear functional on the Banach space H is of the above form thus establishing the third part of the theorem at the same time; the proof is based on the following classical result, an elementary and succinct proof of which we give next.

LEMMA VIII-3-15. *The dual of the Banach space $L^1_{l^2(\mathbf{N})}$ is the space $L^\infty_{l^2(\mathbf{N})}$ under the pairing $(f, g) = E(\sum_{\mathbf{N}} f_n g_n)$.*

PROOF. Indeed, if $f \in L^1_{l^2(\mathbf{N})}$ and $g \in L^\infty_{l^2(\mathbf{N})}$, we can write

$$E\left(\sum_{\mathbf{N}} |f_n|\,|g_n|\right) \leqslant E\left(\sqrt{\left(\sum_{\mathbf{N}} f_n^2\right)}\sqrt{\left(\sum_{\mathbf{N}} g_n^2\right)}\right) \leqslant \left\|\sqrt{\left(\sum_{\mathbf{N}} f_n^2\right)}\right\|_{L^1} \left\|\sqrt{\left(\sum_{\mathbf{N}} g_n^2\right)}\right\|_{L^\infty},$$

and it follows that the map $f \to E(\sum_{\mathbf{N}} f_n g_n)$ is well-defined, linear and continuous with norm $\leqslant \|g\|$ on the space $L^1_{l^2(\mathbf{N})}$, whatever the $g \in L^\infty_{l^2(\mathbf{N})}$. Conversely, if Φ is a continuous linear functional on $L^1_{l^2(\mathbf{N})}$, then for every fixed $n \in \mathbf{N}$ there exists a function $g_n \in L^\infty_{\mathbf{R}}$ such that

$$\Phi(\underbrace{0, 0, \ldots, 0}_{n}, \mathrm{h}, 0 \ldots) = E(hg_n)$$

for all $h \in L^1_{\mathbf{R}}$, since $L^\infty_{\mathbf{R}}$ is the dual of the space $L^1_{\mathbf{R}}$. We then deduce easily that for every $f \in L^1_{l^2(\mathbf{N})}$,

$$E\left(\sum_{m=0}^{n} f_m g_m\right) = \Phi((f_0, f_1, \ldots, f_n, 0, 0 \ldots)) \to \Phi(f)$$

when $n \to \infty$. In order to pass to the limit on the right-hand side let us note that this identity implies that

$$E\left(h \sum_{m=0}^{n} g_m^2\right) = \Phi((hg_0, \ldots, hg_n, 0, 0 \ldots)) \leqslant \|\Phi\| \, E\left(h \sqrt{\left(\sum_{m=0}^{n} g_m^2\right)}\right)$$

for every positive h in $L^1_{\mathbf{R}}$, which is only possible if $\sum_{m=0}^{n} g_m^2 \leqslant \|\Phi\| \sqrt{(\sum_{m=0}^{n} g_m^2)}$ and thus if $\sum_{m=0}^{n} g_m^2 \leqslant \|\Phi\|^2$; hence $g = (g_n, n \in \mathbf{N})$ is an element of $L^\infty_{l^2(\mathbf{N})}$ of norm $\|g\| = \|\sqrt{(\sum_{\mathbf{N}} g_n^2)}\|_\infty \leqslant \|\Phi\|$ and

$$E\left(\sum_{\mathbf{N}} f_n g_n\right) = \lim_{n \to \infty} E\left(\sum_{m=0}^{n} f_m g_m\right) = \Phi(f).$$

The lemma is thus proved. ■

Let Δ' be the adjoint of the continuous linear mapping $\Delta : H \to L^1_{l^2(\mathbf{N})}$ introduced in the first paragraph of this proof; to every element $g = (g_n, n \in \mathbf{N})$ of the dual $L^\infty_{l^2(\mathbf{N})}$ of $L^1_{l^2(\mathbf{N})}$ it associates the continuous linear functional on H defined by

$$\Delta' g(X) = (\Delta X, g) = E\left(\sum_{\mathbf{N}} (X_{n+1} - X_n) g_n\right) \qquad (X \in H).$$

This mapping $\Delta' : L^\infty_{l^2(\mathbf{N})} \to H'$ is linear and norm-decreasing ($\|\Delta' g\|_{H'} \leqslant \|g\|_\infty$) as Δ possesses this property. Furthermore, since Δ preserves the norm, the Hahn–Banach theorem shows that every continuous linear functional Φ on H extends via Δ to a continuous linear functional on $L^1_{l^2(\mathbf{N})}$ of the same norm; in other words, the mapping Δ' is surjective and such that

$$\|\Phi\|_{H'} = \inf(\|g\|_\infty : g \in L^\infty_{l^2(\mathbf{N})} \text{ and } \Delta' g = \Phi)$$

for every $\Phi \in H'$. Using the Hahn–Banach theorem, we have thus established the existence for every functional $\Phi \in H'$ of a sequence $(g_n, n \in \mathbf{N})$ of bounded real-valued r.v.'s with $\|\sqrt{(\sum_{\mathbf{N}} g_n^2\|_\infty} = \|\Phi\|$ and $\Phi(X) = E(\sum_{\mathbf{N}} (X_{n+1} - X_n) g_n)$ for every martingale $X \in H$. Next let us introduce the martingale $Y = (Y_n, n \in \mathbf{N})$ defined by

$$Y_0 = 0, \qquad Y_{n+1} - Y_n = E^{\mathscr{B}_{n+1}} g_n - E^{\mathscr{B}_n} g_n \qquad (n \in \mathbf{N}),$$

as the unique martingale for which $E((X_{n+1} - X_n) g_n = E((X_{n+1} - X_n)(Y_{n+1} - Y_n))$ whatever the $n \in \mathbf{N}$ and $X \in H$. We will show that this martingale Y belongs to the space BMO and that its norm satisfies the inequality

$$\|Y\|_{\text{BMO}} \leqslant 2\left\|\sqrt{\left(\sum_{\mathbf{N}} g_n^2\right)}\right\|_\infty = 2\|\Phi\|.$$

The proposition will then be fully proved.

The definition of the martingale $(Y_n, n \in \mathbf{N})$ implies that on the one hand

$$(Y_{n+1} - Y_n)^2 \leqslant 2(E^{\mathscr{B}_{n+1}} g_n)^2 + 2(E^{\mathscr{B}} g_n)^2 \leqslant 2E^{\mathscr{B}_{n+1}}(g_n^2) + 2E^{\mathscr{B}_n}(g_n^2)$$

and on the other hand, putting $g_n^* = E^{\mathscr{B}_{n+1}} g_n$, that

$$E^{\mathscr{B}_n}((Y_{n+1} - Y_n)^2) = E^{\mathscr{B}_n}((g_n^* - E^{\mathscr{B}_n} g_n^*)^2) \leqslant E^{\mathscr{B}_n}(g_n^{*2}) \leqslant E^{\mathscr{B}_n}(g_n^2).$$

We then deduce that for every $m \in \mathbf{N}$,

$$\begin{aligned}
E^{\mathscr{B}_{m+1}}\left(\sum_{n \geqslant m} (Y_{n+1} - Y_n)^2\right) &= (Y_{m+1} - Y_m)^2 + \sum_{n \geqslant m+1} E^{\mathscr{B}_{m+1}}((Y_{n+1} - Y_n)^2) \\
&\leqslant 2E^{\mathscr{B}_{m+1}}(g_m^2) + 2E^{\mathscr{B}_m}(g_m^2) + \sum_{n \geqslant m+1} E^{\mathscr{B}_{m+1}} g_n^2 \\
&\leqslant 2E^{\mathscr{B}_{m+1}}\left(\sum_{n \geqslant m} g_n^2\right) + 2E^{\mathscr{B}_m}(g_m^2) \\
&\leqslant 4\Big\|\sum_{\mathbf{N}} g_n^2\Big\|_\infty,
\end{aligned}$$

which simultaneously proves that the martingale Y is bounded in L^2, since

$$\sup_{\mathbf{N}} E(Y_n^2) = E\left(\sum_{\mathbf{N}} (Y_{n+1} - Y_n)^2\right) \leqslant 4\Big\|\sum_{\mathbf{N}} g_n^2\Big\|_\infty,$$

and that it belongs to the space BMO with

$$\begin{aligned}
\|Y\|_{\text{BMO}}^2 &= \Big\|\sup_{\mathbf{N}} E^{\mathscr{B}_{m+1}}((Y_\infty - Y_m)^2)\Big\|_\infty \\
&= \left\|\sup_{\mathbf{N}} E^{\mathscr{B}_{m+1}}\left(\sum_{n \geqslant m} (Y_{n+1} - Y_n)^2\right)\right\|_\infty \\
&\leqslant 4\left\|\sum_{\mathbf{N}} g_n^2\right\|_\infty = 4\|g\|^2. \blacksquare
\end{aligned}$$

The following remarkable proposition is a consequence of the preceding results and some precise calculations.

PROPOSITION VIII-3-16. *On the set of integrable martingales* $(X_n, n \in \mathbf{N})$ *the two expressions* $E(\sqrt{V})$ *and* $E(\sup_{\mathbf{N}} |X_n|)$ *are finite or infinite together; more precisely, they satisfy the pair of inequalities*

$$E(\sup_{\mathbf{N}} |X_n|) \leqslant (\sqrt{12})\, E(\sqrt{V}), \qquad E(\sqrt{V}) \leqslant 5E(\sup_{\mathbf{N}} |X_n|).$$

PROOF. To prove the first inequality, we will use the following lemma, which extends the result of Corollary VIII-1-5 to the case $p = \infty$.

LEMMA VIII-3-17. *For every sequence* $(Z_n,\ n \in \mathbf{N})$ *of r.r.v.'s such that* $\sum_{\mathbf{N}} |Z_n| \in L^\infty$, *the r.v.* $Y = \sum_{\mathbf{N}} E^{\mathscr{B}_n}(Z_n)$ *defines a martingale* $(Y_n = E^{\mathscr{B}_n}(Y), n \in \mathbf{N})$ *belonging to the space* BMO *and*

$$\| Y \|_{\mathrm{BMO}} \leqslant (\sqrt{6}) \left\| \sum_{\mathbf{N}} |Z_n| \right\|_\infty .$$

PROOF. Firstly let us note that the series $Y = \sum_{\mathbf{N}} E^{\mathscr{B}_n}(Z_n)$ converges in L^2, since by Corollary VIII-1-5

$$\left\| \sum_{\mathbf{N}} E^{\mathscr{B}_n}(|Z_n|) \right\|_2 \leqslant 2 \left\| \sum_{\mathbf{N}} |Z_n| \right\|_2 \leqslant 2 \left\| \sum_{\mathbf{N}} |Z_n| \right\|_\infty < \infty .$$

The martingale $(Y_n, n \in \mathbf{N})$ associated with the square-integrable r.v. Y then satisfies the identity

$$Y - Y_n = U_n - E^{\mathscr{B}_n}(U_n),$$

where $U_n = \sum_{m>n} E^{\mathscr{B}_m}(Z_m)$ $(n \in \mathbf{N})$; this implies that

$$(Y - Y_n)^2 \leqslant 2U_n^2 + 2(E^{\mathscr{B}_n} U_n)^2$$

and hence that

$$E^{\mathscr{B}_{n+1}}((Y - Y_n)^2) \leqslant 2E^{\mathscr{B}_{n+1}}(U_n^2) + 2(E^{\mathscr{B}_n} U_n)^2 .$$

But since $E^{\mathscr{B}_n} U_n = \sum_{m>n} E^{\mathscr{B}_n}(Z_m)$ in L^2, it is clear that

$$|E^{\mathscr{B}_n} U_n| \leqslant \left\| \sum_{\mathbf{N}} |Z_m| \right\|_\infty ,$$

whilst the following calculation gives an upper bound for $E^{\mathscr{B}_{n+1}}(U_n^2)$:

$$E^{\mathscr{B}_{n+1}}(U_n^2) \leqslant 2E^{\mathscr{B}_{n+1}}\left(\sum_{n<l\leqslant m} E^{\mathscr{B}_l}(|Z_l|)\, E^{\mathscr{B}_m}(|Z_m|)\right)$$

$$= 2E^{\mathscr{B}_{n+1}}\left(\sum_{l>n} E^{\mathscr{B}_l}(|Z_l|) \sum_{m\geqslant l} |Z_m|\right)$$

$$\leqslant 2\left\|\sum_{\mathbf{N}} |Z_m|\right\|_\infty E^{\mathscr{B}_{n+1}}\left(\sum_{l>n} E^{\mathscr{B}_l}(|Z_l|)\right)$$

$$\leqslant 2\left\|\sum_{\mathbf{N}} |Z_m|\right\|_\infty E^{\mathscr{B}_{n+1}}\left(\sum_{l>n} |Z_l|\right)$$

$$\leqslant 2\left\|\sum_{\mathbf{N}} |Z_m|\right\|_\infty^2 .$$

It follows that

$$E^{\mathscr{B}_{n+1}}((Y - Y_n)^2) \leqslant 6\left\|\sum_{\mathbf{N}} |Z_m|\right\|_\infty^2 \qquad (n \in \mathbf{N})$$

and the lemma is established. ■

To prove the first inequality of the proposition, write for every fixed $n \in \mathbf{N}$,

$$\sup_{m\leqslant n} |X_m| = \sum_{m=0}^{n} X_m Z_m,$$

where $Z_m = \operatorname{sign}(X_m)\, 1_{\{\nu=m\}}$, and ν is the first index l such that $|X_l| = \sup_{m\leqslant n} |X_m|$. The r.v. ν is not a stopping time and the r.v.'s Z_m are not adapted, but $\sum_{m=0}^{n} |Z_m| = 1$. Then

$$E(\sup_{m\leqslant n} |X_m|) = E\left(\sum_{m=0}^{n} X_m Z_m\right) = E\left(X_n \sum_{m=0}^{n} E^{\mathscr{B}_m}(Z_m)\right)$$

$$\leqslant E(\sqrt{V})\left\|\sum_{m=0}^{n} E^{\mathscr{B}_m}(Z_m)\right\|_H'$$

by definition of the norm on the Banach space H; by Theorem VIII-3-14 and the preceding lemma we have

$$\left\|\sum_{m=0}^{n} E^{\mathscr{B}_m}(Z_m)\right\|_{H}' \leqslant (\sqrt{2})\left\|\sum_{m=0}^{n} E^{\mathscr{B}_m}(Z_m)\right\|_{\mathrm{BMO}} \leqslant \sqrt{12}$$

since $\sum_{m=0}^{n} |Z_m| = 1$. To see that $E(\sup_{\mathbf{N}} |X_n|) \leqslant (\sqrt{12}) E(\sqrt{V})$ it only remains to let $n \uparrow \infty$.

To prove the second inequality of the proposition, next we use Schwarz's inequality and the elementary formula

$$V_n = X_n^2 - 2 \sum_{m<n} (X_{m+1} - X_m) X_m$$

already encountered earlier to write

$$\begin{aligned}[E(\sqrt{V_n})]^2 &\leqslant E(S_n) E(V_n S_n^{-1}) \\ &\leqslant E(S_n) \left\{ E(S_n) + 2 \left| E\left(\sum_{m<n} (X_{m+1} - X_m) X_m S_n^{-1} \right) \right| \right\},\end{aligned}$$

where we have put $S_n = \sup_{m \leqslant n} |X_m|$. This being so, we will show that

$$\left| E\left(\sum_{m<n} (X_{m+1} - X_m) X_m S_n^{-1} \right) \right| \leqslant 2E(\sqrt{V_n});$$

it will then follow that

$$[E(\sqrt{V_n})]^2 \leqslant E(S_n)[E(S_n) + 4E(\sqrt{V_n})],$$

and, since the inequality $v^2 \leqslant s(s + 4v)$ $(s, \ v \in \mathbf{R}_+)$ can also be written $(v - 2s)^2 \leqslant 5s^2$ and hence implies that $v \leqslant (2 + \sqrt{5})s$, we will have established that

$$E(\sqrt{V_n}) \leqslant (2 + \sqrt{5}) E(S_n),$$

which, by a passage to the limit, will give the second inequality of the proposition with a slightly better constant!

Let $(Y_m, m \in \mathbf{N})$ be the martingale defined for a fixed $n \in \mathbf{N}$ by $Y_0 = 0$,

$$\begin{aligned}Y_{m+1} - Y_m &= E^{\mathscr{B}_{m+1}}(X_m S_n^{-1}) - E^{\mathscr{B}_m}(X_m S_n^{-1}) \\ &= X_m \{E^{\mathscr{B}_{m+1}}(S_n^{-1}) - E^{\mathscr{B}_m}(S_n^{-1})\}\end{aligned}$$

if $1 \leqslant m < n$, and $Y_m = Y_n$ for $m \geqslant n$; then we have, after some rearrangements,

$$\begin{aligned}\left| E\left(\sum_{m<n} (X_{m+1} - X_m) X_m S_n^{-1} \right) \right| &= \left| E\left(\sum_{m<n} (X_{m+1} - X_m)(Y_{m+1} - Y_m) \right) \right| \\ &\leqslant E(\sqrt{V_n}) \| Y \|_H'.\end{aligned}$$

But if $m < n$,

$$E^{\mathscr{B}_m}((Y_{m+1} - Y_m)^2) = E^{\mathscr{B}_m}(X_m^2\{(E^{\mathscr{B}_{m+1}}(S_n^{-1}))^2 - (E^{\mathscr{B}_m}(S_n^{-1}))^2\})$$

and consequently

$$E^{\mathscr{B}_{l+1}}\left(\sum_{m>l}(Y_{m+1} - Y_m)^2\right) = E^{\mathscr{B}_{l+1}}\left(\sum_{l<m<n} X_m^2\{(E^{\mathscr{B}_{m+1}}(S_n^{-1}))^2 - (E^{\mathscr{B}_m}(S_n^{-1}))^2\}\right)$$

$$\leqslant E^{\mathscr{B}_{l+1}}(S_n^2\, E^{\mathscr{B}_n}(S_n^{-1})) = E^{\mathscr{B}_{l+1}}(1) = 1,$$

if $l < m$, whilst

$$|\,Y_{l+1} - Y_l| \leqslant |X_l| \max\{E^{\mathscr{B}_{l+1}}(S_n^{-1}), E^{\mathscr{B}_l}(S_n^{-1})\} \leqslant 1$$

since $|X_l|\; S_n^{-1} \leqslant 1$. Thus we have shown that

$$\|\,Y\|_{\mathrm{BMO}}^2 = \left\|\sup_{l<n} E^{\mathscr{B}_{l+1}}\left(\sum_{m\geqslant l}(Y_{m+1} - Y_m)^2\right)\right\|_\infty \leqslant 2,$$

and hence, by the inequalities of Theorem VIII-3-14, $\|Y\|_H' \leqslant (\sqrt{2})\|Y\|_{\mathrm{BMO}} \leqslant 2$. ■

COROLLARY VIII-3-18. *For every integrable martingale $(X_n, n \in \mathbf{N})$ the following inequalities hold for $1 < p < \infty$:*

$$\|\sqrt{V}\|_p \leqslant 10p\|\sup_{\mathbf{N}}|X_n|\,\|_p, \qquad \|\sup|X_n|\,\|_p \leqslant (p\sqrt{12})\|\sqrt{V}\|_p.$$

Recall that by Proposition IV-2-8 the r.v. $\sup_{\mathbf{N}}|X_n|$ belongs to L^p if and only if $\sup_{\mathbf{N}}\|X_n\|_p < \infty$ $(1 < p < \infty)$.

PROOF. The formulae

$$A_0 = 0, \qquad A_{m+1} = \sup_{n\leqslant m}|X_n| \quad (m \in \mathbf{N})$$

define an increasing process whose potential satisfies the inequality

$$E^{\mathscr{B}_m}(A_\infty - A_m) \leqslant E^{\mathscr{B}_m}(\sup_{n\geqslant m}|X_n - X_{m-1}|) \qquad (m \in \mathbf{N}^*);$$

indeed this inequality follows easily from

$$A_\infty = \sup_{\mathbf{N}} |X_n| \leqslant \sup_{n<m} |X_n| + \sup_{n \geqslant m} |X_n - X_{m-1}|$$

$$= A_m + \sup_{n \geqslant m} |X_n - X_{m-1}|.$$

Then the preceding proposition applied to the integrable martingale $(X_n - X_{m-1}, n \geqslant m)$ adapted to the family $(\mathscr{B}_n, n \geqslant m)$ of σ-fields shows that

$$E^{\mathscr{B}_m}(\sup_{n \geqslant m} |X_n - X_{m-1}|) \leqslant (\sqrt{12})\, E^{\mathscr{B}_m}(\sqrt{V}) \qquad (m \in \mathbf{N}^*),$$

since the quadratic variation of the martingale $(X_n - X_{m-1}, n \geqslant m)$ is dominated by that of $(X_n, n \in \mathbf{N})$. We then find that

$$E^{\mathscr{B}_m}(A_\infty - A_m) \leqslant (\sqrt{12})\, E^{\mathscr{B}_m}(\sqrt{V})$$

for every integer $m \geqslant 1$ and also for $m = 0$ (by direct application of the preceding proposition); Proposition VIII-1-4 then implies that $\|A_\infty\|_p \leqslant (p\sqrt{12})\|V_p\|$ for every $p \in [1, \infty[$, which establishes the second inequality of this corollary.

The first inequality is obtained similarly by considering the increasing process defined by

$$B_0 = 0, \qquad B_{n+1} = \sqrt{V_n} \quad (n \in \mathbf{N}).$$

The inequality

$$\sqrt{V} - \sqrt{V_{m-1}} \leqslant \sqrt{\left(\sum_{n \geqslant m} (X_n - X_{n-1})^2\right)} \qquad (m \in \mathbf{N}^*)$$

in which the right-hand side denotes the quadratic variation of the martingale $(X_n - X_{m-1}, n \geqslant m)$ then implies, with the first formula of the preceding proposition, that

$$E^{\mathscr{B}_m}(B_\infty - B_m) \leqslant E^{\mathscr{B}_m}\left(\sqrt{\left(\sum_{n \geqslant m} (X_n - X_{n-1})^2\right)}\right)$$

$$\leqslant 5E^{\mathscr{B}_m}(\sup_{n \geqslant m} |X_n - X_{m-1}|)$$

for every integer $m \geqslant 1$. Consequently

$$E^{\mathscr{B}_m}(B_\infty - B_m) \leqslant 10E^{\mathscr{B}_m}(\sup_{\mathbf{N}} |X_n|),$$

and a second application of Proposition VIII-1-4 leads to the first inequality of the corollary. ■

The space of integrable martingales $(X_n, n \in \mathbf{N})$ such that $E(\sup_{\mathbf{N}} |X_n|) < \infty$ is a Banach space for the norm defined by this expectation, as is easy to verify directly. The preceding proposition shows that, up to equivalence of norms, this space coincides with the space H of Theorem VIII-3-14; by that theorem the strong dual of the Banach space of martingales such that $E(\sup_{\mathbf{N}} |X_n|) < \infty$ is, up to norm equivalence the space BMO. The next corollary deduces a remarkable representation for martingales in the space BMO from this fact.

COROLLARY VIII-3-19. *The martingales $Y = (Y_n, n \in \mathbf{N})$ of the space* BMO *coincide with those which admit a representation of the form*

$$Y_n = E^{\mathscr{B}_n}(Y),$$

where $Y = \sum_{\bar{\mathbf{N}}} E^{\mathscr{B}_n}(Z_n)$, for a sequence $(Z_n, n \in \bar{\mathbf{N}})$ of real-valued r.v.'s such that $\sum_{\bar{\mathbf{N}}} |Z_n| \in L^\infty$. Also, the norm of every martingale $Y \in$ BMO considered as a linear functional on the space H equipped with the norm $E(\sup_{\mathbf{N}} |X_n|)$, satisfies the identity

$$\sup_{X \in H} \frac{(X, Y)}{E(\sup_{\mathbf{N}} |X_n|)} = \min \left\| \sum_{\bar{\mathbf{N}}} |Z_n| \right\|_\infty,$$

the minimum on the right-hand side being taken over all possible such representations of the martingale Y, and always being attained.

PROOF. The linear mapping Θ defined by $\Theta X = (X_n, n \in \bar{\mathbf{N}})$, where $X_\infty = \lim_{n \to \infty} X_n$, of the Banach space H equipped with the norm $E(\sup_{\mathbf{N}} |X_n|)$ into the Banach space $L^1_{C(\bar{\mathbf{N}})}$ of integrable functions with values in the space $C(\bar{\mathbf{N}})$ of continuous functions on the compact space $\bar{\mathbf{N}} = \mathbf{N} \cup \{\infty\}$ clearly preserves the norm; the Hahn–Banach theorem then shows that its adjoint mapping Θ' which maps the dual $L^\infty_{l^1(\bar{\mathbf{N}})}$ of $L^1_{C(\bar{\mathbf{N}})}$ into the space BMO is a surjection, and that the norm of every element $Y \in$ BMO calculated as a linear functional on $(H, E(\sup_{\mathbf{N}} |X_n|)$ is equal to the minimum of the norms $\|g\|$ of elements $g \in L^\infty_{l^1(\bar{\mathbf{N}})}$ such that $\Theta' g = Y$.

By the definition of the mapping Θ and the duality of the space H and BMO, on the one hand, and of $L^1_{C(\bar{\mathbf{N}})}$ and $L^\infty_{l^1(\bar{\mathbf{N}})}$ on the other, we must have

$$\lim_{n \to \infty} E(X_n Y_n) = E\left(\sum_{\bar{\mathbf{N}}} X_n g_n \right) \quad \text{for every } X \in H$$

when $Y = \Theta' g$. But

$$E\left(\sum_{\bar{\mathbf{N}}} X_n g_n\right) = \sum_{\bar{\mathbf{N}}} E(X_n g_n) = \sum_{\bar{\mathbf{N}}} E(X_\infty E^{\mathscr{B}_n}(g_n)),$$

whilst Lemma VIII-3-17 above shows that the series $\sum_{\bar{\mathbf{N}}} E^{\mathscr{B}_n}(g_n)$ defines a martingale in BMO; we have thus identified the mapping Θ' and shown that the equality $Y = \Theta' g$ can also be written

$$Y = \sum_{\bar{\mathbf{N}}} E^{\mathscr{B}_n}(g_n)$$

when $g \in L^\infty_{l^1(\bar{\mathbf{N}})}$. The corollary is then proved. ∎

VIII-4. Martingale transforms

The notion of martingale transform generalises that of stopped martingale which had many applications in the previous sections. It was first studied systematically by Burkholder.

Definition VIII-4-20. Given two sequences $(U_n, n \in \mathbf{N})$ and $(X_n, n \in \mathbf{N})$ of r.r.v.'s, we will define a new sequence $((U * X)_n, n \in \mathbf{N})$ by putting

$$(U * X)_0 = U_0 X_0,$$

$$(U * X)_{n+1} - (U * X)_n = U_{n+1}(X_{n+1} - X_n) \quad \text{if } n \in \mathbf{N}.$$

The sequence $((U * X)_n, n \in \mathbf{N})$ is called the *transform of the sequence* $(X_n, n \in \mathbf{N})$ *by the sequence* $(U_n, n \in \mathbf{N})$.

It is not hard to see, for example, that if $(X_n, n \in \mathbf{N})$ is a *martingale* in L^1 and if $(U_n, n \in \mathbf{N})$ is a *predictable sequence* in L^∞, then $((U * X)_n, n \in \mathbf{N})$ is also a martingale in L^1 because

$$E^{\mathscr{B}_n}((U * X)_{n+1} - (U * X)_n) = E^{\mathscr{B}_n}(U_{n+1}(X_{n+1} - X_n))$$
$$= U_{n+1} E^{\mathscr{B}_n}(X_{n+1} - X_n) = 0.$$

In particular, we can associate with each stopping time ν a predictable sequence $(U_n, n \in \mathbf{N})$ by putting $U_n = 1_{\{\nu \geqslant n\}}$ $(n \in \mathbf{N})$, and an easy check shows that for this sequence,

$$(U * X)_n = X_{\nu \wedge n} \qquad (n \in \mathbf{N}).$$

On the other hand, if $(X_n, n \in \mathbf{N})$ is a supermartingale and if $(U_n, n \in \mathbf{N})$ is a predictable sequence, the sequence $((U * X)_n, n \in \mathbf{N})$ can in general only be a supermartingale if the r.r.v.'s U_n are positive. It is thus all the more remarkable that without supposing the positivity of the r.r.v.'s U_n we still have the following result.

PROPOSITION VIII-4-21. *Let $(X_n, n \in \mathbf{N})$ be a finite positive supermartingale and let $(U_n, n \in \mathbf{N})$ be a predictable sequence such that $|U_n| \leq 1$ a.s. for all $n \in \mathbf{N}$. Then the sequence $((U * X)_n, n \in \mathbf{N})$ converges a.s. to a finite limit and satisfies the inequality*

$$P^{\mathscr{B}_0}(\sup_{\mathbf{N}} |(U * X)_n| > c) \leq 9c^{-1} X_0 \quad \textit{for every } c \in \mathbf{R}_+.$$

REMARKS. (1) The sequence $((U * X)_n, n \in \mathbf{N})$ of the proposition also satisfies inequalities analogous to those of Dubins (Section II-2).

(2) The hypothesis $|U_n| \leq 1$ a.s. $(n \in \mathbf{N})$ is not indispensable for the a.s. convergence of the sequence $((U^* X)_n, n \in \mathbf{N})$; for an arbitrary sequence this convergence is always true a.s. on the event $\{\sup_{\mathbf{N}} |U_n| < \infty\}$ if the other hypotheses of the proposition are retained. Indeed, if v_a denotes the first integer n such that $|U_{n+1}| > a$, the preceding proposition applied to the sequence $(a^{-1} U_{(v_a-1) \wedge n}, n \in \mathbf{N})$ dominated by 1 in absolute value, already shows that the sequence $((U * X)_n, n \in \mathbf{N})$ converges a.s. on the event

$$\{v_a = \infty\} = \{\sup_{\mathbf{N}} |U_n| \leq a\};$$

it then remains to let $a \uparrow \infty$.

(3) Every adapted sequence $(Z_n, n \in \mathbf{N})$ of integrable r.r.v.'s can be written in the form $Z_n = (U * X)_n$, where $(U_n, n \in \mathbf{N})$ is a predictable sequence taking only the values ± 1 and where $(X_n, n \in \mathbf{N})$ is a supermartingale; indeed, it suffices to put

$$U_0 = \operatorname{sign}(Z_0), \qquad U_n = \operatorname{sign}(E^{\mathscr{B}_{n-1}}(Z_n - Z_{n-1})) \quad \text{if } n \geq 1,$$

$$X_0 = |Z_0|, \; X_n - X_{n-1} = U_n(Z_n - Z_{n-1}).$$

It is therefore clear that the assumption of positivity of the supermartingale $(X_n, n \in \mathbf{N})$ in the above proposition is essential for the convergence of the sequence $((U * X)_n, n \in \mathbf{N})$. [This remark is obvious since a supermartingale which is not positive does not converge in general!]

PROOF. (a) Suppose firstly that the supermartingale $(X_n, n \in \mathbf{N})$ is dominated by a constant c and write down the Doob decomposition $X_n = M_n - A_n$. It is easy to check that

$$(U*X)_n = (U*M)_n - (U*A)_n \qquad (n \in \mathbf{N});$$

in order to establish the a.s. convergence of the transformed sequence $((U*X)_n, n \in \mathbf{N})$ to a finite limit, we will show that both the sequences $((U*M)_n, n \in \mathbf{N})$ and $((U*A)_n, n \in \mathbf{N})$ possess this property.

On the one hand, the r.v.'s $((U*A)_n, n \in \mathbf{N})$ are the partial sums of an a.s. absolutely summable series since

$$|(U*A)_{n+1} - (U*A)_n| = |U_n|(A_{n+1} - A_n) \leqslant A_{n+1} - A_n$$

by the definitions and since $E^{\mathscr{B}_0}(A_\infty) \leqslant c$ by Proposition VIII-**1**-1; the limit $\lim_{n\to\infty}(U*A)_n$ hence exists a.s. and this limit is moreover bounded in absolute value by A_∞.

On the other hand, the martingale $((U*M)_n, n \in \mathbf{N})$ is bounded in L^2. Indeed, the inequalities

$$(U*M)_0^2 \leqslant U_0^2 M_0^2 \leqslant M_0^2,$$

$$[(U*M)_{n+1} - (U*M)_n]^2 = U_{n+1}^2(M_{n+1} - M_n)^2 \leqslant (M_{n+1} - M_n)^2$$

imply that

$$E^{\mathscr{B}_0}((U*M)_p^2) = E^{\mathscr{B}_0}\Big((U*M)_0^2 + \sum_{n<p} [(U*M)_{n+1} - (U*M)_n]^2\Big)$$

$$\leqslant E^{\mathscr{B}_0}\Big(M_0^2 + \sum_{n<p} (M_{n+1} - M_n)^2\Big) = E^{\mathscr{B}_0}(M_p^2)$$

for every $p \in \mathbf{N}$. By Corollary VIII-**3**-12 we thus have

$$\lim_p\uparrow E^{\mathscr{B}_0}((U*M)_p^2) \leqslant 2cX_0 \leqslant 2c^2;$$

the martingale $((U*M)_p, p \in \mathbf{N})$ which is therefore bounded in L^2 converges to an a.s. finite limit.

The foregoing suffices to establish the a.s. convergence of the sequence $((U*X)_n, n \in \mathbf{N})$ when $n \uparrow \infty$. Further,

$$\begin{aligned} P^{\mathscr{B}_0}(\sup_{\mathbf{N}} |(U*X)_n| \geqslant c) &\leqslant P^{\mathscr{B}_0}(\sup_{\mathbf{N}} |(U*A)_n| \geqslant \tfrac{1}{4}c) + P^{\mathscr{B}_0}(\sup_{\mathbf{N}} |(U*M)_n| > \tfrac{3}{4}c) \\ &\leqslant P^{\mathscr{B}_0}(A_\infty \geqslant \tfrac{1}{4}c) + P^{\mathscr{B}_0}(\sup_{\mathbf{N}} |(U*M)_n| > \tfrac{3}{4}c) \\ &\leqslant 4c^{-1} E^{\mathscr{B}_0}(A_\infty) + \left(\frac{4}{3}c^{-1}\right)^2 \lim_n \uparrow E^{\mathscr{B}_0}((U*M)_n^2) \\ &\leqslant 4c^{-1} X_0 + \left(\frac{4}{3}c^{-1}\right)^2 2cX_0 \leqslant 8c^{-1} X_0. \end{aligned}$$

(b) We now go on to the general case. If $(X_n, n \in \mathbf{N})$ is a positive supermartingale, the sequence $(X_n^{(c)} = \min(X_n, c), n \in \mathbf{N})$ is a positive supermartingale bounded above by c, for every fixed $c \in \mathbf{R}_+$. Further the two martingales coincide on $\{\sup_{\mathbf{N}} X_n \leqslant c\}$ and hence the same is true for the two transformed sequences $((U*X)_n, n \in \mathbf{N})$ and $((U*X^{(c)})_n, n \in \mathbf{N})$.

On the other hand,

$$P^{\mathscr{B}_0}(\sup_{\mathbf{N}} X_n > c) \leqslant c^{-1} X_0 \downarrow 0 \quad \text{when } c \uparrow \infty$$

since by hypothesis $X_0 < \infty$ a.s.; the a.s. convergence of the sequence

$$((U*X^{(c)})_n, n \in \mathbf{N}) \qquad (c \in \mathbf{R}_+)$$

that we have proved in (a) above implies therefore that the sequence $((U * X)_n, n \in \mathbf{N})$ converges a.s. when $n \uparrow \infty$. Further,

$$\begin{aligned} &P^{\mathscr{B}_0}(\sup_{\mathbf{N}} |(U*X)_n| \geqslant c) \\ &\leqslant P^{\mathscr{B}_0}(\sup_{\mathbf{N}} X_n > c) + P^{\mathscr{B}_0}(\sup_{\mathbf{N}} X_n \leqslant c, \sup |(U*X^{(c)})_n| \geqslant c) \\ &\leqslant c^{-1} X_0 + 8c^{-1} \min(X_0, c) \leqslant 9c^{-1} X_0. \blacksquare \end{aligned}$$

COROLLARY VIII-4-22. *For every L^1-bounded martingale $(X_n, n \in \mathbf{N})$ and every predictable sequence $(U_n, n \in \mathbf{N})$ such that $|U_n| \leqslant 1$ a.s. $(n \in \mathbf{N})$, the martingale $((U*X)_n, n \in \mathbf{N})$ converges a.s. to a finite limit and satisfies the inequality*

$$P(\sup_{\mathbf{N}} |(U*X)_n| > c) \leqslant 18c^{-1} \sup_{\mathbf{N}} E(|X_n|) \quad \textit{for every } c \in \mathbf{R}_+.$$

PROOF. The condition $\sup_{\mathbf{N}} E(|X_n|) < \infty$ permits writing the martingale $(X_n, n \in \mathbf{N})$ as the difference of two positive integrable martingales, say

$X_n = X_n' - X_n''(n \in \mathbf{N})$ (see the proof of Theorem IV-1-2); we have

$$E(X_0') = \lim_n \uparrow E(X_n^+), \qquad E(X_0'') = \lim_n E(X_n^-),$$

so that

$$E(X_0' + X_0'') = \lim_n \uparrow E(|X_n|).$$

It then suffices to apply the preceding proposition to these two positive martingales and the predictable sequence $(U_n, n \in \mathbf{N})$ to find that firstly the sequence $(U*X)_n = (U*X')_n - (U*X'')_n$ converges a.s. to a finite limit when, $n \uparrow \infty$ and secondly that

$$\begin{aligned} P(\sup_{\mathbf{N}} |(U*X)_n| > c) &\leqslant P(\sup_{\mathbf{N}} |(U*X')_n| > \tfrac{1}{2}c) + P(\sup_{\mathbf{N}} |(U*X'')_n| > \tfrac{1}{2}c) \\ &\leqslant 18c^{-1} E(X_0' + X_0'') = 18c^{-1} \sup_{\mathbf{N}} E(|X_n|). \blacksquare \end{aligned}$$

Contrary to what one might expect, the hypotheses of the above corollary do not imply that $\sup_{\mathbf{N}} E(|(U*X)_n|) < \infty$ or that the a.s. limit $\lim_{n\to\infty}(U*X)_n$ is integrable.

VIII-5. Exercises

VIII-1. (Multiplicative decomposition of positive supermartingales.) Let $(X_n, n \in \mathbf{N})$ be a finite positive supermartingale defined on the space $[\Omega, \mathscr{A}, P; (\mathscr{B}_n, n \in \mathbf{N})]$. Show that the formula

$$\nu = \begin{cases} \inf(n : E^{\mathscr{B}_n}(X_{n+1}) = 0), \\ +\infty \qquad \text{if } E^{\mathscr{B}_n}(X_{n+1}) > 0 \text{ for every } n \in \mathbf{N}, \end{cases}$$

defines a stopping time such that $X_n = 0$ on $\{\nu < n\}$. Then show that there exists a unique finite positive martingale $(M_n, n \in \mathbf{N})$ and a unique increasing process $(A_n, n \in \mathbf{N})$ such that

$$X_n = \frac{M_n}{1 + A_n} \quad \text{on } \{\nu \geqslant n\} \qquad (n \in \mathbf{N})$$

and that $M_n = M_{\nu \wedge n}$, $A_n = A_{\nu \wedge n}$ for every $n \in \mathbf{N}$.

VIII-2. A real-valued r.v. Z defined on a probability space $(\Omega, \mathscr{A}, P)$ is said to be *symmetric* if the two r.v.'s Z and $-Z$ have the same probability distribution. On the other hand, let us denote by $\varepsilon : \mathbf{R} \to \mathbf{R}$ the Borel-measurable function defined by

$\varepsilon(x) = +1$, 0 or -1 according as $x > 0$, $x = 0$ or $x < 0$ respectively; note that $|x| = \varepsilon(x)x$ and $x = \varepsilon(x)|x|$ for every $x \in \mathbf{R}$. If Z is a symmetric r.v. and if $p = P(Z = 0)$, it is clear that $P(\varepsilon(Z) = \pm 1) = \frac{1}{2}(1 - p)$.

Let $(Z_n, n \in \mathbf{N}^*)$ be a sequence of independent and identically distributed symmetric r.r.v.'s defined on $(\Omega, \mathscr{A}, P)$; for every $n \in \mathbf{N}$ denote by $\mathscr{B}_n$ the sub-σ-field of $\mathscr{A}$ generated by $Z_1, Z_2, \ldots, Z_n$ and $|Z_{n+1}|$. Then show that on $[\Omega, \mathscr{A}, P; (\mathscr{B}_n, n \in \mathbf{N})]$ the sequence $(X_n = \sum_{m=1}^{n} \varepsilon(Z_m), n \in \mathbf{N})$ is a martingale, that the sequence $(|Z_n|, n \in \mathbf{N})$ is predictable and that $\sum_{m=1}^{n} Z_m = (|Z| * X)_n$ for every $n \in \mathbf{N}$.

VIII-3. Assuming only the a.s. convergence theorem for L^2-bounded martingales and using the result of Corollary VIII-3-12, prove the a.s. convergence of positive supermartingales. [Note that a sequence $(X_n, n \in \mathbf{N})$ of positive r.r.v.'s converges a.s. whenever every sequence $(\min(X_n, c), n \in \mathbf{N})$ converges a.s. $(c \in \mathbf{R}_+)$.]

VIII-4. Let $(X_n, n \in \mathbf{N})$ be a square-integrable martingale whose associated increasing process is denoted by $(A_n, n \in \mathbf{N})$. For every real $\alpha > 0$, let $(Y_m^{(\alpha)}, n \in \mathbf{N})$ be the martingale transform $A^\alpha * X$ defined by

$$Y_{n+1}^{(\alpha)} - Y_n^{(\alpha)} = A_{n+1}^\alpha (X_{n+1} - X_n) \qquad (n \in \mathbf{N}).$$

Show that the increasing process $(B_n^{(\alpha)}, n \in \mathbf{N})$ associated with the square of the martingale $(Y_n^{(\alpha)}, n \in \mathbf{N})$ is such that

$$B_n^{(\alpha)} \geqslant (2\alpha + 1)^{-1} A_n^{2\alpha+1},$$

and on the other hand establish that

$$|Y_n^{(\alpha)}| \leqslant 2A_n^\alpha \sup_{m \leqslant n} |X_n|.$$

The equality $E(B_n^{(\alpha)}) = E[(Y_n^{(\alpha)})^2]$ then implies that

$$(2\alpha + 1)^{-1} E(A_n^{2\alpha+1}) \leqslant 4\, E(A_n^{2\alpha} \sup_{m \leqslant n} |X_m|^2).$$

Deduce from this that

$$\|A_\infty\|_p \leqslant 4p \|\sup_{\mathbf{N}} |X_n|^2\|_p$$

for every $p > 1$. [By another method, Corollary VIII-1-6 gives a slightly better result.]

If X is a martingale such that $E(A_\infty^p) < \infty$ for a fixed real number $p \in \left]0, 1\right]$, show that the martingale transform $Y = A^{(p-1)/2} * X$ is such that $\sup_{\mathbf{N}} E(Y_n^2) \leqslant p^{-1} E(A_\infty^p)$ and hence deduce the existence of a constant c_p independent of X such that

$$E(\sup_{\mathbf{N}} |X_n|^{2p}) \leqslant c_p\, E(A_\infty^p).$$

[Deduce from the identity $X = A^{(1-p)/2} * Y$ that $|X_n| \leqslant 2A_n^{(1-p)/2} \sup_{m \leqslant n} |Y_m|$ for every $n \in \mathbf{N}$.]

VIII-5. Let $\nu : \Omega \to \mathbf{N}$ be a geometric r.v. defined on a probability space $(\Omega, \mathscr{A}, P)$; we have $P(\nu = n) = a/(1+a)^{n+1}$ for every $n \in \mathbf{N}$ and a fixed real number $a > 0$. If $\mathscr{B}_n$ denotes the sub-σ-field of $\mathscr{A}$ generated by the r.v. $\nu \wedge n\, (n \in \mathbf{N})$ show that the sequence

$$X_n = (\nu \wedge n)\, a - 1_{\{\nu < n\}} \qquad (n \in \mathbf{N})$$

is a martingale with zero mean and increasing process $A_n = (\nu \wedge n)a\ (n \in \mathbf{N})$ for the sequence $(\mathscr{B}_n, n \in \mathbf{N})$. Show that

$$E(\sup_{p \leqslant n} |X_p|) \leqslant E((\nu \wedge n)\, a + 1_{\{\nu < n\}}) = 2P(\nu < n) = 2[1 - (1+a)^{-n}]$$

and that

$$E(\sqrt{A_n}) \geqslant \sqrt{(an)}\, P(\nu \geqslant n) = (1+a)^{-n} \sqrt{(an)}$$

Then deduce that there cannot exist any finite constant C such that

$$E(\sqrt{A_n}) \leqslant C\, E(\sup_{p \leqslant n} |X_p|)$$

for every $n \in \mathbf{N}$ and every real number $a > 0$.

VIII-6. On the interval $\Omega = [-\frac{1}{2}, +\frac{1}{2}]$ equipped with Lebesgue measure, let us consider the trivial σ-field $\mathscr{B}_0 = \{\emptyset, \Omega\}$, the σ-field $\mathscr{B}_1$ of symmetric Borel subsets of Ω, and the σ-fields $\mathscr{B}_2, \mathscr{B}_3, \ldots$ all equal to the Borel σ-fields of Ω. We also take two Borel measurable functions f, g on Ω, the first being odd and integrable, and the second even and square integrable, such that the function fg is not integrable; such a choice is clearly possible. Then show that the sequence $(X_0 = X_1 = 0,\ X_2 = X_3 = \ldots = f)$ is a martingale such that

$$E(\sqrt{A_\infty}) = \int |f(x)| \mathrm{d}x < \infty,$$

whilst the sequence $(Y_0 = 0,\ Y_1 = Y_2 = \ldots = g)$ is a martingale such that

$$E^{\mathscr{B}_0}(B_\infty) = \int g^2(x)\, \mathrm{d}x < \infty,$$

$$E^{\mathscr{B}_n}(B_\infty - B_n) = 0 \text{ if } n \geqslant 1;$$

nevertheless the products $X_n Y_n (n \geqslant 2)$ are not integrable even though

$$\sum_{\mathbf{N}} [(X_{n+1} - X_n)(Y_{n+1} - Y_n)] = 0.$$

VIII-7. If $(B_n, n \in \mathbf{N})$ is an increasing (not necessarily adapted) sequence of positive r.r.v.'s such that $B_\infty = \lim\uparrow_n B_n$ is integrable, and if $(A_n, n \in \mathbf{N})$ denotes the increasing process with the same potential, so that $E^{\mathscr{B}_n}(A_\infty - A_n) = E^{\mathscr{B}_n}(B_\infty - B_n)$ $(n \in \mathbf{N})$, show that for every real $p \in]0, 1]$,

$$[E(B_\infty^p)]^{1/p} \leqslant p[E(A_\infty^p)]^{1/p}.$$

[Extend the argument of the proof of Proposition VIII-**2**-13(c), where B is the quadratic variation of a martingale and $p = \frac{1}{2}$. Hence deduce that Corollary VIII-**1**-5 remains true for all $p \in]0, 1]$ subject to the direction of the inequality being reversed!]

APPENDIX

ON THE USE OF YOUNG'S FUNCTIONS IN THE THEORY OF MARTINGALES

A-1. Young's functions

Let $\phi : \mathbf{R}_+ \to \mathbf{R}_+$ be an increasing left-continuous function which is zero at the origin. The indefinite integral of this function, say

$$\Phi(t) = \int_0^t \phi(s)\,\mathrm{d}s,$$

is then a continuous convex increasing mapping of $\mathbf{R}_+$ into $\mathbf{R}_+$ which is zero at the origin. Conversely, every mapping $\Phi : \mathbf{R}_+ \to \mathbf{R}_+$ having these properties is of the above form for a unique function ϕ: the function ϕ is the left derivative of Φ at every point $t > 0$.

We will suppose for the moment that $\lim\uparrow_{+\infty} \phi(s) = +\infty$ or, what is clearly equivalent, that $\lim\uparrow_{+\infty} t^{-1}\Phi(t) = +\infty$. Then let us denote by $\psi : \mathbf{R}_+ \to \mathbf{R}_+$ the increasing left-continuous function which is zero at the origin, inverse to the function ϕ; by definition,

$$\psi(u) = \sup(s : \phi(s) < u) \quad \text{for every } u > 0.$$

The function ψ is finite as $\lim\uparrow_{+\infty} \phi(s) = +\infty$ and it likewise has a limit at infinity $\lim\uparrow_{+\infty} \psi(u)$ equal to $+\infty$. It is easy to check that conversely, beginning with the function ψ, the function ϕ is given by

$$\phi(s) = \sup(u : \psi(u) < s) \quad \text{for every } s > 0.$$

It is also not difficult to establish that the indefinite integral Ψ of the function ψ, say

$$\Psi(v) = \int_0^v \psi(u)\,\mathrm{d}u \qquad (v \in \mathbf{R}_+),$$

is such that

$$tv \leqslant \Phi(t) + \Psi(v) \qquad (t, v \in \mathbf{R}_+)$$

and that further, if $t, v \in \mathbf{R}_+$,

$$tv = \Phi(t) + \Psi(v) \Leftrightarrow v \in [\phi(t), \phi(t+0)] \Leftrightarrow t \in [\psi(v), \psi(v+0)].$$

These properties imply the two formulae

$$\Psi(v) = \sup_{t \in \mathbf{R}_+} (tv - \Phi(t)), \qquad \Phi(t) = \sup_{v \in \mathbf{R}_+} (tv - \Psi(v)).$$

All of these results have simple geometric interpretations in R_+^2 which we leave the reader to provide for himself.

We also note that the function ϕ (resp. ψ) is continuous on $\mathbf{R}_+$ if and only if its inverse function ψ (resp. ϕ) is strictly increasing. Consequently, if ϕ is both continuous and strictly increasing, then the same is true for ψ; the two indefinite integrals Φ and Ψ are then differentiable at every point of $\mathbf{R}_+$ (and not only left and right differentiable).

A pair of such functions (Φ, Ψ) is called a pair of Young's functions and the function Ψ (resp. Φ) is said to be conjugate (or complementary) to Φ (resp. Ψ). Here are two important examples:

(a) If we put $\Phi_\alpha(t) = (1+\alpha)^{-1} t^{1+\alpha}$ on $\mathbf{R}_+$ for every real number $\alpha \in]0, \infty[$, the functions Φ_α and $\Phi_{1/\alpha}$ form a pair of Young functions (note that $\phi_\alpha(t) = t^\alpha$). In particular the function $\Phi_1(t) = \frac{1}{2}t^2$ is its own conjugate.

(b) To the pair $\phi(s) = e^s - 1$ $(s \in \mathbf{R}_+)$, $\psi(u) = \log(1+u)$ $(u \in \mathbf{R}_+)$ of functions corresponds the pair of Young functions

$$\Phi(t) = e^t - 1 - t, \; \Psi(v) = (1+v)\log(1+v) - v \qquad (t, v \in \mathbf{R}_+).$$

It frequently arises in analysis that one would like to impose the growth condition $\sup_{t>0}(\Phi(at)/\Phi(t)) < \infty$ for some $a > 1$ on a Young function. For example the functions $\Phi(t) = (1+\alpha)^{-1} t^{1+\alpha}$ $(\alpha > 0)$ satisfy this condition, whereas the exponential function $\Phi(t) = e^t - 1 - t$ does not.

LEMMA A-1-1. *For every Young function Φ the following conditions are equivalent:*

(a) $\sup_{t>0}(\Phi(at)/\Phi(t)) < \infty$ *for some* $a > 1$,
(b) $\sup_{t>0}(\phi(at)/\phi(t)) < \infty$ *for some* $a > 1$,
(c) $\sup_{t>0}\{t\phi(t)/\Phi(t)\} < \infty$.

They imply the existence of a finite constant A such that $\Psi \circ \phi \leqslant A\Phi$ on $\mathbf{R}_+$, where Ψ denotes the Young function conjugate to Φ; more generally, for every $\varepsilon > 0$

there exists a constant $C_\varepsilon > 0$ *such that* $C_\varepsilon \Psi(\phi(s)/C_\varepsilon) \leqslant \varepsilon\Phi(s)$ *for every* $s \in \mathbf{R}_+$, *and then*

$$u\phi(v) \leqslant C_\varepsilon \Phi(u) + \varepsilon\Phi(v) \qquad (u, v \in \mathbf{R}_+).$$

PROOF. We show successively that (b) ⇒ (a) ⇒ (c) ⇒ (b). Firstly if $\phi(at) \leqslant A\phi(t)$ for every $t \in \mathbf{R}_+$, we also have

$$\Phi(at) = a\int_0^t \phi(as)\,\mathrm{d}s \leqslant Aa\Phi(t) \qquad (t \in \mathbf{R}_+).$$

Hence $(b) \Rightarrow$ (a).

Next, if $\Phi(at) \leqslant A'\Phi(t)$ for every $t \in \mathbf{R}_+$ $(a > 1)$, we have

$$(a-1)\,t\,\phi(t) \leqslant \int_t^{at} \phi(s)\,\mathrm{d}s \leqslant \Phi(at) \leqslant A'\Phi(t)$$

since ϕ is increasing; hence (a) ⇒ (c).

Finally, if $t\phi(t) \leqslant A''\Phi(t)$ for every $t \in \mathbf{R}_+$, we can write

$$t\phi(t) \leqslant A''\,\Phi(t) = A''\left[\int_0^{\alpha t} \phi(s)\,\mathrm{d}s + \int_{\alpha t}^t \phi(s)\,\mathrm{d}s\right]$$

$$\leqslant A''\,[\alpha t\phi(\alpha t) + (1-\alpha)\,t\phi(t)]$$

for every $\alpha \in\,]0, 1[$; if α is chosen so that $A''(1-\alpha) < 1$, we see that $\phi(t) \leqslant B\phi(\alpha t)$ for a finite constant B. It remains to put $a = \alpha^{-1}$ and the implication (c) ⇒ (b) is proved.

Since $\Psi(\phi(t)) = t\phi(t) - \Phi(t)$ for every $t \in \mathbf{R}_+$, it is clear that condition (c) implies that $\sup_{t>0}(\Psi(\phi(t))/(\Phi(t)) < \infty$. More generally, let us suppose that $\phi(at) \leqslant A\phi(t)$ for every $t \in \mathbf{R}_+$, for some $a > 1$; then by iteration

$$\phi(t) \leqslant A^n\,\phi(a^{-n}t) \quad \text{for all } n \in \mathbf{N}$$

and upon applying the function ψ to both sides of this inequality and dividing by A^n we find that

$$\psi(A^{-n}\,\phi(t)) \leqslant a^{-n}\,t \quad (t \in \mathbf{R}_+, n \in \mathbf{N}).$$

Using the inequality $\Psi(v) \leqslant v\psi(v)$, this allows us to write

$$A^n\,\Psi(A^{-n}\,\phi(t)) \leqslant \phi(t)\,\psi(A^{-n}\,\phi(t)) \leqslant a^{-n}\,t\,\phi(t) \leqslant a^{-n}\,B\Phi(t)$$

since the hypothesis made implies that $B = \sup_{t>0}(t\phi(t)/\Phi(t)) < \infty$. It then remains to put $C_\varepsilon = A^n$ for every $\varepsilon > 0$, choosing $n \in \mathbf{N}$ so that $a^{-n}B < \varepsilon$. From the inequality $C_\varepsilon \Psi(\phi(t)/C_\varepsilon) \leqslant \varepsilon\Phi(t)$ thus proved for every $t \in \mathbf{R}_+$ and Young's inequality it follows that

$$u\phi(v) = C_\varepsilon u\,\phi(v)/C_\varepsilon \leqslant C_\varepsilon\{\Phi(u) + \Psi(\phi(v)/C_\varepsilon)\} \leqslant C_\varepsilon\,\Phi(u) + \varepsilon\Psi(v).$$

for every $u,v \in \mathbf{R}_+$. ■

A-2. Orlicz spaces

Throughout this section we suppose given a probability space $(\Omega, \mathscr{A}, P)$ and a pair (Φ, Ψ) of Young functions. When $\Phi(t) = (1+\alpha)^{-1}t^{1+\alpha}$ on $\mathbf{R}_+$ $(0 < \alpha < \infty)$, the Banach space L^Φ of the following proposition coincides with the usual space $L^{1+\alpha}(\Omega, \mathscr{A}, P)$.

PROPOSITION A-2-2. *The set $L^\Phi(\Omega, \mathscr{A}, P)$ of equivalence classes of real-valued random variables defined on $(\Omega, \mathscr{A}, P)$ for which there exists at least one real number $a > 0$ such that $E(\Phi(a^{-1}|X|)) \leqslant 1$, is a vector subspace of $L^1(\Omega, \mathscr{A}, P)$ containing $L^\infty(\Omega, \mathscr{A}, P)$. Furthermore, the formula*

$$\|X\|_\Phi = \inf(a : a > 0,\ E(\Phi(a^{-1}|X|)) \leqslant 1)$$

defines a norm on L^Φ and there exist two constants c_1, $c_\infty > 0$ such that $c_1\|X\|_1 \leqslant \|X\|_\Phi \leqslant c_\infty\|X\|_\infty$ for every r.v. X of L^Φ. The normed vector space L^Φ is complete and this Banach space is called an Orlicz space.

For every pair $X \in L^\Phi$, $Y \in L^\Psi$ the product r.v. XY is integrable and in fact satisfies the inequality $\|XY\|_1 \leqslant 2\|X\|_\Phi\|Y\|_\Psi$. For every $Y \in L^\Psi$, the mapping $X \to E(XY)$ is therefore a continuous linear functional on L^Φ; further, the formula

$$\|Y\|'_\Phi = \sup_X(\|X\|_\Phi^{-1}\,E(XY) : X \neq 0)$$
$$= \sup_X(E(XY) : E(\Phi(|X|)) \leqslant 1)\ \ (Y \in L^\Psi)$$

defines a norm on L^Ψ equivalent to the norm $\|\cdot\|_\Psi$ of this space and such that, more precisely,

$$\|Y\|_\Psi \leqslant \|Y\|'_\Phi \leqslant 2\|Y\|_\Psi$$

for all $Y \in L^\Psi$.

PROOF. It will be convenient to define $\|X\|_\Phi$ for every r.r.v. X by putting $\|X\|_\Phi = \infty$ when there exists no real a such that $E(\Phi(a^{-1}|X|)) \leqslant 1$. The definition of $\|X\|_\Phi$ immediately implies that $\|X\|_\Phi = 0$ if and only if $X = 0$ a.s. and that $\|cX\|_\Phi = |c|\,\|X\|_\Phi$ for every r.r.v. X and constant $c \in \mathbf{R}$. The triangle inequality

$$\|X+Y\|_\Phi \leqslant \|X\|_\Phi + \|Y\|_\Phi$$

is easily deduced for every pair X, Y from the fact that the function Φ is convex and increasing; indeed, in the case where $\|X\|_\Phi, \|Y\|_\Phi \in \,]0, \infty[$ (in other cases there is nothing to prove) we have

$$\Phi\left(\frac{|X+Y|}{\|X\|_\Phi + \|Y\|_\Phi}\right) \leqslant \Phi\left(\frac{|X|+|Y|}{\|X\|_\Phi + \|Y\|_\Phi}\right)$$

$$\leqslant \frac{\|X\|_\Phi}{\|X\|_\Phi + \|Y\|_\Phi}\,\Phi\left(\frac{|X|}{\|X\|_\Phi}\right) + \frac{\|Y\|_\Phi}{\|X\|_\Phi + \|Y\|_\Phi}\,\Phi\left(\frac{|Y|}{\|Y\|_\Phi}\right)$$

and taking expectations we find that

$$E\left(\Phi\left(\frac{|X+Y|}{\|X\|_\Phi + \|Y\|_\Phi}\right)\right) \leqslant 1$$

i.e. $\|X+Y\|_\Phi \leqslant \|X\|_\Phi + \|Y\|_\Phi$. It is therefore clear that L^Φ is a vector space normed by $\|\cdot\|_\Phi$.

We go on to establish the inclusions $L^\infty \subset L^\Phi \subset L^1$ and the existence of two constants c_1, c_∞ such that $c_1\|X\|_1 \leqslant \|X\|_\Phi \leqslant c_\infty\|X\|_\infty$ for every r.r.v. X. To this end let us first choose a real number $u_0 > 0$ such that $\phi(u_0) > 0$; using the inequality

$$\Phi(u) = \int_0^u \phi(s)\,\mathrm{d}s \geqslant (u-u_0)^+\,\phi(u_0) \qquad (u \in \mathbf{R}_+)$$

for every r.r.v. $X \neq 0$ in L^Φ, let us write

$$1 \geqslant E\left(\Phi\left(\frac{|X|}{\|X\|_\Phi}\right)\right) \geqslant \phi(u_0)\,E\left(\left(\frac{|X|}{\|X\|_\Phi} - u_0\right)^+\right);$$

it follows that

$$\frac{E(|X|)}{\|X\|_\Phi} \leqslant E\left(\left(\frac{|X|}{\|X\|_\Phi} - u_0\right)^+\right) + u_0 \leqslant \left(\frac{1}{\phi(u_0)} + u_0\right) < \infty.$$

The thesis concerning L^1 is then established. On the other hand, if u_1 is a real number >0 such that $\Phi(u_1) \leqslant 1$, the inequality $\Phi(a^{-1}|X|) \leqslant 1$ is valid a.s. whenever $\|X\|_\infty \leqslant au_1$; consequently $\|X\|_\infty \geqslant u_1\|X\|_\Phi$.

To show that the normed vector space L^Φ is complete, one proceeds as in the case of the L^p spaces. If $(X_n, n \in \mathbf{N})$ is a Cauchy sequence in L^Φ, extract a subsequence $(X_{n_p}, p \in \mathbf{N})$ such that

$$\sum_{p \in \mathbf{N}} \|X_{n_{p+1}} - X_{n_p}\|_\Phi < \infty;$$

the series $\sum_{p\in\mathbf{N}} \|X_{n_{p+1}} - X_{n_p}\|_1$ is hence also convergent and the limit $X = \lim_{p\to\infty} X_{n_p}$ then exists a.s. since the sum r.v. $\sum_{p\in\mathbf{N}} |X_{n_{p+1}} - X_{n_p}|$ is integrable and hence a.s. finite. But Fatou's lemma applied to the definition of the norm in L^Φ easily gives

$$\|X\|_\Phi \leqslant \liminf_{p\to\infty} \|X_{n_p}\|_\Phi < \infty,$$

so that $X \in L^\Phi$; also

$$\|X - X_{n_q}\|_\Phi \leqslant \liminf_{p\to\infty} \|X_{n_p} - X_{n_q}\|_\Phi \to 0 \quad \text{when } q\uparrow\infty,$$

so that $X_{n_q} \to X$ in L^Φ as $q\uparrow\infty$. Finally the triangle inequality allows us to show that $X_n \to X$ in L^Φ when $n\uparrow\infty$ since

$$\|X - X_n\|_\Phi \leqslant \|X - X_{n_q}\|_\Phi + \|X_n - X_{n_q}\|_\Phi \to 0 \quad \text{if } n, n_q\uparrow\infty.$$

If X and Y are two non-zero elements of L^Φ and L^Ψ respectively, the fundamental inequality $uv \leqslant \Phi(u) + \Psi(v)$ applied to the r.v.'s $|X|/\|X\|_\Phi$ and $|Y|/\|Y\|_\Psi$ gives

$$\begin{aligned} E(|X|\,|Y|) &\leqslant \|X\|_\Phi \|Y\|_\Psi \left\{ E\left(\Phi\left(\frac{|X|}{\|X\|_\Phi}\right)\right) + E\left(\Psi\left(\frac{|Y|}{\|Y\|_\Psi}\right)\right)\right\} \\ &\leqslant 2\|X\|_\Phi \|Y\|_\Psi. \end{aligned}$$

Hence $XY \in L^1$ and the inequality in the proposition concerning $\|XY\|_1$ is proved. Furthermore, the linear functional defined by every element Y of L^Ψ is such that

$$\|Y\|'_\Phi = \sup_X (E(XY) : \|X\|_\Phi \leqslant 1) \leqslant 2\|Y\|_\Psi. \blacksquare$$

REMARK. When the function Φ satisfies the growth condition of Lemma A-1-1, $\Phi(at) \leqslant A\Phi(t)$ $(t \in \mathbf{R}_+)$ for some $a > 1$ and $A < \infty$, the vector space L^Φ coincides with the set of equivalence classes of r.v.'s X such that $E(\Phi(|X|)) < \infty$. Indeed, if $X \in L^\Phi$, there exists an integer $n > 1$ such that $\|X\|_\Phi \leqslant a^n$, and then the inequality $E(\Phi(a^{-n}|X|)) \leqslant 1$ implies that

$$E(\Phi(|X|)) \leqslant A^n < \infty.$$

If $E[\Phi(|X|)] < \infty$, the general inequality $\|X\|_\Phi \leqslant \max(1, E(\Phi(|X|)))$ proved above implies conversely that $X \in L^\Phi$. On the other hand, in the case of the exponential function $\Phi(t) = e^t - 1 - t$, there exist r.v.'s X belonging to L^Φ such that $E(\Phi(|X|)) = +\infty$.

PROPOSITION A-2-3. *For every sub-σ-field $\mathscr{B}$ in the probability space $(\Omega, \mathscr{A}, P)$, the restriction of the conditional expectation $E^\mathscr{B}$ to the Orlicz space* L $(\Omega, \mathscr{A}, P)$ *defines an idempotent positive linear operator of norm* 1.

PROOF. Everything reduces to showing that $\|E^\mathscr{B}(X)\|_\Phi \leqslant \|X\|_\Phi$ if $X \in L^\Phi$. To this end we use the convexity inequality

$$\Phi(u) \geqslant \Phi(v) + (u - v)\,\phi(v)$$

valid for every $u,v \in \mathbf{R}_+$ (which is easy to establish from the monotonicity of the function ϕ) to write

$$\Phi(Z) \geqslant \Phi(E^\mathscr{B}(Z)) + [Z - E^\mathscr{B}(Z)]\,\phi[E^\mathscr{B}(Z)]$$

for any positive r.v. Z whose conditional expectation $E^\mathscr{B}(Z)$ is a.s. finite; this inequality implies that $E^\mathscr{B}(\Phi(Z)) \geqslant \Phi(E^\mathscr{B}(Z))$ by taking conditional expectations of both sides and hence

$$E(\Phi(Z)) \geqslant E(\Phi(E^\mathscr{B}(Z)).$$

If $0 < \|X\|_\Phi < \infty$, we apply this inequality to the positive r.v. $Z = |X|/\|X\|_\Phi$ and find that

$$E\left(\Phi\left(\frac{|E^\mathscr{B}(X)|}{\|X\|_\Phi}\right)\right) \leqslant E(\Phi(E^\mathscr{B}(X))) \leqslant 1$$

and hence that $\|E^\mathscr{B}(X)\|_\Phi \leqslant \|X\|_\Phi$. ■

A-3. Applications to the theory of martingales

We begin with a result generalising Propositions IV-**2**-7, IV-**2**-8 and IV-**2**-10, proved in essentially the same way as these results.

PROPOSITION A-**3**-4. *Every martingale* $(X_n, n \in \mathbf{N})$ *defined on a space* $[\Omega, \mathscr{A}, P; (\mathscr{B}_n, n \in \mathbf{N})]$ *such that* $\sup_{\mathbf{N}} \|X_n\|_\Phi < \infty$ *is regular and converges in* L^Φ *to its a.s. limit* $X_\infty = \lim_{n \to \infty} X_n$. *Furthermore, if we denote by* $\xi : \mathbf{R}_+ \to \mathbf{R}_+$ *the increasing function defined by*

$$\xi(u) = u\phi(u) - \Phi(u) \equiv \Psi[\phi(u)],$$

the r.v. $S = \sup_{\mathbf{N}} |X_n|$ *satisfies the inequality*

$$E\left(\xi\left(\frac{S}{\rho\|X_\infty\|_\Phi}\right)\right) \leqslant \frac{1}{\rho - 1}$$

for every constant $\rho > 1$.

The preceding inequality reduces to the inequality of Proposition IV-**2**-8 when $\Phi(t) = t^p$ $(1 < p < \infty)$, provided we put $\rho = p/(p-1)$; it reduces to an inequality equivalent to that of Proposition IV-**2**-10 when

$$\Phi(t) = t \log^+ t,$$

in which case $\xi(t) \sim t$ at infinity; in any case the increasing function ξ is continuous and increases to $+\infty$ as $t \uparrow \infty$; in the case of the function $\Phi(t) = t\log^+(t)$, it nonetheless increases at infinity more slowly than this function, namely, as fast as t only. But whenever the assumption $\sup_{t > 0}(\Psi(bt)/\Psi(t)) = B < \infty$, where $b > 1$, is satisfied by the conjugated function, Lemma A-**1**-1 shows that $\xi(t) = \Psi(\phi(t)) \geqslant (b-1)B^{-1}\Phi(t)$; in this case the preceding proposition therefore implies that

$$\|S\|_\Phi \leqslant c\|X_\infty\|_\Phi$$

for some constant c not depending on the function Φ (put $c = 1 + (b-1)^{-1}B$).

PROOF. The function $x \to x^{-1}\Phi(x)$ is increasing on $\mathbf{R}_+$; hence every real-valued r.v. $Z > 0$ satisfies the inequality

$$\int_{\{Z \geqslant a\}} Z \,\mathrm{d}P \leqslant \frac{a}{\Phi(a)} E(\Phi(Z)) \quad \text{for every } a > 0.$$

Let us apply this result to the r.v.'s $\sigma^{-1}|X_n|$ ($n \in \mathbf{N}$) associated with the martingale of the proposition, the constant σ being defined by $\sigma = \sup_{\mathbf{N}}(\|X_n\|_\Phi) < \infty$ (we can clearly suppose that $\sigma > 0$). Since for every $n \in \mathbf{N}$ we have $E(\Phi(\sigma^{-1}|X_n|)) \leqslant 1$ by the definition of the norms $\|X_n\|_\Phi$, we obtain

$$\sup_{\mathbf{N}} \int_{\{|X|>b\}} |X_n| \, dP \leqslant \frac{b}{\Phi(\sigma^{-1} b)} \qquad (b > 0)$$

(putting $b = a\sigma$), and it remains to let $b \uparrow \infty$ to see that the right-hand side tends to zero and that the martingale $(X_n, n \in \mathbf{N})$ is uniformly integrable.

By Proposition IV-**2**-3 we have established that the martingale $(X_n, n \in \mathbf{N})$ is regular; it thus converges a.s. and in L^1 to an integrable r.v. X_∞ such that $X_n = E^{\mathscr{B}_n}(X_\infty)$ for every $n \in \mathbf{N}$. But then

$$\Phi(\sigma^{-1}|X_\infty|) = \lim_{n \to \infty} \text{a.s.}\ \Phi(\sigma^{-1}|X_n|),$$

and Fatou's lemma therefore shows that $E(\Phi(\sigma^{-1}|X_\infty|)) \leqslant 1$; in other words, $X_\infty \in L^\Phi$ and $\|X_\infty\|_\Phi \leqslant \sigma$. The martingales $(X_n, n \in \mathbf{N})$ for which $\sup_{\mathbf{N}} \|X_n\|_\Phi < \infty$ are therefore exactly those of the form $(X_n = E^{\mathscr{B}_n}(X), n \in \mathbf{N})$ for some $X \in L^\Phi$. It is then easy to show, beginning as in the proof of Proposition II-**2**-11 with the case of a bounded r.v. X, that $E^{\mathscr{B}_n}(X) \to E^{\mathscr{B}_\infty}(X)$ in L^Φ as $n \to \infty$ whenever $X \in L^\Phi$.

To prove the inequality of the proposition, we begin with the inequality of Lemma IV-**2**-9 applied to the sub-martingale $(|X_n|, n \in \mathbf{N})$

$$au\, P(a^{-1} S > u) \leqslant E(|X_\infty| \, 1_{\{a^{-1}S > u\}}) \quad (n \in \mathbf{N}, a, u > 0).$$

Integrating both sides in u with respect to the positive measure $d\phi(u)$, Fubini's theorem gives

$$aE\left(\int_0^{a^{-1}S} u \, d\phi(u)\right) \leqslant E(|X_\infty| \, \phi(a^{-1} S)) \qquad (a > 0).$$

Next we apply the basic inequality $v\phi(t) \leqslant \Phi(v) + \Psi(\phi(t))$ to the positive real-valued r.v.'s $b^{-1}|X_\infty|$ and $a^{-1} S$, where b denotes a constant >0; it follows that

$$E(|X_\infty| \phi(a^{-1} S)) \leqslant b\{E(\Phi(b^{-1}|X_\infty|)) + E(\Psi \circ \phi(a^{-1} S))\} \qquad (a, b > 0)$$

But an integration by parts shows that

$$\int_0^t u \, d\phi(u) = t\phi(t) - \Phi(t) \equiv \Psi(\phi(t)) \qquad (t \in \mathbf{R}_+)$$

and if we denote by ξ this increasing function of t, the two inequalities obtained above show that

$$aE(\xi(a^{-1}S)) \leqslant bE(\Phi(b^{-1}|X_\infty|)) + bE(\xi(a^{-1}S))$$

for any constants $a,b > 0$. Let us put $b = \|X_\infty\|_\Phi$ and $a = \rho b$, where $\rho > 1$; recalling the definition of the norm $\|X_\infty\|$, the preceding inequality can be rewritten

$$(\rho - 1)E\left(\xi\left(\frac{S}{\rho\|X_\infty\|_\Phi}\right)\right) \leqslant 1,$$

at least if the preceding expectation is finite. The formula of the proposition is thus proved when $X_\infty \in L^\infty$ for in this case the r.v.

$$S = \sup_{\mathbf{N}} |E^{\mathscr{B}_n}(X_\infty)|$$

also belongs to L^∞. If $X_\infty \in L^\Phi$ is positive but not bounded, we approximate X_∞ by the increasing sequence $(\min(X_\infty, p), p \in \mathbf{N})$, and obtain the desired inequality by passing to the limit since

$$\sup_{\mathbf{N}} E^{\mathscr{B}_n}(\min(X_\infty, p)) \uparrow S \quad \text{when } p \uparrow \infty.$$

Finally, the case where $X_\infty \in L^\Phi$ is arbitrary is immediately reduced to the positive case since $\sup_{\mathbf{N}}|E^{\mathscr{B}_n}(X_\infty)| \leqslant \sup_{\mathbf{N}} E^{\mathscr{B}_n}(|X_\infty|)$, whilst

$$\|X_\infty\|_\Phi = \|\,|X_\infty|\,\|_\Phi. \blacksquare$$

Remark. It is not hard to show that for every uniformly integrable martingale $(X_n, n \in \mathbf{N})$ there exists at least one Young function Φ such that $\sup_{\mathbf{N}} E(\Phi(|X_n|)) \leqslant 1$. ■

Next we generalise Proposition VIII-1-4.

Proposition A-3-5. *Let Φ be a Young function such that*

$$\alpha = \sup_{t>0} \frac{t\phi(t)}{\Phi(t)} < \infty$$

(cf. Lemma A-1-1). Let Z be a positive r.v. in $L^\Phi(\Omega, \mathscr{A}, P)$. Every increasing process $(A_n, n \in \mathbf{N})$ whose potential $(X_n = E^{\mathscr{B}_n}(A_\infty - A_n), n \in \mathbf{N})$ is dominated by the positive martingale $(E^{\mathscr{B}_n}(Z), n \in \mathbf{N})$ then satisfies the inequality

$$\|A_\infty\|_\Phi \leqslant \alpha\|Z\|_\Phi.$$

In particular the function $\Phi(t)=t^p$, where $1<p<\infty$, satisfies the assumption with $\alpha=p$; the preceding proposition thus generalises Proposition VIII-1-4 with the best possible constant in the inequality. On the other hand, it is clear that the preceding proposition has the same applications as Proposition VIII-1-4.

PROOF. We begin with the inequality

$$E((A_\infty-au)^+)\leqslant E(Z\,1_{\{A_\infty>au\}})\qquad (a,u>0)$$

satisfied by the limiting r.v. $A_\infty=\lim\uparrow_{\mathbf{N}}A_n$ of the increasing process and the r.v. Z (see the proof of Proposition VIII-1-4). We integrate out u both sides of this inequality with respect to the positive measure $d\phi(u)$; by Fubini's theorem we thus find that

$$aE\left(\int_0^\infty (a^{-1}A_\infty-u)^+\,\mathrm{d}\phi(u)\right)\leqslant E(Z\phi(a^{-1}A_\infty)).$$

But on the one hand

$$\int_0^\infty (z-u)^+\,\mathrm{d}\phi(u)=\int_0^z \phi(u)\,\mathrm{d}u=\Phi(z)\qquad (z\in\mathbf{R}_+),$$

and on the other hand the basic inequality

$$v\phi(t)\leqslant\Phi(v)+\Psi(\phi(t))\qquad (v,t\in\mathbf{R}_+)$$

implies that for every $b>0$,

$$E(Z\phi(a^{-1}A_\infty))\leqslant b\{E(\Phi(b^{-1}Z))+E(\Psi\circ\phi(a^{-1}A_\infty))\}.$$

Comparing the two preceding inequalities we therefore find that

$$aE(\Phi(a^{-1}A_\infty))\leqslant bE(\Phi(b^{-1}Z))+bE(\Psi\circ\phi(a^{-1}A_\infty))$$

for any constants $a,b>0$.

Since the r.v. A_∞ appears in both sides of this last inequality, the latter can only be interesting if the function $\Psi\circ\phi$ does not grow faster at infinity than the function Φ. Under the assumption

$$\alpha=\sup_{\mathbf{R}_+}\frac{t\phi(t)}{\Phi(t)}<\infty$$

which implies that

$$\psi(\phi(t)) \equiv t\phi(t) - \Phi(t) \leqslant (\alpha - 1)\Phi(t)$$

for every $t \in \mathbf{R}_+$, the above inequality can be rewritten

$$[a - b(\alpha - 1)]\, E(\Phi(a^{-1} A_\infty)) \leqslant bE(\Phi(b^{-1} Z)).$$

Taking $b = \|Z\|_\Phi$ and $a = b\alpha$ we thus find that

$$E\left(\Phi\left(\frac{A_\infty}{\alpha\|Z\|_\Phi}\right)\right) \leqslant 1,$$

which clearly shows that $\|A_\infty\|_\Phi \leqslant \alpha\|Z\|_\Phi$.

The preceding argument is not entirely rigorous since we do not know in advance that $E(\Psi \circ \phi(a^{-1} A_\infty)) < \infty$. Nonetheless, this condition is always fulfilled if the r.v. A_∞ is bounded, and the preceding inequality is therefore established in this case; the general case follows by applying the inequality to the increasing processes $(\min(A_n, p), n \in \mathbf{N})$ and their potentials, then letting $p \uparrow \infty$. The passage to the limit can be made since the potentials of the truncated increasing processes are all dominated by the potential $(X_n, n \in \mathbf{N})$. ■

REFERENCES

[1] A. F. Abrahamse, A comparison between the Martin boundary theory and the theory of likelihood ratios, *Ann. Math. Statist.* **41** (1970) 1064–1067.

[2] G. Alexits, *Convergence Problems of Orthogonal Series* (Pergamon Press, Oxford, 1961) ix + 350 pp.

[3] G. Alexits and A. Sharma, On the convergence of multiplicatively orthogonal series, *Acta Math. Acad. Sci. Hung.* **22** (1971) 257–266.

[4] C. Alloin, Martingales progressives, *Cahiers Centre Etudes Rech. Opér.* **12** (1970) 201–210.

[5] T. Ando, Contractive projections in L_p-spaces, *Pacific J. Math.* **17** (1966) 391–405.

[6] T. Ando and I. Amemiya, Almost everywhere convergence of prediction sequence in L^p $(1 < p < \infty)$, *Z. Wahrscheinlichkeitstheorie Verw. Gebiete* **4** (1965) 113–120.

[7] D. G. Austin, A sample function property of martingales, *Ann. Math. Statist.* **37** (1966) 1396–1397.

[8] L. Báez-Duarte, Another look at the martingale theorem, *J. Math. Anal. Appl.* **23** (1968) 551–557.

[9] L. Báez-Duarte, On the convergence of martingale transforms, *Z. Wahrscheinlichkeitstheorie Verw. Gebiete* **19** (1971) 319–322.

[10] L. Báez-Duarte, An a.e. divergent martingale that converges in probability, *J. Math. Anal. Appl.* **36** (1971) 149–150.

[11] R. R. Bahadur, Statistics and subfields, *Ann. Math. Statist.* **26** (1955) 490–497.

[12] R. R. Bahadur and P. J. Bickel, Substitution in conditional expectation, *Ann. Math. Statist.* **39** (1968) 377–378.

[13] B. B. Bhattacharya, Reverse submartingale and some functions of order statistics, *Ann. Math. Statist.* **41** (1970) 2155–2157.

[14] P. J. Bickel and J. A. Yahav, On an A.P.O. rule in sequential estimation with quadratic loss, *Ann. Math. Statist.* **40** (1969) 417–426.

[15] P. J. Bickel and J. A. Yahav, Some contributions to the asymptotic theory of Bayes solutions, *Z. Wahrscheinlichkeitstheorie Verw. Gebiete.* **11** (1969) 257–276.

[16] P. Billingsley, The Lindeberg–Lévy theorem for martingales, *Proc. Am. Math. Soc.* **12** (1961) 788–792.

[17] P. Billingsley, *Convergence of Probability Measures* (Wiley, New York, 1968).

[18] E. Bishop, An upcrossing inequality with applications, *Michigan Math. J.* **13** (1966) 1–13.

[19] D. Blackwell, Discounted dynamic programming, *Ann. Math. Statist.* **36** (1965) 226–235.

[20] D. Blackwell and L. E. Dubins, A converse to the dominated convergence theorem, *Illinois J. Math.* **7** (1963) 508–514.

[21] D. Blackwell and D. Freedman, A remark on the coin-tossing game. *Ann. Math. Statist.* **35** (1964) 1345–1347.

[22] D. BLACKWELL and C. RYLL-NARDZEWSKI, Non-existence of everywhere proper conditional distributions, *Ann. Math. Statist.* **34** (1963) 223–225.

[23] L. H. BLAKE, A generalisation of martingales and two consequent convergence theorems, *Pacific J. Math.* **35** (1970) 279–283.

[24] L. H. BLAKE, A note concerning the L_1 convergence of a class of games which become fairer with time, *Glasgow Math. J.* **13** (1972) 39–41.

[25] L. H. BLAKE, Further results concerning games which become fairer with time, *J. London Math. Soc.* (2) **6** (1973) 311–316.

[26] S. BOCHNER, Partial ordering in the theory of martingales, *Ann. Math.* **62** (1955) 162–169.

[27] E. S. BOYLAN, Equiconvergence of martingales, *Ann. Math. Statist.* **42** (1971) 552–559.

[28] L. BREIMAN, First exit times from a square root boundary, in: L. M. Le Cam and J. Neyman, eds., *Proc. 5th Berkeley Symp. on Mathematical Statistics and Probability* II **2** (Univ. of California Press, Berkeley, Calif., 1967) 9–16.

[29] L. BREIMAN, *Probability* (Addison-Wesley, Reading, Mass., 1968).

[30] E. BRIEM, A. GUICHARDET, and NGUYEN-XUAN-LOC, Les martingales généralisées, *C.R. Acad. Sci. Paris* (A) **270** (1970) 373–375.

[31] B. M. BROWN, Moments of a stopping rule related to the central limit theorem, *Ann. Math. Statist.* **40** (1969) 1236–1249.

[32] B. M. BROWN, Martingale central limit theorems, *Ann. Math. Statist.* **42** (1971) 59–66.

[33] B. M. BROWN, A general three-series theorem, *Proc. Am. Math. Soc.* **28** (1971) 573–577; errata see [145].

[34] B. M. BROWN and G. K. EAGLESON, Martingale convergence to infinitely divisible laws with finite variances, *Trans. Am. Math. Soc.* **162** (1971) 449–453.

[35] R. S. BUCY, Stability and positive supermartingales, *J. Differential Equations* **1** (1965) 151–155.

[36] H. BÜHLMANN, L^2-martingales and orthogonal decomposition, *Z. Wahrscheinlichkeitstheorie Verw. Gebiete* **1** (1963) 394–414.

[37] D. L. BURKHOLDER, Successive conditional expectations of an integrable function, *Ann. Math. Statist.* **33** (1962) 887–893.

[38] D. L. BURKHOLDER, Maximal inequalities as necessary conditions for almost everywhere convergence, *Z. Wahrscheinlichkeitstheorie Verw. Gebiete.* **3** (1964) 75–88.

[39] D. L. BURKHOLDER, Martingale transforms, *Ann. Math. Statist.* **37** (1966) 1494–1504.

[40] D. L. BURKHOLDER, Independent sequences with the Stein property, *Ann. Math. Statist.* **39** (1968) 1282–1288.

[41] D. L. BURKHOLDER, Martingale Inequalities, in: *Martingales*, Lecture Notes in Math. **190** (Springer, 1970) 1–8.

[42] D. L. BURKHOLDER, Inequalities for operators on martingales, in: *Proc. Intern. Congr. Mathematicians*, Nice, 1970 Vol. 2 (Gauthier-Villars, Paris, 1971) 551–557.

[43] D. L. BURKHOLDER and R. F. GUNDY, Extrapolation and interpolation of quasi-linear operators on martingales, *Acta Math.* **124** (1970) 250–304.

[44] D. L. BURKHOLDER, B. J. DAVIS and R. F. GUNDY, Integral inequalities for convex functions of operators on martingales, in: L. M. Le Cam et al., eds., *Proc 6th Berkeley Symp. on Mathematical Statistics and Probability*, II (Univ. of California Press, Berkeley, Calif., 1972) 223–240.

[45] D. L. BURKHOLDER, Distribution function inequalities for martingales, *Ann. Probab.* **1** (1973) 19–42.

[46] R. CAIROLI, Une inegalité pour martingales à indices multiples et ses applications, in: *Sém. Probabilités IV, Univ. de Strasbourg*, Lecture Notes in Math. **124** (Springer, Berlin, 1970) 1–27.

[47] R. CAIROLI, Décomposition de processus à indices double, in: *Sém. de Probabilités V, Univ. de Strasbourg*, Lecture Notes in Math. **191** (Springer, Berlin, 1970) 37–57.

[48] R. CAIROLI, Une théorème de convergence pour martingales à indices multiples, *C.R. Acad. Sci. Paris* (A) **269** (1969) 587–589.

[49] Ch. CASTAING, Sur les multi-applications mesurables, *Rev. Francaise Inform. Rech. Oper.* **1**, (1967) 91–126.

[50] S. D. CHATTERJI, Martingales of Banach-valued random variables, *Bull. Am. Math. Soc.* **66** (1960) 395–398.

[51] S. D. CHATTERJI, A note on the convergence of Banach-space valued martingales, *Math. Ann.* **153** (1964) 142–149.

[52] S. D. CHATTERJI, Comments on the martingale convergence theorem, in: *Symp. on Probability Methods in Analysis* (*Loutraki*), Lecture Notes in Math. **31** (Springer, Berlin, 1967) 55–61.

[53] S. D. CHATTERJI, Martingale convergence and the Radon–Nikodym theorem in Banach spaces, *Math. Scand.* **22** (1968) 21–41.

[54] S. D. CHATTERJI, Differentiation along algebras, *Manuscript Math.* **4** (1971) 213–224.

[55] S. D. CHATTERJI, Les martingales et leurs applications analytiques, in: J. L. Bretagnolle et al., *Ecole d'Été de Probabilités: Processus Stochastiques*, Lecture Notes in Math. **307** (Springer, Berlin, 1973) 27–164.

[56] D. CHAZAN, A note on the convergence of submartingales, *Ann. Math. Statist.* **35** (1964) 1811–1814.

[57] H. CHERNOFF, A note on risk and maximal regular generalised submartingales in stopping problems, *Ann. Math. Statist.* **38** (1967) 606–607.

[58] H. CHERNOFF, Optimal stochastic control, *Sankhyā* (A) **30** (1968) 221–252.

[59] G. Y. H. CHI, Conditional expectations and submartingale sequences of random Schwartz distributions, *J. Multivariate Anal.* **3** (1973) 71–92.

[60] C. CHING-SUNG, Φ-pseudo-entropie du processus et regularité des martingales, *C.R. Acad. Sci. Paris* (A) **274** (1972) 104–107.

[61] C. CHING-SUNG, Φ-pseudo-entropie du processus à l'indice multiple et regularité des martingales, *C.R. Acad. Sci. Paris* (A) **274** (1972) 206–207.

[62] J. CHOVER, On Strassen's version of the loglog law, *Z. Wahrscheinlichkeitstheorie Verw. Gebiete* **8** (1967) 83–90.

[63] Y. S. CHOW, A martingale inequality and the law of large numbers, *Proc. Am. Math. Soc.* **11** (1960) 107–111.

[64] Y. S. CHOW, Martingales in a σ-finite measure space indexed by directed sets, *Trans. Am. Math. Soc.* **97** (1960) 254–285.

[65] Y. S. CHOW, Convergence theorems of martingales, *Z. Wahrscheinlichkeitstheorie Verw. Gebiete* **1** (1962) 340–346.

[66] Y. S. CHOW, A martingale convergence theorem of Ward's type, *Illinois J. Math.* **9** (1965) 569–576.

[67] Y. S. CHOW, Local convergence of martingales and the law of large numbers, *Ann. Math. Statist.* **36** (1965) 552–558.

[68] Y. S. CHOW, On the expected value of a stopped submartingale, *Ann. Math. Statist.* **38** (1967) 608–609.

[69] Y. S. Chow, On a strong law of large numbers for martingales, *Ann. Math. Statist.* **38** (1967) 610.
[70] Y. S. Chow, Convergence of sums of squares of martingale differences, *Ann. Math. Statist.* **39** (1968) 123–133.
[71] Y. S. Chow, Martingale extensions of a theorem of Marcinkiewiez and Zygmund, *Ann. Math. Statist.* **40** (1969) 427–433.
[72] Y. S. Chow and H. Robbins, A martingale system theorem and applications, in: J. Neyman, ed., *Proc.* 4*th Berkeley Symp. on Mathematical Statistics and Probability*, Vol. 1 (Univ. of California Press, Berkeley, Calif., 1961) 93–104.
[73] Y. S. Chow and H. Robbins, On sums of independent random variables with infinite first moments and "fair" games, *Proc. Nat. Acad. Sci. U.S.A.* **47** (1961) 330–335.
[74] Y. S. Chow and H. Robbins, On optimal stopping rules, *Z. Wahrscheinlichkeitstheorie Verw. Gebiete* **2** (1963) 33–49.
[75] Y. S. Chow and H. Robbins, On optimal stopping rules for s_n/n, *Illinois J. Math.* **9** (1965) 444–454.
[76] Y. S. Chow and H. Robbins, A class of optimal stopping problems, in: L. M. Le Cam and J. Neyman, eds., *Proc.* 5th *Berkeley Symp. on Mathematical Statistics and Probability*, Vol. 1 (Univ. of California Press, Berkeley, Calif., 1967) 419–426.
[77] Y. S. Chow, H. Robbins and D. Siegmund, *Great Expectations: The Theory of Optimal Stopping* (Houghton Mifflin, Boston, Mass., 1971) 141 pp.
[78] Y. S. Chow, H. Robbins and H. Teicher, Moments of randomly stopped sums, *Ann. Math. Statist.* **36** (1965) 789–799.
[79] Y. S. Chow and W. F. Stout, On the expected value of a stopped stochastic sequence, *Ann. Math. Statist.* **40** (1969) 456–461.
[80] Y. S. Chow and W. J. Studden, On the monotonicity of $E_p(S_t/t)$, *Ann. Math. Statist.* **39** (1968) 1755.
[81] Y. S. Chow and W. J. Studden, Monotonicity of the variance under truncation and variations of Jensen's inequality, *Ann. Math. Statist.* **40** (1969) 1106–1108.
[82] K. L. Chung and J. L. Doob, Fields, optionality and measurability, *Am. J. Math.* **87** (1965) 397–424.
[83] A. Cornea and G. Licea, General optional sampling of supermartingales, *Rev. Roumaine Math. Pures Appl.* **10** (1965) 1349–1367.
[84] Ph. Courrège and P. Priouret, Temps d'arrêt d'une function aléatoire. Théoremes de décomposition, *Publ. Inst. Statist. Univ. Paris* **14** (1965) 245–274, 275–377.
[85] E. Csáki, An iterated logarithm law for semimartingales and its application to empirical distribution function, *Studia Sci. Math. Hung.* **3** (1968) 287–292.
[86] M. Csörgö, On the strong law of large numbers and the central limit theorem for martingales, *Trans. Am. Math. Soc.* **131** (1968) 259–275; addendum: *ibid.* **136** (1969) 545–6.
[87] I. Cuculescu, Martingales on von Neumann algebras, *J. Multivariate Analysis* **1** (1971) 17–27.
[88] E. C. Curtis, A potential theory for supermartingales, *Ann. Math. Statist.* **39** (1968) 802–814.
[89] D. A. Darling and H. Robbins, Iterated logarithm inequalities, *Proc. Nat. Acad. Sci. U.S.A.* **57** (1967) 1188–1192.
[90] D. A. Darling and H. Robbins, Inequalities for sequences of sample means, *Proc. Nat. Acad. Sci. U.S.A.* **57** (1967) 1577–1580.
[91] D. A. Darling and H. Robbins, Confidence sequences for mean, variance and median, *Proc. Nat. Acad. Sci. U.S.A.* **58** (1967) 66–68.

[92] D. A. Darling and H. Robbins, Some further remarks on inequalities for sample sums, *Proc. Nat. Acad. Sci. U.S.A.* **60** (1968) 1175–1182.
[93] J.-P. Daurès, Version multivoque du théorème de Doob, *C.R. Acad. Sci. Paris* (A) **275** (1972) 527–530.
[94] B. Davis, Comparison tests for the convergence of martingales, *Ann. Math. Statist.* **39** (1968) 2141–2144.
[95] B. Davis, A comparison test for martingale inequalities, *Ann. Math. Statist.* **40** (1969) 505–508.
[96] B. Davis, Divergence properties of some martingale transforms, *Ann. Math. Statist.* **40** (1969) 1852–1854.
[97] B. Davis, On the integrability of the martingale square function, *Israel J. Math.* **8** (1970) 187–190.
[98] B. Davis, Stopping rules for S_n/n and the class $L \log L$ *Z. Wahrscheinlichkeitstheorie Verw. Gebiete* **17** (1971) 147–150.
[99] C. Dellacherie, Une représentation intégrale des surmartingales à temps discret, *Publ. Inst. Statist. Univ. Paris* **17** (1968) 1–18.
[100] S. W. Dharmadhikari and K. Jogdeo, Bounds on moments of certain random variables, *Ann. Math. Statist.* **40** (1969) 1506–1509.
[101] S. W. Dharmadhikari, V. Fabian and K. Jogdeo, Bounds on the moments of martingales, *Ann. Math. Statist.* **39** (1968) 1719–1723.
[102] J. Dieudonné, Sur un théorème de Jessen, *Fund. Math.* **37** (1950) 242–248.
[103] H. Dinges, Inequalities leading to a proof of the classical martingale-convergence theorem, in: *Martingales*, Lecture Note in Math. **190** (Springer, Berlin, 1970) 9–12.
[104] J. L. Doob, Regularity properties of certain families of chance variables, *Trans. Am. Math. Soc.* **47** (1940) 455–486.
[105] J. L. Doob, Applications of the theory of martingales, *Colloq. Intern. du Centre Nat. Rech. Sci., Paris* (1949) 23–27.
[106] J. L. Doob, *Stochastic Processes* (Wiley, New York, 1953) 654 pp.
[107] J. L. Doob, Discrete potential theory and boundaries, *J. Math. Mech.* **8** (1959) 433–458.
[108] J. L. Doob, Notes on martingale theory, in: J. Neyman, ed., *Proc. 4th Berkeley Symp. on Mathematical Statistics and Probability*, Vol. 2 (Univ. of California Press, Berkeley, Calif., 1960) 95–102.
[109] J. L. Doob, Generalised sweeping-out and probability, *J. Functional Analysis* **2** (1968) 207–225.
[110] J. L. Doob, What is a martingale?, *Am. Math. Monthly* **76** (1971) 451–463.
[111] J. L. Doob, J. L. Snell and R. E. Williamson, Applications of boundary theory to sums of independent random variables, in: I. Olkin et al., eds., *Contributions to Probability and Statistics*, Essays in Honour of Harold Hotelling (Stanford Univ. Stanford, Calif., 1960) 182–197.
[112] R. G. Douglas, Contractive projections on an $\mathscr{L}_1$ space, *Pacific J. Math.* **15** (1965) 443–462.
[113] R. Drogin, An invariance principle for martingales, *Ann. Math. Statist.* **43** (1972) 602–620.
[114] L. E. Dubins, Rises and upcrossings of non-negative martingales, *Illinois J. Math.* **6** (1962) 226–241.
[115] L. E. Dubins, A note on upcrossings of semimartingales, *Ann. Math. Statist.* **37** (1966) 728.
[116] L. E. Dubins, Rises of non-negative semimartingales, *Illinois J. Math.* **12** (1968) 649–653.
[117] L. E. Dubins, On a theorem of Skorohod, *Ann. Math. Statist.* **39** (1968) 2094–2097.

[118] L. E. Dubins, Sharp bounds for the total variance of uniformly bounded semimartingales, *Ann. Math. Statist.* **43** (1972) 1559–1665.

[119] L. E. Dubins, Some upcrossing inequalities for uniformly bounded martingales, *Ist. Alta Mat., Symp. Math.* **9** (1972) 169–177.

[120] L. E. Dubins and D. A. Freedman, A sharper form of the Borel–Cantelli lemma and the strong law, *Ann. Math. Statist.* **36** (1965) 800–807.

[121] L. E. Dubins and D. A. Freedman, On the expected value of a stopped martingale, *Ann. Math. Statist.* **37** (1966) 1505–1509.

[122] L. E. Dubins and G. Schwarz, On extremal martingale distributions, in: L. M. Le Cam and J. Neyman, eds., *Proc. 5th Berkeley Symp. on Mathematical Statistics and Probability*, II, Vol. 1 (Univ. of California Press, Berkeley, Calif., 1967) 295–300.

[123] L. E. Dubins, E. Lester and H. Teicher, Optimal stopping when the future is discounted, *Ann. Math. Statist.* **38** (1967) 601–605.

[124] A. Dvoretzky, Existence and properties of certain optimal stopping rules, in: L. M. Le Cam and J. Neyman, eds., *Proc. 5th Berkeley Symp. on Mathematical Statistics and Probability*, Vol. 1 (Univ. of California Press, Berkeley, Calif., 1967) 441–452.

[125] A. Dvoretzky, Asymptotic normality of sums of dependent random variables, in: L. M. Le Cam et al., eds., *Proc. 6th Berkeley Symp. on Mathematical Statistics and Probability*, II, Vol. 1 (Univ. of California Press, Berkeley, Calif., 1972) 513–535.

[126] E. B. Dynkin, The optimum choice of the instant of stopping a Markov process, *Dokl. Akad. Nauk SSSR* **150** (1963) 238–240 (in Russian; English Transl.: *Soviet Math. Dokl.* **4**, 627–629).

[127] E. B. Dynkin, Controlled random sequences, *Teor. Verojatnost. Primenen.* **10** (1965) 3–18(in Russian; English Transl.: *Theor. Probab. Appl.* **10**, 1–14).

[128] E. B. Dynkin, Sufficient statistics for the optimal stopping problem, *Teor. Verojatnost. Primenen.* **13** (1968) 150–152 (in Russian; English Transl.: *Theor. Probab. Appl.* **13**, 152–153).

[129] E. B. Dynkin, A game variant of a problem on optimal stopping, *Dokl. Akad. Nauk SSSR* **185** (1969) 16–19 (in Russian; English Transl. *Soviet Math. Dokl.* **10**, 270–274).

[130] E. B. Dynkin and A. A. Yushkevich, *Markov processes. Theorems and Problems* (Nauka, Moscow, 1967) (in Russian; English Transl. by J. S. Wood: Plenum Press, New York, 1969) 237 pp.

[131] D. Eustice, Orthogonal series and probability, *Proc. Am. Math. Soc.* **18** (1967) 465–471.

[132] A. G. Fakeev, On necessary conditions for optimality in sequential decision problems, *Teor. Verojatnost. Primenen.* **14** (1969) 742–746 (in Russian; English Transl.: *Theor. Probab. Appl.* **14**, 710–713.

[133] Ch. Fefferman, Characterisations of bounded mean oscillation, *Bull. Am. Math. Soc.* **77** (1971) 587–588.

[134] Ch. Fefferman, and E. M. Stein, H^p-spaces of several variables, *Acta Math.* **129** (1972) 137–194.

[135] D. L. Fisk, Quasi-martingales, *Trans. Am. Math. Soc.* **120** (1965) 369–389.

[136] H. Föllmer, The exit measure of a supermartingale, *Z. Wahrscheinlichkeitstheorie Verw. Gebiete* **21** (1972) 154–166.

[137] H. Föllmer, On the representation of semimartingales, *Ann. Probab.* **1** (1973) 580–589.

[138] E. B. Frid, The optimal stopping rule for a two-person Markov chain with opposing interests, *Teor. Verojatnost. Primenen.* **14** (1969) 746–749 (in Russian; English Transl.: *Theor. Probab. Appl.* **14**, 713–716).

[139] A. GARSIA, *Topics in Almost Everywhere Convergence* (Markham, Chicago, Ill., 1970).
[140] A. M. GARSIA, On a convex function inequality for martingales, *Ann. Probab.* **1** (1973) 171–174.
[141] A. M. GARSIA, The Burgess Davis inequalities via Fefferman's inequality, *Ark. Mat.* **11** (1973) 229–237.
[142] A. M. GARSIA, *Martingale Inequalities*, Seminar notes on recent progress (Addison Wesley/Benjamin, New York, 1973) 184 pp.
[143] A. GARSIA, The B. Davis inequalities via Fefferman's inequality, *Ark Mat.* **11** (1973) 229–237.
[144] R. K. GETOOR and M. J. SHARPE, Conformal martingales, *Invent. Math.* **16** (1972) 271–308.
[145] D. GILAT, On the non existence of a three series condition for series of non independent random variables, *Ann. Math. Statist.* **42** (1971) 409.
[146] D. GILAT, Convergence in distribution, convergence in probability and almost sure convergence of discrete martingales, *Ann. Math. Statist.* **43** (1972) 1374–1379.
[147] M. I. GORDIN, The central limit theorem for stationary processes, *Dokl. Akad. Nauk SSSR* **188** (1969) 739–741 (in Russian; English Transl.: *Soviet Math. Dokl.* **10**, 1174–1176).
[148] L. GORDON, An equivalent to the martingale square function inequality, *Ann. Math. Statist.* **43** (1972) 1927–1934.
[149] B. I. GRIGELIONIS and A. N. SHIRYAEV, Criteria of "truncation" for the optimal stopping time in sequential analysis, *Teor. Verojatnost. Primenen.* **10** (1965) 601–613 (in Russian; English Transl.: *Theor. Probab. Appl.* **10**, 541–552).
[150] B. I. GRIGELIONIS, and A. N. SHIRYAEV, On Stefan's problem and optimal stopping rules for Markov processes, *Teor. Verojatnost. Primenen.* **11** (1966) 631–654 (in Russian; English Transl.: *Theor. Probab. Appl.* **11**, 541–558).
[151] R. F. GUNDY, Martingale theory and pointwise convergence of certain orthogonal series, *Trans. Am. Math. Soc.* **124** (1966) 228–248.
[152] R. F. GUNDY, The martingale version of a theorem of Marcinkiewicz and Zygmund, *Ann. Math. Statist.* **38** (1967) 725–734.
[153] R. F. GUNDY, A decomposition theorem for L^1-bounded martingales, *Ann. Math. Statist.* **39** (1968) 134–138.
[154] R. F. GUNDY, On the class L log L, martingales and singular integrals, *Studia Math.* **33** (1969) 109–118.
[155] R. F. GUNDY and D. SIEGMUND, On a stopping rule and the central limit theorem, *Ann. Math. Statist.* 38 (1967) 1915–1917.
[156] E. J. HANNAN and C. C. HEYDE, On limit theorems for quadratic functions of discrete time series, *Ann. Math. Statist.* **43** (1972) 2058–2066.
[157] C. A. HAYES and C. Y. PAUC, *Derivation and Martingales*, Ergebn. Math. Grenzgeb., Band 49 (Springer, Berlin, 1970) 203 pp.
[158] L. L. HELMS, Mean convergence of martingales, *Trans. Am. Math. Soc.* **87** (1958) 439–446.
[159] D. J. HERBERT, Jr., Generalised balayage and a Radon–Nikodym theorem, *Proc. Am. Math. Soc.* **26** (1970) 165–167.
[160] C. S. HERZ, H_p-spaces of martingales, $0 < p < 1$, *Z. Wahrscheinlichkeitstheorie Verw. Gebiete* **28** (1974) 189–205.
[161] C. S. HERZ, Bounded mean oscillation and regulated martingales, *Trans. Am. Math. Soc.* **192** (1974).
[162] C. C. HEYDE, An iterated logarithm result for martingales and its application in estimation theory for autoregressive processes, *J. Appl. Probab.* **10** (1973) 146–157.

[163] C. C. HEYDE and B. M. BROWN, On the departure from normality of a certain class of martingales, *Ann. Math. Statist.* **41** (1970) 2161–2165.

[164] C. C. HEYDE and D. J. SCOTT, Invariance principles for the law of the iterated logarithm for martingales and processes with stationary increments, *Ann. Probab.* **1** (1973) 428–436.

[165] G. A. HUNT, *Martingales et Processus de Markov* (Dunod, Paris, 1966) 147 pp.

[166] G. A. HUNT, Markoff chains and Martin boundaries, *Illinois J. Math.* **4** (1960) 313–340.

[167] A. IONESCU TULCEA, and C. IONESCU TULCEA, Abstract ergodic theorems, *Trans. Am. Math. Soc.* **107** (1963) 107–124.

[168] R. ISAAC, A proof of the martingale convergence theorem, *Proc. Am. Math. Soc.* **16** (1965) 842–844.

[169] D. ISAACSON, Uniform integrability of square integrable martingales, *Ann. Math. Statist.* **43** (1972) 688–689.

[170] K. ITO, Lectures on Stochastic Processes, Tata Institute, Bombay (1963).

[171] K. ITO and H. P. MCKEAN, Jr., *Diffusion Processes and Their Sample Paths* (Springer, Berlin, 1965) 321 pp.

[172] B. JAMISON and S. OREY, An optional stopping theorem, *Ann. Math. Statist.* **40** (1969) 677–678.

[173] M. JERISON, Martingale formulation of ergodic theorems, *Proc. Am. Math. Soc.* **10** (1959) 531–539.

[174] S. JOHANSEN and J. KARUSH, On the semimartingale convergence theorem, *Ann. Math. Statist.* **37** (1966) 690–694.

[175] F. JOHN and L. NIRENBERG, On functions of bounded mean oscillation, *Commun. Pure Appl. Math.* **14** (1961) 415–426.

[176] P. E. KOPP, A ratio limit theorem for contraction projections and applications, *Glasgow Math. J.* **14** (1973) 80–85.

[177] K. KRICKEBERG, Convergence of martingales with a directed index set, *Trans. Am. Math. Soc.* **83** (1956) 313–337.

[178] K. KRICKEBERG, Stochastische Konvergenz von Semimartingalen, *Math. Z.* **66** (1957) 470–486.

[179] K. KRICKEBERG, Semimartingales à base filtrante decroissante, in: *Le calcul des probabilités et ses applications*, Colloq. Intern. Centre Nat. Rech. Sci. **87** (Paris, 1959).

[180] K. KRICKEBERG, Seminar on martingales, Mat. Inst. Aarhus Univ. (1959) 58 pp.

[181] K. KRICKEBERG, Absteigende Semimartingale mit filtrierenden Parameterbereich, *Abh. Math. Sem. Univ. Hamburg* **24** (1960) 109–125.

[182] K. KRICKEBERG, Convergence of conditional expectation operators, *Teor. Verojatnost. Primenen.* **9** (1964) 595–607 (in Russian; English Transl.: *Theor. Probab. Appl.* **9**, 538–549).

[183] K. KRICKEBERG, Notwendige Konvergenzbedingungen bei Martingalen und verwandten Prozessen, in: *Trans.* 2nd *Prague Conf. on Information Theory, Statistical Decision Functions and Random Processes* (Czech. Acad. Sci., Prague, 1969) 279–305.

[184] K. KRICKEBERG and C. PAUC, Martingales et dérivation, *Bull. Soc. Math. France* **91** (1963) 455–543.

[185] N. V. KRYLOV, The construction of an optimal strategy for a finite controlled chain *Teor. Verojatnost. Primenen.* **10** (1965) 51–60 (in Russian; English Transl.: *Theor. Probab. Appl.* **10**, 45–54).

[186] Ch. W. LAMB, A short proof of the martingale convergence theorem, *Proc. Am. Math. Soc.* **38** (1973) 215–217.

[187] Ch. W. LAMB, A ratio limit theorem for approximate martingales, *Can. J. Math.* **25** (1973) 772–779.

[188] D. LANDERS and L. ROGGE, A generalised martingale theorem, *Z. Wahrscheinlichkeitstheorie Verw. Gebiete.* **23** (1972) 289–292.

[189] L. LE CAM, Sufficiency and approximate sufficiency, *Ann. Math. Statist.* **35** (1964) 1419–1455.

[190] P. LÉVY, *Theorie de l'Addition des Variables Aléatoires*, 2nd ed. (Gauthier-Villars, Paris, 1954) 387 pp.

[191] P. LÉVY, *Processus Stochastique et Mouvement Brownien*, Followed by a note by M. Loève, 2nd ed. (Gauthier-Villars, Paris, 1965) 439 pp.

[192] G. LICEA, On supermartingales with partially ordered parameter set, *Teor. Verojatnost. Primenen.* **14** (1969) 135–137; see also: *Theor. Probab. Appl.* **14**, 135–137.

[193] R. M. LOYNES, The central limit theorem for backward martingales, *Z. Wahrscheinlichkeitstheorie Verw. Gebiete* **13** (1969) 1–8.

[194] R. M. LOYNES, An invariance principle for reversed martingales, *Proc. Am. Math. Soc.* **25** (1970) 56–64.

[195] J. B. MACQUEEN, A linear extension of the martingale theorem, *Ann. Probab.* **1** (1973) 263–271.

[196] D. MAHARAM, On a theorem of von Neumann, *Proc. Am. Math. Soc.* **9** (1958) 987–994.

[197] D. MAHARAM, On two theorems of Jessen, *Proc. Am. Math, Soc.* **9** (1958) 995–999.

[198] B. MAISSONNEUVE, Quelques martingales remarquables associées à une martingale continue, *Publ. Inst. Statist. Univ. Paris*, **17** (1968) 13–28.

[199] J. MARCINKIEWICZ and A. ZYGMUND, Quelques théorèmes sur les fonctions indépendantes, *Studia Math.* **7** (1938) 104–120.

[200] N. F. G. MARTIN, Uniform convergence of families of martingales, *Ann. Math. Statist.* **40** (1969) 1071–1074.

[201] B. J. MCCABE and L. A. SHEPP, On the supremum of S_n/n, *Ann. Math. Statist.* **41** (1970) 2166–2168.

[202] H. P. MCKEAN, Jr., *Stochastic Integrals* (Academic Press, New York, 1969) 140 pp.

[203] M. MÉTIVIER, Limites projectives de mesures, martingales. Applications, *Ann. Mat. Pura Appl.* **63** (1963) 225–352.

[204] M. MÉTIVIER, Convergence de martingales à valeurs vectorielles, *Bull. Soc. Math. Grèce* **5** (1964) 54–74.

[205] M. MÉTIVIER, Martingales à valeurs vectorielles; application à la dérivation des mesures vectorielles, *Ann. Inst. Fourier* (*Grenoble*) **17** (1967) 175–298.

[206] M. MÉTIVIER, Martingales à valeurs vectorielles. Applications à la dérivation, in: *Symp. on Probability Methods in Analysis*, Lecture Notes in Math. **31** (Springer, Berlin, 1967) 239–255.

[207] M. MÉTIVIER, Martingales faibles et martingales fortes, *C.R. Acad. Sci. Paris* (A) **261** (1965) 3723–3726.

[208] P. A. MEYER, *Probabilités et Potentiels* (Hermann, Paris; English Transl.: *Probability and Potentials*, Blaisdell, Waltham, Mass., 1966, 266 pp.)

[209] P. A. MEYER, Une majoration du processus croissant naturel associé à une surmartingale, in: *Sém. de Probabilités II, Univ. of Strasbourg*, Lecture Notes in Math. **51** (Springer, Berlin, 1968) 166–170.

[210] P. A. MEYER, Les inégalités de Burkholder en théorie des martingales d'après Gundy, in: *Sém. de Probabilités III, Univ. of Strasbourg*, Lecture Notes in Math. **88** (Springer, Berlin, 1969) 163–174.

[211] P. A. MEYER, Quelques inégalités sur les martingales, in: *Sém. de Probabilités IV, Univ. of Strasbourg*, Lecture Notes in Math. **124** (Springer, Berlin, 1970) 162–169.

[212] P. A. Meyer, Sur un article de Dubins, in: *Sém. de Probabilités V, Univ. of Strasbourg*, Lecture Notes in Math. **191** (Springer, Berlin, 1970) 170–176.
[213] P. A. Meyer, *Martingales and Stochastic Integrals I*, Lecture Notes in Math. **284** (Springer, Berlin, 1972) 89 pp.
[214] P. W. Millar, Martingale integrals, *Trans. Am. Math. Soc.* **133** (1968) 145–166.
[215] P. W. Millar, Transforms of stochastic processes, *Ann. Math. Statist.* **39** (1968) 372–376.
[216] P. W. Millar, Martingales with independent increments, *Ann. Math. Statist.* **40** (1969) 1033–1041.
[217] H. D. Miller, A generalisation of Wald's identity with applications to random walks, *Ann. Math. Statist.* **32** (1961) 549–560.
[218] Shu-Teh Chen Moy, Measure extensions and the martingale convergence theorem, *Proc. Am. Math. Soc.* **4** (1953) 902–907.
[219] P. A. Nelson, A class of orthogonal series related to martingales, *Ann. Math. Statist.* **41** (1970) 1684–1694.
[220] J. Neveu, Deux remarques sur la théorie des martingales, *Z. Wahrscheinlichkeitstheorie Verw. Gebiete* **3** (1964) 122–127.
[221] J. Neveu, Relations entre la théorie des martingales et la théorie ergodique, *Ann. Inst. Fourier (Grenoble)* **15** (1965) 31–42.
[222] J. Neveu, *Bases Mathématiques du Calcul des Probabilités* (Masson, Paris, 1964) 203 pp. (2nd ed., 1970, 213 pp.; English Transl. of 1st ed. by A. Feinstein: *Mathematical Foundations of the Calculus of Probability*, Holden-Day, San Francisco, Calif., 1965, 223 pp.).
[223] J. Neveu, *Processus Aléatoires Gaussiens* (Presses de l'Univ. de Montréal, Montréal, 1968) 225 pp.
[224] J. Neveu, Convergence presque sûre de martingales multivoques, *Ann. Inst. Henri Poincaré* (B) **8** (1972) 1–7.
[225] J. Neveu, Un lemme élémentaire de la théorie des martingales, *Period. Math. Hung.* **2** (1972) 291–294.
[226] R. A. Olshen and D. O. Siegmund, Some first passage problems for $S_n/n^{1/2}$ *Ann. Math. Statist.* **40** (1969) 648–652.
[227] M. Ph. Olson, On a characterisation of conditional probability, *Pacific J. Math.* **15** (1965) 971–983.
[228] S. Orey, *F*-processes, in: L. M. Le Cam and J. Neyman, eds., *Proc. 5th Berkeley Symp. on Mathematical Statistics and Probability*, II, Vol. 1 (Univ. of California Press, Berkeley, Calif., 1965) 301–314.
[229] D. Ornstein, On the pointwise behaviour of iterates of a self-adjoint operator, *J. Math. Mech.* **18** (1968/9) 473–490.
[230] R. Panzone, Alternative proofs for certain upcrossing inequalities, *Ann. Math. Statist.* **38** (1967) 735–741.
[231] V. V. Petrov, On the strong law of large numbers, *Teor. Verojatnost. Primenen.* **14** (1969) 193–202 (in Russian; English Transl.: *Theor. Probab. Appl.* **14**, 183–192.
[232] J. Pfanzagl, Characterisations of conditional expectations, *Ann. Math. Statist.* **38** (1967) 415–421.
[233] J. Pfanzagl, On the existence of regular conditional probabilities, *Z. Wahrscheinlichkeitstheorie Verw. Gebiete* **11** (1969) 244–256.
[234] Z. P. Pop-Stojanovič, Decomposition of Banach-valued quasimartingales, *Math. Systems Theory* **5** (1971) 344–348.
[235] Z. P. Pop-Stojanovič, Riesz decomposition for weak Banach-valued quasi-martingales, *Ann. Math. Statist.* **43** (1972) 1020–1026.
[236] C. J. Preston, On the convergence of multiplicatively orthogonal series, *Proc. Am. Math. Soc.* **28** (1971) 453–455.

[237] J. J. Price, Orthonormal sets with non-negative Dirichlet kernels, *Trans. Am. Math. Soc.* **95** (1960) 256–262.
[238] J. J. Price and R. E. Zink, On sets of completeness for families of Haar functions, *Trans. Am. Math. Soc.* **119** (1965) 262–269.
[239] K. M. Rao, Quasimartingales, *Math. Scand.* **24** (1969) 79–92.
[240] M. M. Rao, Conditional expectations and closed projections, *Indag. Math.* **27** (1965) 100–112.
[241] M. M. Rao, Interpolation ergodicity and martingales, *J. Math. Mech.* **16** (1966) 543–567.
[242] M. M. Rao, Opérateurs de moyennes conditionelles, *C.R. Acad. Sci. Paris* (A) **268** (1969) 795–797.
[243] M. M. Rao, Contractive projections and prediction operators, *Bull. Am. Math. Soc.* **75** (1969) 1369–1373.
[244] M. M. Rao, Generalised martingales, in: *Contributions to Ergodic and Probability Theory*, Lecture Notes in Math. **160** (Springer, Berlin, 1970) 241–261.
[245] M. Rao, Doob decomposition and Burkholder inequalities, in: *Sém. de Probabilités VI, Univ. of Strasbourg*, 1970/1971, Lecture Notes in Math. **258** (Springer, Berlin, 1972) 198–201.
[246] J. P. Raoult, Généralisation de la notion de sous-martingale: asymptosous-martingales. Définition et théorèmes de comvergence en moyenne, *C.R. Acad. Sci. Paris* (A) **263** (1966) 738–741.
[247] J. P. Raoult, Asympto-martingales et continguité, *C.R. Acad. Sci. Paris* (A) **264** (1967) 329–332.
[248] S. N. Ray, Bounds on the maximum size of a Bayes sequential procedure, *Ann. Math. Statist.* **36** (1965) 859–878.
[249] M. A. Rieffel, The Radon–Nikodym theorem for the Bochner integral, *Trans. Am. Math. Soc.* **131** (1968) 466–487.
[250] H. Robbins, Statistical methods related to the law of the iterated logarithm, *Ann. Math. Statist.* **41** (1970) 1397–1409.
[251] H. Robbins and E. Samuel, An extension of a lemma of Wald, *J. Appl. Probab.* **3** (1966) 272–273.
[252] H. Robbins and D. Siegmund, Probability distributions related to the law of the iterated logarithm, *Proc. Nat. Acad. Sci. U.S.A.* **62** (1969) 11–13.
[253] H. Robbins and D. Siegmund, Boundary crossing probabilities for the Wiener process and sample sums, *Ann. Math. Statist.* **41** (1970) 1410–1429.
[254] H. Robbins and D. Siegmund, A convergence theorem for non-negative almost supermartingales and some applications, in: J. S. Rustagi, ed., *Optimizing Methods in Statistics* (Academic Press, New York, 1971) 233–258.
[255] H. Robbins, D. Siegmund and J. Wendel, The limiting distribution of the last time $s_n \geqslant n\varepsilon$, *Proc. Nat. Acad. Sci. U.S.A.* **61** (1968) 1228–1230.
[256] G. C. Rota, On the representation of averaging operators, *Rend. Sem. Mat. Univ. Padova* **30** (1960) 52–64.
[257] G. C. Rota, Une theorie unifiée des martingales et des moyennes ergodiques, *C.R. Acad. Sci. Paris.* (A) **252** (1961) 2064–2066.
[258] J. de Sam Lazaro and P. A. Meyer, Méthodes de martingales et théorie des flots, *Z. Wahrscheinlichkeitstheorie Verw. Gebiete* **18** (1971) 116–140.
[259] F. S. Scalora, Abstract martingale convergence theorems, *Pacific J. Math.* **11** (1961) 347–374.
[260] D. J. Scott, An invariance principle for reversed martingales, *Z. Wahrscheinlichkeitstheorie Verw. Gebiete* **20** (1971) 9–27.
[261] D. J. Scott, Central limit theorems for martingales and for processes with stationary increments using a Skorohod representation approach, *Adv. Appl. Probab.* **5** (1973) 119–137.

[262] L. A. SHEPP, Explicit solutions to some problems of optimal stopping, *Ann. Math. Statist.* **40** (1969) 993–1010.

[263] A. N. SHYRAEV, On the theory of decision functions and control by the process of observation of partial data, in: *Trans. 3rd Prague Conf. on Information Theory, Statistical Decision Functions and Random Processes* (in Russian). (Czech. Acad. Sci, 1964) 657–681.

[264] A. N. SHYRAEV, Sequential analysis and controlled random processes, *Kibernetika* (*Prague*) **3** (1965) 1–24.

[265] D. A. SIEGMUND, Some one-sided stopping rules, *Ann. Math. Statist.* **38** (1967) 1641–1646.

[266] D. O. SIEGMUND, Some problems in the theory of optimal stopping rules, *Ann. Math. Statist.* **38** (1967) 1627–1640.

[267] D. O. SIEGMUND, On the moments of the maximum of normed partial sums, *Ann. Math. Statist.* **40** (1969) 527–531.

[268] D. O. SIEGMUND, The variance of one-sided stopping rules. *Ann. Math. Statist.* **40** (1969) 1074–1077.

[269] D. O. SIEGMUND, G. SIMONS and P. FEDER, Existence of optimal stopping rules for rewards related to S_n/n, *Ann. Math. Statist.* **39** (1968) 1228–1235.

[270] G. SIMONS, A martingale decomposition theorem, *Ann. Math. Statist.* **41** (1970) 1102–1104.

[271] A. V. SKOROHOD, *Studies in the Theory of Stochastic Processes* (Kiev Univ. Press, Kiev, 1961) (in Russian; English Transl. by Scripta Technica: Addison-Wesley, Reading, Mass., 1965, 199 pp.).

[272] J. L. SNELL, Applications of martingale systems theorems, *Trans. Am. Math. Soc.* **73** (1952) 293–312.

[273] J. L. SOX, JR. and W. J. HARRINGTON, A class of complete orthogonal sequences of step functions, *Trans. Am. Math. Soc.* **157** (1971) 129–136.

[274] N. STARR, Operator limit theorems, *Trans. Am. Math. Soc.* **121** (1966) 90–115.

[275] W. L. STEIGER, A best possible Kolmogoroff-type inequality for martingales and a characteristic property, *Ann. Math. Statist.* **40** (1969) 764–769.

[276] W. L. STEIGER, Bernstein's inequality for martingales, *Z. Wahrscheinlichkeitstheorie Verw. Gebiete* **16** (1970) 104–106.

[277] W. F. STOUT, Some results on the complete and almost sure convergence of linear combinations of independent random variables and martingale differences, *Ann. Math. Statist.* **39** (1968) 1549–1562.

[278] W. F. STOUT, The Hartman–Wintner law of the iterated logarithm for martingales, *Ann. Math. Statist.* **41** (1970) 2158–2160.

[279] W. F. STOUT, A martingale analogue of Kolmogorov's law of the iterated logarithm, *Z. Wahrscheinlichkeitstheorie Verw. Gebiete* **15** (1970) 279–290.

[280] W. F. STOUT, Maximal inequalities and the law of the iterated logarithm, *Ann. Probab.* **1** (1973) 322–328.

[281] V. STRASSEN, Almost sure behaviour of sums of independent random variables and martingales, in: L. M. Le Cam and J. Neyman, eds., *Proc. 5th Berkeley Symp. on Mathematical Statistics and Probability*, II, Vol. 1 (Univ. of California Press, Berkeley, Calif., 1965) 315–343.

[282] R. E. STRAUCH, Negative dynamic programming, *Ann. Math. Statist.* **37** (1966) 871–890.

[283] R. E. STRAUCH, Measurable gambling houses, *Trans. Am. Math. Soc.* **126** (1967) 64–72.

[284] D. W. STROOK, Applications of Fefferman–Stein type interpolation to probability theory and analysis, *Commun. Pure Appl. Math.* **20** (1973) 477–495.

[285] W. D. SUDDERTH, On the existence of good stationary strategies, *Trans. Am. Math. Soc.* **135** (1969) 399–414.

[286] H. M. Taylor, Optimal stopping in a Markov process, *Ann. Math. Statist.* **39** (1968) 1333–1344.
[287] H. Teicher and J. Wolfowitz, Existence of optimal stopping rules for linear and quadratic rewards, *Z. Wahrscheinlichkeitstheorie Verw. Gebiete* **5** (1966) 361–368, **9** (1968) 357.
[288] J. J. Uhl, Jr., The Radon–Nikodym theorem and the mean convergence of Banach space valued martingales, *Proc. Am. Math. Soc.* **21** (1969) 139–144.
[289] J. J. Uhl, Jr., The range of a vector-valued measure, *Proc. Am. Math. Soc.* **23** (1969) 158–164.
[290] J. J. Uhl, Jr., Applications of Radon–Nikodym theorems to martingales of vector-valued functions, *Bull. Am. Math. Soc.* **75** (1969) 840–842.
[291] M. Valadier, Sur l'integration des ensembles convexes compacts en dimension infinie, *C.R. Acad. Sci. Paris* (A) **266** (1968) 14–16.
[292] M. Valadier, Multi-applications mesurables à valeurs convexes compactes, *J. Math. Pures Appl.* **50** (1971) 265–297.
[293] B. van Cutsem, Espérances conditionelles d'une multiapplication à valeurs convexes compactes, *C.R. Acad. Sci. Paris* (A) **269** (1969) 212–214.
[294] B. van Cutsem, Martingales de multiapplications à valeurs convexes compactes, *C.R. Acad. Sci Paris* (A) **269** (1969) 429–432.
[295] B. van Cutsem, Thèse, Univ. of Grenoble (1971).
[296] J. Ville, *Étude critique de la notion de collectif*, Monographies des Probabilités, Fasc. **3** (Gauthier-Villars, Paris, 1939).
[297] B. von Bahr and C.-G. Esseen, Inequalities for the rth absolute moment of a sum of random variables, $1 \leqslant r \leqslant 2$, *Ann. Math. Statist.* **36** (1965) 299–303.
[298] H. Walk, Convergence properties of martingale transforms, *Ann. Math. Statist.* **41** (1970) 706–709.
[299] L. H. Walker, Regarding stopping rules for brownian motion and random walks, *Bull. Am. Math. Soc.* **75** (1969) 46–50.
[300] P. Whittle, Refinements of Kolmogorov's inequality, *Teor. Verojatnost. Primenen.* **14** (1969) 315–317; see also: *Theor. Probab. Appl.* **14**, 310–311.
[301] A. Zygmund, *Trigonometric Series*, Vols. I, II, 2nd ed. (Cambridge Univ. Press, London, 1959).

INDEX OF TERMINOLOGY AND NOTATION

Terminology

Notation

We will denote $\mathbf{R}$ (resp. $\mathbf{Q}$) the real line (resp. the rationals), by $\mathbf{R}_+$ (resp. $\overline{\mathbf{R}}_+$) the half-line of reals $\geqslant 0$ (resp. completed with $+\infty$), by $\mathbf{N}$ (resp. $\mathbf{N}^*$) the half-line of integers $\geqslant 0$ (resp. $\geqslant 1$). The usual abbreviations r.v., a.s. denote, respectively; random variable, almost surely.

The symbol ■ indicates the end of a proof.